Grain Legumes: Evolution and Genetic Resources

T0275743

Grain Legumes

Evolution and Genetic Resources

J. Smartt

Department of Biology, University of Southampton

CAMBRIDGE UNIVERSITY PRESS

Cambridge

New York Port Chester Melbourne Sydney

CAMBRIDGE UNIVERSITY PRESS
Cambridge, New York, Melbourne, Madrid, Cape Town, Singapore, São Paulo

Cambridge University Press
The Edinburgh Building, Cambridge CB2 8RU, UK

Published in the United States of America by Cambridge University Press, New York

www.cambridge.org
Information on this title: www.cambridge.org/9780521307970

© Cambridge University Press 1990

This publication is in copyright. Subject to statutory exception
and to the provisions of relevant collective licensing agreements,
no reproduction of any part may take place without the written
permission of Cambridge University Press.

First published 1990
This digitally printed version 2008

A catalogue record for this publication is available from the British Library

Library of Congress Cataloguing in Publication data

Smartt, J.
Grain legumes: evolution and genetic resources / J. Smartt.
p. cm.
Bibliography: p. 333
Includes index.
ISBN 0-521-30797-X
1. Legumes. 2. Legumes – Evolution. 3. Legumes – Germplasm
resources. I. Title.
SB177.L45S58 1990
633.3 – dc20 89–7181 CIP

ISBN 978-0-521-30797-0 hardback
ISBN 978-0-521-05052-4 paperback

Contents

Preface

Research on grain legumes is now coming more to the forefront in the world as a whole than at any time in the past. This is reflected in the number of monographic treatments of individual legume crops which have appeared in the past decade. Forty years ago one would have been able to unearth rather little information on grain legumes generally and very little indeed specifically on tropical grain legumes. In present circumstances it still can be argued that the market is not as yet oversupplied with literature on the legumes. Be this as it may, it nevertheless behoves any author setting forth his labours before the public to define his aims and objectives as clearly as possible for the benefit of the potential reader. Most writers on technical subjects perhaps feel the urge to write the book they would themselves need if they were embarking on work in the field it covered. There has been a tendency in all this new writing for treatments to become very detailed and specialised. The rise of the multi-authored tome has been irresistible, with all this convergent thought; perhaps there is scope and a need for thinking on a broader front, in a more lateral vein.

My own contact with legumes as research material goes back to my student days, some thirty-five years ago. I found them to be problematic but fascinating then as I still do. They can be rewarding to work with but the rewards are often hard-won. A comparative approach such as that taken in this work may help to give encouragement to those who have come to believe, for example, that the producer of legumes must inevitably resign himself to low yields. Certainly the problem of low yields appears to be very intractable to the breeders of many grain legume species, but so it did to those working forty years ago in East Africa on the groundnut scheme. Yet within the past two decades some of the highest grain legume yields recorded have been produced, not in North America, but in Africa with groundnuts. We now have available the genetic resources and the plant breeding expertise to bring about substantial improvement in the productivity of many of these crops. In some ways perhaps we are spoiled for choice, but the availability of good passport data in modern collections can be of very material assistance. Unfortunately there is still much valuable material in the older collections for which these data were not collected; some material has been incorrectly named at the outset or through curatorial errors. A very good case could be made for working over old collections and adequately evaluating and classifying the material they contain, much of which may no longer be extant elsewhere.

There is of course a broader context for work on grain legumes and that is in the utilisation of other members of the family in sustaining agricultural production. All too frequently the role of wild legume species in restoring land to productivity under systems of shifting cultivation is forgotten. It is difficult to imagine how, for example, the *chitemene* system of agriculture in Central Africa could be sustained without the legume species that provide the material for the periodic burns. It is likely that there will be an increasing role for herbaceous, shrub and tree legume species in the restoration of biological productivity to former rainforest areas and man-made deserts. If agricultural production becomes possible in any of these restored areas then grain legumes may have a place, permitting exploitation without despoliation.

It is hoped that this work will be of interest to a range of people interested in legumes in the context of specialised agricultural disciplines, plant breeding, crop protection, agronomy and soil science, and also to those interested in the nutritional and industrial uses of grain legumes. I should be most gratified if palaeoethnobotanists, archaeologists, anthropologists and students of higher plant evolution also found this work to be of interest. Most of all, I hope that it will be in the context of sustainable utilisation and conservation of the environment and its genetic resources that my efforts will be of most value.

Acknowledgements

It is a great pleasure to acknowledge the encouragement, moral support and material assistance received which has enabled the present book to be produced. This help has been received over a period of more than thirty years. I should like to place on record by debt of gratitute to Magnus Halcrow, Stephen Hoyle and Charles Ducker, who provided enormous stimulation and gave great encouragement to me in my days in Northern Rhodesia, and my indebtedness to many mentors and friends in the United States, especially Dr Walton C. Gregory, Dr Margaret Pfluge Gregory, the late Dr Ben W. Smith and Dr Ray O. Hammons. Acknowledgement of the help and encouragement given by the late Professor Sir J. B. Hutchinson, F.R.S., Professor J. G. Hawkes and Professor N. W. Simmonds F.R.S.E. since my resettlement in Britain is more than appropriate, as well as those others, too numerous to mention individually, who have made working with legumes a source of considerable satisfaction to me.

I was most gratified at the readiness of many friends and correspondents who made available unstintingly much of the material which illustrates this volume. I was frankly overwhelmed by their generosity and I should like to assure them all of my gratitude and appreciation, namely Drs D. A. Bond, D. Debouck, J. A. Duke, P. Gates, H. S. Gentry, P. Gepts, J. S. Gladstones, W. C. and M. P. Gregory, F. N. Hepper; Mrs I. Herklots; Drs M. Hopf, T. Hymowitz, L. Kaplan, T. N. Khan, E. Klozová, F. Kupicha; Professors G. Ladizinsky, R. Maréchal, J. W. Purseglove; Drs J. P. Moss, C. E. Simpson, A. K. Smith, H. T. Stalker, R. J. Summerfield, J. E. M. Valls; Professor L. J. G. van der Maesen, Drs B. Verdcourt, J. G. Waines, E. Westphal and D. Zohary. The courtesy of the following publishers in permitting reproduction of illustrative material is also acknowledged gratefully: APRES (The American Peanut Research and Education Society); CIAT; The Crop Science Society; FAO; ICARDA; ICRISAT; Le Jardin Botanique de Genève; le Jardin Botanique National de Belgique; Kluwer Academic Publishers; Longman; Macmillan Magazines; PBI Ltd, Cambridge; Plenum;

PUDOC; The Royal Botanic Garden, Edinburgh; The Royal Botanic Gardens, Kew; and the Systematics Association. The writer is also indebted to editors of the following journals for permission to publish illustrative material: *Biologia Plantarum*, *Botanical Gazette*, *Economic Botany*, *Euphytica*, *Science*, *Theoretical and Applied Genetics* and the *Wageningen Papers*.

Table 1.2. *Taxonomic affinities of grain legumes**

Tribe	Sub-tribe	Species
Aeschynomeneae		*Arachis hypogaea* L. *A. villosulicarpa* Hoehne
Cicereae		*Cicer arietinum* L.
Vicieae		*Lens culinaris* Med. *Pisum sativum* L. *Vicia faba* L. *Lathyrus sativus* L.
Genisteae	Lupininae	*Lupinus albus* L. *L. luteus* L. *L. angustifolius* L. *L. mutabilis* Sweet. *L. cosentinii* Guss.
Phaseoleae	Erythrininae	*Mucuna* spp. (velvet beans)
	Diocleinae	*Canavalia ensiformis* (L.) DC. *C. gladiata* (Jacq.) DC. *Pachyrrhizus erosus* (L.) Urban *P. tuberosus* (Lam.) Spreng. *Calopogonium mucunoides* Desv.
	Glycininae	*Pueraria phaseoloides* (Roxb.) Benth. *Glycine max* (L.) Merr.
	Clitoriinae	*Centrosema pubescens* Benth. *Clitoria ternatea* L.
	Phaseolinae	*Psophocarpus tetragonolobus* (L.) DC. *Lablab purpureus* (L.) Sweet *Macrotyloma geocarpa* (Harms) Maréchal & Baudet *M. uniflorum* (Lamb.) Verdc. *Vigna aconitifolia* (Jacq.) Maréchal *V. angularis* (Willd.) Ohwi & Ohashi *V. mungo* (L.) Hepper *V. radiata* (L.) Wilczek *V. subterranea* (L.) Verdc. *V. umbellata* (Thunb.) Ohwi & Ohashi *V. unguiculata* (L.) Walp. *Phaseolus acutifolius* A. Gray *P. coccineus* L. *P. lunatus* L. *P. polyanthus* Greenm. *P. vulgaris* L.
	Cajaninae	*Cajanus cajan* (L.) Millsp.
Indigoferae		*Cyamopsis tetragonoloba* (L.) Taubert
Crotalarieae		*Crotalaria juncea* L.

industrially.

The ecological role of species of a number of genera in this sub-family is an exceedingly important one. The most important of these in Africa are the genera *Brachystegia*, *Isoberlinia* and *Julbernardia*, which serve to maintain the productivity of many sub-Saharan ecosystems. The maintenance of many indigenous systems of shifting cultivation is dependent on the ability of these trees to regenerate and restore soil fertility to an exploitable level after a cycle of cropping.

1.2. Sub-family Mimosoideae

The economic exploitation of this sub-family has many parallels with that of the Caesalpinioideae. There are a number of outstanding ornamental species, good examples of which are the Saman tree (*Samanea saman* Merr.) and the mimosas (*Acacia* spp.) of horticulture and floristry. A number of *Acacia* species are also sources of tannin (e.g. *A. mearnsii* de Wild), gum arabic and similar mucilages (*A. senegal* (L.) Willd.), dyes (*A. catechu* Willd.), perfume (*A. farnesiana* (L.) Willd.) and a high-quality wood (*A. seyal* Del.). Other useful products are the edible fruits of *Inga laurina* (Sw.) Willd. and the pods, seeds and leaves of *Parkia filicoides* Welw., the African locust bean. Other species find use as browse and fodder plants; these include *Leucaena glauca* (L.) Benth. and species of *Albizia*, *Prosopis* and *Pithecellobium*.

Ecologically the *Acacia* species are very important in the semi-arid tropics. They are an important constituent of savanna vegetation. Here they probably play a highly significant role in the nitrogen economy of the ecosystem, comparable to that of *Brachystegia*, *Isoberlinia* and *Julbernardia* species in woodland.

1.3. Sub-family Papilionoideae

This sub-family shows perhaps the widest range of morphological diversity within the three sub-families of the Leguminosae. It includes all the pulse crops and the leguminous oilseeds, and the major forage species such as the clovers (*Trifolium* spp.), the medicks (and lucerne or alfalfa) (*Medicago* spp.) and representatives of many other genera. A wide range of other uses are found for members of this group in producing dyes (*Indigofera* spp.), fibres (*Crotalaria*), insecticides (*Derris*, *Lonchocarpus*), spices (*Trigonella*) and flavourings (*Glycyrrhiza*). In addition to a number of extremely handsome woody species of the genera *Erythrina*, *Laburnum* and *Wisteria*, there are several shrubby and herbaceous genera such as *Cytisus*, *Lupinus* and *Lathyrus*, to name but three, that find extensive use as horticultural ornamentals. Rather specialist uses for papilionate legume species include use as shade trees for tropical plantation crops such as cocoa (*Gliricidia*) and cover crops to prevent soil erosion such as the kudzu, *Pueraria phaseoloides* (Roxb.) Benth., pioneer crops in land reclamation (*Lupinus* spp.) and green manures (*Crotalaria* spp.).

The ecological role of papilionate legumes is at least as great as and arguably greater than that of the other sub-families. The role of genera in the Trifolieae has been particularly well studied in enhancing soil nitrogen status. The results of the introduction of clovers together with selected effective strains of *Rhizobium* to the Antipodes are especially remarkable. Skilful management of the symbiotic *Trifolium/Rhizobium* nitrogen fixation system has very substantially reduced the requirement for artificial nitrogenous fertilisers in Australia and New Zealand (G. D. Hill, personal communication).

The legumes which are most directly exploited by man can conveniently be called grain legumes; these include pulses, the seeds of which are exploited directly as food, and the oilseeds which can also be exploited directly as food but which contain 20–50% extractable lipids, Pulses and oilseeds are not randomly distributed throughout the sub-family. By far the majority belong to the Phaseoleae; several important pulses are members of the Vicieae and the related tribe Cicereae; the tribe Aeschynomeneae contains one important grain legume, the groundnut; while the Genisteae furnishes one genus (*Lupinus*) to the catalogue of grain legumes. The taxonomic affinities of grain legumes are summarised in Table 1.2. It can be seen that taxonomically they comprise a small number of groups of rather closely related species.

A particular difficulty arises with the grain legumes and other cultivated plants concerning those species which essentially exist in both wild and domesticated populations. Separate codes of nomenclature exist for wild plants and those which have been domesticated but it is only recently that the problem has been addressed of those species which are found at the present time both in the wild and in cultivation. The nature of the problem is determined by the type of species concept adopted, whether morphological or biological. Morphologically there is in most cases little difficulty in distinguishing wild and domesticated populations and on morphological grounds alone there would be little or no difficulty for the taxonomist in producing separate diagnoses for such wild and cultivated populations. However, the application of the biological species concept to this situation reveals serious practical limitations to the morphological species concept. It becomes very clear that quite extreme morphological divergence can occur between wild and domesticated populations without generating any genetic isolating mechanisms. This can raise quite serious agricultural problems in producing seed for a crop like sugar beet where, if the seed crop is produced within pollinating distance of wild beets, crossing can occur and produce weedy beets as contaminants of the seed crop.

Obviously it is sensible to recognise the fact that morphological divergence has occurred between wild and cultivated populations within the same biological species. The biological species concept is effectively based on a gene pool concept of the species, defined in terms of actual or potential gene flow between its component individuals. There is no problem in recognising segments of a biological species on morphological grounds. This is most conveniently done at the sub-species level. In many cases a wild and a domesticated sub-species can be recognised. A difficulty can arise in a species which has an extensive geographic range in which populations, within the wild component of the species, have achieved a level of differentiation that might well merit recognition at the sub-species level. It would probably be imprudent to recommend a uniform approach to such situations, but rather one should determine each case on its merits. This notwithstanding it is desirable that the affinities of wild and domesticated components of a biological species be recognised in their nomenclature and that this be done in as consistent a fashion as is practicable.

At first sight it is perhaps surprising that more intensive use of the seeds produced by legumes has not been made by man since, without exception, they are very rich in protein. The seed of the vast majority of legumes is, however, generally well protected against predation by man and many other organisms. It is in fact only those which are poorly protected or whose protective devices are easily circumvented that can be exploited. The vast majority of legume seeds are apparently protected by a range of toxic and anti-metabolite materials. These range from highly toxic alkaloids (*Laburnum anagyroides* Med.) to non-protein amino acids, lectins, protease inhibitors and cyanogenic compounds. In species which are exploited as pulses these compounds are either absent, removed or neutralised in the course of preparation. It is possible to regard the majority of legumes which produce seeds containing various toxic materials as representing a potential protein resource which might be exploited when appropriate technologies are developed for removing, neutralising or eliminating such toxic materials from seeds.

2 The role of grain legumes in the human economy

If one were writing a school report on grain legumes the observation 'could do better' might very well be apposite, with all that this remark implies. The major preoccupation of many engaged in legume research is to find out why these crops in their performance so often fall far short of our expectations. There is some encouraging evidence that, in favourable conditions, some grain legumes at least can in fact perform very well indeed. The nadir of grain legume performance, as far as many people are concerned, is undoubtedly the notorious fiasco of the Overseas Food Corporation's East African Groundnut Scheme of the late 1940s and early 1950s. Curiously enough, it is also the groundnut which has given the clearest indication of what might lie in the realms of future grain legume production. The highest recorded groundnut yields (9.6 t ha^{-1} pods, equivalent to 6.4 t ha^{-1} kernels) have been obtained in Zimbabwe with the cultivar Makulu Red (Hildebrand and Smartt, 1980). Interestingly, this cultivar was not the product of a long-drawn-out and expensive breeding programme but obtained as a single plant selection from a Bolivian landrace 'maní pintado'. Agricultural improvement depends as much on recognising opportunities such as this and exploiting them efficiently, as on the execution of complex research programmes. Unfortunately it is the more grandiose schemes, not the most cost-effective, which make the headlines and attract the attention of the politicians. It is ironic that the media, which have had many field days castigating almost all concerned with the Groundnut Scheme, have never seen fit to cover the rehabilitation of the crop's reputation in Africa.

2.1. The legume problem

It is worth while to attempt an analysis of the broad nature of what might be called the legume problem. How is it that crops, seemingly with everything in their favour in theory, in practice prove to be so unpredictable and unreliable? This unpredictability was well appreciated in North America when the very useful book *The peanut – the unpredictable legume* was published in 1951. Unpredictability is perhaps less characteristic of the groundnut than of other legumes at the present time. A stronger contender for the title of 'the unpredictable legume' now might well be the faba bean. Duke (1981) reports yield levels of 6600 kg ha^{-1} in the USA and 3000 kg ha^{-1} in the UK. If such yields could be obtained reliably, production of this crop would offer very good prospects of satisfactory economic returns. In the UK the faba bean is, however, an unreliable cropper; satisfactory yields in one season may be followed by uneconomic yields in the next. The reasons for this instability of yield are not clear, except that they are somehow related to climatic vagaries in the growing season. In poor seasons, characterised by cool, overcast and wet weather, disappointing yields are frequently obtained. This has been ascribed to poor pollination; it could equally and perhaps better be ascribed to the effects of low temperatures on the efficiency of the *Rhizobium* symbiosis itself (Sprent, 1979).

The legume–*Rhizobium* symbiosis

Agronomists and agricultural scientists often fail to understand that the benefits of the legume–*Rhizobium* symbiosis are obtained only at a price. Instead of considering just the environmental needs of the crops, those of the *Rhizobium* bacterium itself must also be met and favourable conditions provided for efficient nitrogen fixation by the symbiotic partnership. Among the ill-informed there is a tendency to consider that because legumes generally are self-sufficient for nitrogen they are self-sufficient for other nutrients as well.

The evolutionary history of the legume family sheds some light on the problem. Its early evolution appears to have been in tropical rainforest. In such a nutrient-depleted environment the ability to fix nitrogen would undoubtedly confer a considerable selective advantage. The primitive members of the family were probably arborescent; lianas, climbing and non-climbing herbaceous plants can be regarded as derived from this ancestral form. It is worth noting that the balance between woody and herbaceous legumes is tilted very much more in favour of herbaceous forms as one moves from the tropics. Another trend is also apparent among herbaceous forms, and that is that colonisation of more extreme habitats, such as deserts and their margins, tends to favour the annual rather more than the perennial life form. This imposes a restriction on the general effectiveness of the rhizobial symbiosis in that it has to be re-established for each annual generation. In woody and herbaceous perennials there is in most cases a more or less constant cycle of root-nodule turnover. For most of their life span perennial legumes can be expected to benefit from the symbiotic relationship, while for a much greater proportion of the life span annuals will derive no such benefit. In some studies quoted by Sprent (1979) this certainly seems to be the case; the common bean (*Phaseolus vulgaris*) in Colombia is thought to meet only 50% of its inorganic nitrogen requirements by fixation. This is probably a reasonable estimate of the best that is being achieved, since growing conditions in Colombia are probably very close to the optimum. Moving away from upland tropical areas into the higher temperate latitudes, seasonal temperatures may become an important limiting factor. It is a matter of common observation in the UK that although nodulation may well occur, this is often only apparent when plants are nearly mature. Quite frequently root nodules may be very difficult to find, especially in dwarf-determinate varieties of the common bean. These cultivars appear to be less effective in nitrogen fixation than the longer-term climbing forms. In selecting for more convenient morphology, man may very well have been unconsciously selecting against efficient nitrogen fixation simply by reducing the time span in which it could occur effectively. In northern temperate latitudes the clear preference for dwarf French beans (*Phaseolus vulgaris*) combined with the strong probability of sub-optimal temperatures in late spring and early summer may reduce effective nitrogen fixation to very low levels. *Phaseolus* beans show these effects quite clearly and these observations should encourage us to look more closely at the effectiveness of rhizobial fixation in legumes such as the common pea (*Pisum sativum*) and the faba bean (*Vicia faba*). This applies not only to north temperate regions but elsewhere, where field performance leaves a lot to be desired.

Rhizobial nitrogen fixation has generally been regarded as a free bonus; this it certainly appears not to be. There is undoubtedly a bonus to be derived but it may very well have to be earned. Our expectations should also be more realistic. For example, short-term catch crops of *Phaseolus* beans or cowpeas (*Vigna unguiculata*) ('90 day cowpeas') may do very little indeed to improve soil nitrogen status and may even deplete it. There may also be a temptation to take two successive crops of 90 day cowpeas, rather than a single crop of a longer-term cultivar. From the point of view of the soil nitrogen economy, the latter alternative may be far preferable. These considerations may, in particular circumstances, be outweighed by economic factors, but they should not be overlooked in policy formulation.

Another point which is often lost sight of is that implicit in a plant's supporting a nitrogen-fixing symbiosis is the net energy cost of doing so. This will entail a yield penalty; yield comparisons with cereals are often made, to the disadvantage of the legume, on the basis of crude dry matter production. Comparisons can quite cheerfully be made between dry matter production of maize and rice, say, on the one hand, with groundnuts and soyabeans on the other. This kind of comparison is totally misleading, because grain legumes have commonly twice and sometimes even three times the protein content of cereals. In the case of the legume oilseeds one half to one quarter of the dry matter is lipid, which is more energy-rich than the carbohydrate of the cereals.

It is clearly apparent that what we are lacking most is the necessary skill in managing the symbiosis to the best advantage. This is well shown in the soyabean where this has received the greatest attention from agronomists. It is a reasonable question to ask, what are the limits imposed by symbiotic nitrogen fixation on the production of higher grain legume yields? Are we approaching, or have we actually reached, a yield plateau determined by a ceiling of *Rhizobium*-mediated nitrogen fixation? It is of interest to note that the highest groundnut yields recorded (Smartt, 1978*b*) were in fact achieved with natural nodulation. The gap between average yields and this maximum indicates that the level of management required for effective exploitation of *Rhizobium* symbiosis is not necessarily highly sophisticated. However, where endemic low yields persist in spite of efforts to improve them, the legume–*Rhizobium* relationship might well repay study.

The possibility that the soyabean–*Rhizobium* symbiotic output of fixed nitrogen might be the major limiting factor in restricting or inhibiting further increases in soyabean yield certainly merits investigation and it has received some attention. It should be borne in mind that other important possible limiting factors exist, which could impose ceilings on soyabean yields. The most likely is that of the canopy; compared with that of the groundnut, the soyabean canopy does not appear to be particularly effective in intercepting incident light energy (Duncan *et al.*, 1978). Other things being equal this could explain, in greater part, the differing yield capacities of groundnuts and soyabeans.

Returning to the question of the ability of symbiotic nitrogen fixation to support higher soyabean yields, if there were to be a shortfall of nutrient nitrogen how could this best be met? The studies carried out so far serve to highlight the complexity of the symbiosis. The review by Sprent (1979) is very illuminating in this respect. She notes apparent contradictions in the results obtained from different workers studying this problem. Hardy *et al.* (1971) concluded that nitrogen fixation in the soyabean is geared to demand and that the main period of nitrogen fixation activity is during pod fill; however, other workers claim that marked falls in nodule activity can occur during pod fill. Provided that there are nitrogen reserves in the plant, these can be mobilised and translocated to the sink of developing pods. The soyabean seems to be particularly efficient in this respect; leaves appear at pod maturity to be strongly depleted of reserves whereupon they wither and are shed. Sinclair and de Wit (1975) regard the soyabean as the most self-destructive of legumes. Certainly in this regard it can be considered as exhibiting the highest development of the annual life form; all resources that can be mobilised are mobilised and translocated to the seed. In this regard, at least, the soyabean can be considered to have a highly effective physiology. It appears that, during the pod-filling phase, uptake of nitrate from the soil does not occur. This is, however, possible at other stages of development until the onset of flowering and pod setting. There are also effects of combined soil nitrogen on the establishment of nodules, and on the nitrogen fixation process itself. Even here the effects are not straightforward; a period of nitrogen shortage may well occur after seed reserves are exhausted and before symbiotic fixation becomes effective. Adequate soil nitrogen status may be critical at this stage to maintain vigorous seedling growth and that of the root system and nodules in particular. Where soil nitrogen status is low, seedlings may develop a transient chlorosis at the late seedling stage, which rapidly disappears when effective nitrogen fixation is established.

The establishment of efficient nodulation depends on effective interactions between host and *Rhizobium* genotypes and the environment. This, as can be imagined, produces extremely complex genotype × genotype × environment interactions which are well exemplified in the data from *Vicia faba* of El-Sherbeeny *et al.* (1977*a*, *b*). The effects of environmental factors have been studied individually, and one of the interesting features to emerge is that the process may be most efficiently carried out in terms of the quantity of N fixed per unit of carbon respired (Halliday, 1976) at cooler (but not cold!) temperatures, even for a tropical legume such as the cowpea. The question of diurnal temperature range could be very important. The greater temperature range experienced on the Central African plateau may be a contributory factor in the achievement of the highest recorded groundnut yields there, rather than in the south-eastern USA, where the greatest breeding and agronomic efforts have been made.

While it is inappropriate to elaborate further on the crucial questions relating to rhizobial nitrogen fixation, this brief consideration does underline the initial contention that although much has been expected from grain legumes, little sustained real attention has been given to their particular and peculiar needs. The human race has benefitted more from the legumes than it had the right to expect or it probably deserved, judging by the resources devoted to legume research. It stands to benefit enormously in the future, if it is prepared to devote adequate energy and resources in the study of symbiotic nitrogen fixation in a much wider range of grain legumes and to develop sufficiently ingenious strategies in its exploitation. It is perhaps significant that the major studies have been carried out in Australia, New Zealand and North America on introduced legumes, where appropriate *Rhizobium* strains were absent and had to be introduced. Much broader studies are needed, especially on indigenous species and their associated *Rhizobium* strains, to resolve the perhaps even more complex problems of manipulating established *Rhizobium* populations.

The reviews of Sprent (1979) and those contained in Summerfield and Roberts (1985) are extremely valuable in providing background information on the legume–Rhizobium symbiosis. These are useful in deepening our appreciation of the role of legumes both in cropping cycles and during the restorative non-cropping phase.

Legumes in the context of farming systems

The patterns of agricultural thought which prevail in those of us whose formative years were spent in the 'developed' world are very largely conditioned by a certain permanency in patterns of land use. This may be as arable or pasture land where even short-term fallows are exceptional, and land tends to be kept in a revenue-earning condition continuously. This state of affairs is not self-sustaining and can only be maintained by a constant input of nutrients, largely in the form of artificial fertilisers. The economics of production in subsistence agriculture do not permit such a pattern of permanent agriculture. Virgin land, cleared of its natural vegetation, may only be able to support satisfactory production of crops for a few seasons. Sooner or later productivity falls, and it becomes more cost-effective in energy terms to clear new land than to persist in cultivating the old. The old land is usually allowed to revert to nature. In poorer soil areas it may be necessary to allow climax vegetation to develop, as in the *chitemene* agriculture of Zambia (Trapnell, 1953). Under a very extended fallow the physical condition of the soil and its nutrient status are to a greater or lesser degree restored. The role of wild legumes in this process is highly significant and especially that of the tree species, largely members of the genera *Brachystegia* and *Isoberlinia*, in semi-arid woodland areas of Africa.

Systems of shifting cultivation can maintain populations at an equilibrium level provided population pressures do not increase, which unfortunately happens all too frequently. Any attempt at improving agriculture in the developing world should take as its starting point the indigenous agricultural systems; these have usually stood the test of time which many of the 'schemes' which have supplanted them manifestly have not. The most effective strategy is likely to be one which exploits the advantages of indigenous farming systems, identifies the constraints and bottlenecks limiting production, and devises effective means to reduce or eliminate them. Many 'package' development schemes fail because of total unfamiliarity with the traditions and practices of the intended beneficiaries. These farmers find little or nothing they are attuned to in the proposed new systems and fail to understand or appreciate the operational needs, to the extent of knowing which are the essentials and which are not.

There are three main areas of concern in the improvement of agricultural productivity. The first is soil nutrient status: without adequate provision of nutrients, satisfactory production is not possible. The second aspect is soil physical structure: without a satisfactory soil structure, erosion hazards are increased and conditions may soon become unsuitable for crop growth. Finally there is biotic competition: there may be direct competition between crop plants and weeds; there may be competition between sinks, whether the dry matter production of the crop is exploited by man as intended, or whether it is exploited by pest and pathogen competitors.

The role of legumes is very important in the improvement of soil nutrient status in natural vegetation or fallows. It is no exaggeration to say that agriculture would be impossible in vast areas of Africa, for example, were it not for the restorative capacity of indigenous legume trees. In cultivation it is hoped that legumes perform something of a similar restorative function, in addition to producing exploitable crops. Legumes thus have a dual role to play in many subsistence agricultural systems, in helping to maintain nitrogen status of the soil during the cropping phase and, where fallowing or long recovery periods occur between cropping phases, to restore soil nitrogen status. The wild legumes can be trees, shrubs, lianas or herbaceous plants. Their role is presumed to be of great significance largely on the basis of their abundance in natural regrowth vegetation. Very little work has been published on this role, which would be a logical development of Trapnell's ecological studies in Central Africa (Trapnell, 1853; Trapnell and Clothier, 1957).

While a reasonable balance between population and land area is maintained, productivity of such ecosystems can be maintained; a diversity of crops can be produced, and a satisfactory diet provided. With increased population pressures, resting periods are reduced, productivity declines, diversity of crops is reduced, and the quality of diet suffers, with increasing reliance on crops such as cassava, which can maintain some production even from soils of poor nutrient status.

In cropping systems where a range of crops can be grown satisfactorily, legumes are usually to be found as a major supplement to carbohydrate staple crops. These may be seed crops (cereals and pseudo-cereals), or tuberous crops such as cassava, yams, sweet potatoes and potatoes. In the best subsistence farming systems a wide range of crops can be produced, often in mixed cropping systems. These systems have several advantages, in that the composition of the mixture and the proportions of the constituents can be varied according to the needs of the producer. These mixed cropping systems also tend to reduce pest and disease incidence; the species diversity contributes to this effect, which is also reinforced in many instances by intra-specific genetic diversity. This also tends to promote stability of production in what is ideally an equilibrium situation. The nutrient balance is maintained when those removed from the soil in crops are replaced by those returned to it together with those fixed in it and those released by weathering of soil minerals.

This equilibrium situation has been disturbed by commercialisation of farming. This often leads to the abandonment of traditional mixed cropping systems in favour of monocrops. Monocropping is by no means unknown in Third World traditional farming systems but this is usually restricted to production of small grains such as finger millet and rice (for obvious reasons!). Large-seeded crops such as the groundnut, produced in mixed stands for subsistence, may be grown in pure stands as a cash crop. This almost invariably exacerbates pest and disease problems; the nutrient balance in the soil tends to deteriorate unless corrective measures are taken. The cost of ameliorative inputs may be so high relative to the cash value of the crop that the enterprise may be of dubious value. A credit balance may only be achieved by skimping inputs and effectively engaging in soil mining.

It is at this juncture appropriate to consider (all too briefly) the pattern of grain legume production and consumption in the world as a whole. Certain changes are in train at the moment; in some areas of the world pulse production is under pressure but in others pulses are achieving a somewhat enhanced status. Pulses are regarded as a 'health food'; their high contents of dietary fibre and protein are especially appreciated. The increased cost of animal protein plus an increased popularity of vegetarianism have also encouraged interest in grain legumes. The use of vegetable protein from legumes in the production of meat substitutes (textured vegetable protein) and meat extenders can be expected to increase. The 'vanishing beefsteak' may eventually disappear and not be missed!

2.2. Geographical patterns of pulse production and consumption

Although there are pulse crops available that will grow in most environments in which arable agriculture can be carried out, the pattern of pulse production and consumption is very varied over the world as a whole. This variation can arise from two major causes, biotic and cultural. It is not always possible, however, to distinguish sharply between these two.

In general the greatest stimulus to the production and consumption of pulses is a shortage of animal protein and it is significant that in civilisations of ancient origin the greatest importance of pulses is achieved in those in which supplies of animal protein are or have been limiting. Such civilisations where pulses are a very important component of the diet are to be found in both the Old and the New Worlds, although the underlying causes of animal protein shortage are different. In the New World animal protein was in short supply in pre-Columbian times because very few animals were domesticated, apart from the indigenous cameloids and the guinea pig. This restricted the sources of animal protein largely to game and fish; the most dependable supply of protein throughout the year, therefore, was in all probability protein in the seeds of cereals, cucurbits and legumes. That civilisations can develop and be maintained largely on plant proteins has been shown by the development of the Aztec and Mayan civilisations in which no animals were domesticated for their meat, and the main protein sources for the population as a whole were the staple grain and pulse crops. In the Old World the situation is different and in the ancient civilisations of India, China, Japan and the Far East in general, a different combination of factors has produced a comparable situation to that of the New World. In the Old World it seems that a pastoral phase either pre-existed or co-existed with the early phases of arable agriculture, and so domesticated animals which could be used both as beasts of burden and sources of food (meat and milk) have been a long established feature of agricultural systems.

With the establishment of ancient dynasties and effective maintenance of settled or comparatively settled conditions over periods of several millennia, populations expanded and imposed a high pressure on the productive capacity of the land. This in turn necessitated an efficient use of land in meeting the protein requirements of the population. It is well known that the intermediate processing of plant protein by an animal which is in turn consumed by humans is, in terms of energy use, wasteful. In stable societies which support dense populations, animals are used to exploit only those plant protein resources which are not directly utilisable by man. The use of pigs and poultry, which subsist largely on food they find and scavenge, is typical of this type of agriculture, as contrasted with stock fed protein concentrates in current Western agricultural practice. In India we have a special situation, where religious beliefs accentuate artificially the shortage of animal protein. The veneration of the cow in the Hindu religion as the fount of many benefits – a mother god figure – means that the cow is the source only of milk and dairy products, not of meat. The excessive and uncontrolled populations of sacred cattle probably mean that their food consumption is much more used in maintenance than in production. Productivity of cattle would be higher if the available food supply was used by a smaller selected population.

It is probably true, therefore, to conclude that high production of

Table 2.1. *Supplies of legumes in 34 countries*

Figures are grams per head per day.

Country	Quantity	Country	Quantity	Country	Quantity	Country	Quantity
Germany (West)		Canada		Italy	16	Greece	29
Sweden		Denmark		Israel	16	Ecuador	36
New Zealand		France	8–13	Ceylon	16	Honduras	36
Uruguay		United Kingdom		United States	16	Paraguay	44
Belgium/Luxembourg		South Africa		Portugal	19	Japan	50
Finland				Chile	22	Mexico	51
Switzerland	3–7			Egypt	26	Brazil	68
Ireland				Turkey	27	India	71
Netherlands							
Norway							
Argentina							
Austria							
Australia							

FAO food balance sheet figures from Aykroyd and Doughty (1964). A more detailed breakdown is given in Aykroyd and Doughty (1982).

pulses is characteristic of areas which have supported an ancient civilisation and where, locally at any rate, high population densities have been maintained; population densities of such as order as to have outstripped local supplies of animal protein from domesticated and wild animal sources. Bearing this in mind we can consider in broad outline the pattern of production and use of pulses in the world today. We must also take account of the fact that there is a certain cultural inertia. People's food habits, built up gradually over millennia to meet particular nutritional situations, may well persist after the nutritional situation has changed. Pulses, for example, are widely used in Latin America today even though the Old World domesticated cattle and sheep thrive there. This can be traced back at least in part to the cultural heritage of the Inca, Maya and Aztec upon which a Latin culture was superimposed. In parts of Latin America where there was no fusion of indigenous and European culture, where in fact a European culture replaced the native culture, the pattern of pulse usage is similar to that of a typical Western European community; the best example of this is probably Argentina.

Pattern of pulse consumption as shown by diet surveys

Diet survey data compiled by FAO and presented in Aykroyd and Doughty (1964) give an overall idea of the pattern of legume supply and consumption. Although these figures were compiled over twenty years ago the pattern of consumption is still broadly similar.

These data give us the best available indication of the relative importance of pulses in the diets of different people. The ratio of supplies per head between extremes is 23:1; this gives some measure of the potential role of legumes in meeting the protein needs of the world's population in a situation which might develop in the future where animal protein becomes scarcer and less readily available. A brief survey of the areas mentioned will indicate differences in patterns of use and of the different species employed. Just as there is a cultural inertia in maintaining high- or low-pulse diets, so there is a parallel inertia in keeping to those species which have been hallowed by long and effective use and have become entrenched in a local culture.

The New World

The consumption of pulses in the New World ranges from very low orders in Argentina and Uruguay (3–7 g per head per day) through moderate levels in Chile (22 g), Ecuador and Honduras (36 g) to relatively high levels in Paraguay (44 g), Mexico (51 g) and to a very high level in Brazil (68 g).

In Mexico the most important pulse crop is undoubtedly *Phaseolus vulgaris*, as it is in the Central American republics of Costa Rica, El Salvador, Guatemala and Panama. In the cooler upland areas the indigenous *Phaseolus coccineus* can be grown, but the introduced Old World pulse, the faba bean (*Vicia faba*) is of perhaps equal or greater importance. Other introduced legumes are also grown: peas, lentils and chickpeas. The species *Phaseolus lunatus* and *P. acutifolius* are also to be found in cultivation; the latter species is, however, tending to go out of cultivation. The groundnut is also grown. The actual consumption of *Phaseolus vulgaris*, the most important pulse, shows a mean of 42 g daily with a range of 11–70 g.

A generally similar pattern of pulse use is found in North American Indian communities; the predominant bean is *P. vulgaris* but other pulses, particularly *P. lunatus*, are also eaten. Very high consumption of pulses (*ca.* 70 g daily) are common in the Red Indian communities of Arizona and New Mexico. Seen in the context of low availability and consumption of animal protein and the general diet of maize, cucurbits and pulses, the latter have an extremely important role in local nutrition.

In South America the general pattern of high pulse dependence is maintained with two important exceptions, Uruguay and Argentina. In these two countries principally, and locally elsewhere, beef cattle are produced in large numbers and protein requirements are met from animal sources, because these have been abundant and relatively cheap. Levels of pulse consumption are consequently among the lowest to be found anywhere. In contrast the level of consumption of pulses in Brazil is one of the highest in the world. The most important pulse crop is undoubtedly the kidney bean *P. vulgaris*; this is supplemented by the related American species *P. lunatus* and the introduced pulses, the chickpea (*Cicer arietinum*) and the cowpea (*Vigna unguiculata*). The groundnut (*Arachis *hypogaea*) is also widely grown as a cash crop. Consumption is high everywhere in the country and it is of interest that pulses supplement all three major carbohydrate staple crops, rice in the south, maize in the central portion and cassava in the north. In all probability supplementation is reasonably effective with the rice and maize diets. It is doubtful if beans could supplement efficiently a cassava diet without the addition of some animal protein, such as fish, or plant protein with a complementary amino acid profile.

In Venezuela, consumption of pulses is high. There is an apparent direct relationship between social status and consumption of beans, ranging from an average of 70 g daily among the poorer people to about 50 g among the better off. This represents a very high level of consumption throughout Venezuelan society. *Phaseolus* beans predominate among the pulses, especially *P. vulgaris*. In Colombia, levels of consumption are less and to a large extent are inversely correlated with social status and the consumption of other protein-rich foods, fish, meat and milk. Consumption of both dried and fresh legumes is relatively high in Ecuador. Reported consumptions of dry legumes ranging between 3 and 42 g daily were recorded in one survey; coupled with this were consumptions of up to 140 g daily per head of fresh legumes in areas of high dry pulse consumption. An interesting observation was made in Paraguay that consumption of pulses was higher in rural than in urban areas; in the latter, consumption fell to the order of one half or one quarter of the level of consumption in the country (*ca.* 66 g per head daily). A salient feature of consumption of pulses in Peru is that there is considerable seasonal variation in the amount of pulses consumed daily, ranging from 20 g in February to 70 g in July, a reflection of their availability. In addition, it was pointed out that where fish is freely available lower pulse consumption was noted. As in most of South America, the common bean is the most widespread pulse with the lima bean, faba bean and cowpea in subordinate positions.

It would be of great interest to have more detailed surveys of the pattern of utilisation of pulses in Latin America. A number of very important points would yield useful information if investigated more thoroughly; among them are the effects of the following on the consumption of pulses.

1 Age: how soon pulses and pulse preparations are fed to infants and in what quantities, and how much is fed to children and adolescents.
2 Social class: to what extent the amount of pulse foods taken is affected by the ability to afford more expensive protein-rich foods.

3 The nature of the community: to what extent differences in the amounts of pulses used can be related to the rural or urban situation of communities.
4 Alternative protein sources: to what extent protein intake is self-regulating, i.e. if animal protein sources are available are these in fact used deliberately in replacement of pulse plant proteins, or whether food of animal origin replaces plant material on a basis of calorie equivalence.
5 Seasonal variation in pulse intake: it is reasonable to expect that some variation in the intake of pulses over the year may occur, and perhaps lead to protein imbalance in the diet in the lean season of the year.

Surveys have usually been undertaken independently with different objectives in view and employing different sampling techniques and thus it is difficult to produce an objective picture of how, for example, the effectiveness of pulse utilisation varies from place to place. It is very difficult indeed, particularly in areas of subsistence agriculture, to assess accurately the population of a given large area and the volumes of both production and consumption. Taken all in all, surveys carried out to date give a good indication of the range of variation one is likely to encounter and aspects which obviously would repay further study. The agriculturalist and nutrition worker in the meantime must do what they can with the relatively incomplete and fragmentary evidence available. The situation in the more sophisticated societies is much better and more reliably documented. Much more of the volume of agricultural production passes through trade channels and is recorded by appropriate Government agencies. We can therefore in the case of North America and western Europe take the figures given as accurately reflecting a moderate or a low level of pulse consumption. It is interesting to note that in the United States the level of consumption is high for a developed society. The settlers, after encountering *Phaseolus* beans, took up their culture and both *P. vulgaris* and *P. lunatus* are consumed in quantity as dry beans or fresh. The lower level of consumption in Canada is perhaps a reflection of the less suitable climate for the production of *Phaseolus* beans.

The position of pulses in the Caribbean islands is perhaps rather intermediate between the situation found in the contrasted mainland communities of high (e.g. Brazil) and low (e.g. Argentina) dependence on pulses. They resemble the low-dependence areas sociologically in that the indigenous Amerindian culture has been swept away and a very polyglot society has developed with culture and traditions of no great antiquity. Nevertheless a wide range of pulses is grown and consumed which help to supplement diets based on maize, rice and imported wheat. These include common beans, lima beans, cowpeas, pigeonpeas and the groundnut.

The Old World

The Indian sub-continent is probably the area of greatest dependence on pulses in the Old World, if not in the world as a whole. This arises from both high population pressure on the land and the religious prohibition on the slaughter of cattle for beef among the Hindus, together with a prohibition of meat consumption of any kind for orthodox Hindus. While the overall figure of consumption is high (71 g per head per day), this average derives from a wide range of actual consumption figures. Some groups sampled ate no legumes at all, whereas in others consumption exceeded 85 g daily; the general level most frequently observed was in the range 45–70 g. The pattern of consumption also showed geographical variation; consumption in the South Indian states of Madras and Kerala was of the order of 26 g daily, whereas in North India consumptions were of the order of 240 g daily by labourers engaged in heavy physical work.

There seems to be a direct relationship in some cases in India between social status and the level of pulse consumption. This is so in Bombay City and a part of the State; however, steel-workers in Bihar, regardless of their wage levels, consumed much the same quantity of pulses. There were some areas of Bombay State where milk and milk products were relatively abundant; in these there was some evidence that these animal products replaced pulses, particularly among the well-to-do. The pulses grown in India are predominantly those of the Old World; there has been only a very slow acceptance of the New World pulses *P. vulgaris* and *P. lunatus*, although the New World groundnut has been readily accepted. The pulses actually grown extensively are green gram (*Vigna radiata*), black gram (*V. mungo*), chickpea (*Cicer arietinum*), pigeonpea (*Cajanus cajan*), pea (*Pisum sativum*), lentil (*Lens culinaris*) and khesari dhal (*Lathyrus sativus*).

The position regarding utilisation of pulses is probably little different in Pakistan and India, in spite of the availability of beef in Pakistan. There is, however, a great contrast in the amounts of pulses utilised in the neighbouring states, Burma, Thailand and Malaysia. Surveys made before World War II in Burma found consumption levels of pulses of about 14 g per head per day. In Thailand figures of a similarly low order are reported for 1960. In Malaysia consumption of pulses is low, the predominant legume being green gram; the prime protein sources for these areas of Malaysia are pork, beef, poultry and fish, both fresh and dried. In Indonesia legumes are more extensively used than in Malaysia. The soyabean used in making tempe, soy sauce and curd predominates; green gram is grown extensively and the groundnut is also used.

The Far East (China and Japan) is an area of high pulse utilisation. There are no recent data for consumption or production of pulses on the Chinese mainland but an unofficial estimate of legume supplies is 42 g per head per day, of which 18 g is from the soyabean.

The position in Japan is of considerable interest as it is a rare example of a technologically advanced society with a high level of pulse utilisation. The figure of daily use per head (50 g) is about the same as that of Mexico. A further feature of interest is that the use made of pulses is, if anything, rather higher in urban society than in the country; this is due to harnessing of advanced Japanese technology to the processing and manufacture of highly acceptable and palatable soyabean products. Perhaps it is an augury of the shape of things to come in other societies that the predominant pulse in Japan is the soyabean. Other pulses which are cultivated occupy a relatively minor position in comparison; these are kidney beans and groundnuts. It was shown, for example, in a survey of a population in which the average pulse intake was found to be 70 g daily, that 64 g of this was soyabean in various forms, largely processed. There are of course indications that, by the use of spun protein fibres, synthetic meats can be produced and that these can replace the natural products satifactorily provided that appropriate nutritional factors are added where necessary.

In Australasia, in both advanced and primitive societies, the use made of pulses is very low. Australia and New Zealand have a pattern of pulse use closely similar to that of European countries where these are least used, a reflection perhaps of the relative abundance of meat and dairy products. In New Guinea and the Pacific Islands the use of pulses is low and there is apparently scope for extending their use. Staple crops in this area are frequently starchy roots and fruit (e.g. taro and breadfruit), which might not be effectively supplemented with pulses alone, but together with fish and/or the use of cereals some dietary improvement should be possible. There is very little sound information available from this area on pulse cultivation and utilisation.

Africa, as far as we are concerned, can be considered in two parts, the area South of the Sahara (so-called 'black' Africa) and what may be called North, or Mediterranean, Africa. The pattern of pulse utilisation is reasonably consistent in the North African and Mediterranean area generally and is characterised by moderately intensive use of pulses. In Africa south of the Sahara, the use made of legumes varies enormously from exceedingly intensive use made of the common bean in Eastern Zaire, adjacent parts of Uganda and Rwanda and Burundi, where *P. vulgaris* achieves the rare status of a staple crop, to very low levels elsewhere. In the staple pulse-growing area of Uganda, consumption figures of 100–150 g per head per day are reported. The intake of beans is so large, in fact, that 65% of calorie intake may be in the form of pulses and only 35% as starchy roots, fruits (plantains) and grains. These areas are well watered, on rich volcanic soils with a mild climate due to high elevation; as they are on the equator, 3 or 4 crops can be produced annually. They are certainly exceptional as far as Africa is concerned, but they do show that legumes can fulfil a much more important role than they do at the moment.

The indigenous pulse of Africa is the cowpea. It is widely grown and consumed throughout much of Africa, and it is well suited to some of the relatively poor porous soils which are found there. It seems to be less suitable on heavier soils, particularly under wet conditions where it is a prey to various pathogens such as *Ascochyta phaseolorum*. Africa is one area of the world where it does seem that the consumption of pulses is limited by the ability to produce them. Africans, however, do have, like the Indians, certain traditions which influence their attitude to introduced legumes. The American *Phaseolus* species are readily accepted by the Africans because of their general similarity in form and culinary characteristics to the cowpea. *Phaseolus vulgaris* is a perfectly acceptable pulse in much of Africa, particularly Central and Southern Africa, where it may be preferred to the indigenous cowpea as it stores better. Kidney beans are less attractive to weevils than are cowpeas, and this fact is of some importance where clean storage is not easy. The problem in Central Africa can be summed up by saying that the difficulty is not in growing pulses, but is growing in quantity pulses acceptable to the population. Pulses which can be grown reasonably well in Central Africa are the cowpea (*Vigna unguiculata*), the kidney bean (*Phaseolus vulgaris*), the lima bean (*P. lunatus*), white-seeded dwarf varieties of *P. coccineus*, pigeonpeas (*Cajanus cajan*), soyabean (*Glycine max*) and groundbean (*Vigna subterranea*). The garden pea (*Pisum sativum*) is a cool-season plant; the broad bean might also grow in the cool season under irrigation; the important pulses chickpea (*Cicer arietinum*) and lentil (*Lens culinaris*) are unsuitable for normal field cultivation, but might be useful under dry season irrigation. Green (*V. radiata*) and black (*V. mungo*) grams can be grown satisfactorily but yields are low. Of all these pulses only cowpeas, common beans, perhaps the lima beans and groundbeans are readily acceptable, in addition to the groundnut (*Arachis hypogaea*). Attempts to introduce the soyabean and pigeonpea to the diet have been disappointing in the extreme. Both crops are exceedingly well adapted to conditions in much of Africa, yield well and are agronomically desirable. All this counts for nothing, however, in view of the lack of acceptance. Efforts by Europeans to persuade Africans to eat soyabeans in one form or another inevitably fail when it is appreciated that the Europeans themselves eat them in negligible amounts. In Africa pulses are not prestige foods, and so exotic pulses are much more difficult to introduce than exotic staples, such as wheaten bread, to which some prestige is attached.

There is a great nutritional role in Africa to be fulfilled by the legumes; we have here a situation in which animal protein supplies are being reduced with the progressive extermination of game animals. This is reinforced by the difficulty of extending the production of cattle because of trypanosomiasis (although eventual if inadvertent extermination of game may make this possible) and the inefficient use made of other protein sources, for example fish.

The levels quoted for consumption of legumes per head per day in Africa range from 0 to 150 g, with an average level of about 40 g. This is close to the level reported in 1953 for Southern Rhodesia (Zimbabwe) but considerably above that for South Africa (13 g). Some of these figures are on the generous side; for myself I believe a level of consumption at about the South African level is probably much nearer the mark over much of Africa than the Rhodesian figure. The figures quoted by the excellent FAO publication *Legumes in human nutrition* are probably very generous over-estimates, in some cases, of actual consumption. Much depends on the timing of the sampling: after harvest pulse consumption may be high but may sink to negligible levels after a few months. Be this as it may, there is enormous scope in Africa south of the Sahara for expansion in production and consumption of pulses.

It is reasonable to consider along with the remainder of Africa the rest of the Mediterranean region and Europe. Throughout the region consumption levels are moderate; in North Africa the most popular pulse is probably the broad bean (*Vicia faba*) but others are grown and consumed to a significant extent, e.g. chickpeas, lentils and fenugreek (*Trigonella foenum-graecum*). The latter is highly aromatic and used mostly for flavouring, but in Ethiopia it is consumed traditionally by nursing mothers, who eat larger quantities of pulses than is otherwise customary, to maintain the supply of breast milk.

In the North Mediterranean consumption of pulses is relatively high by European standards, but apparently as the standard of living improves the level of consumption of pulses declines. This has been marked in Italy over the past four decades. Pulses apparently suffer from the classical stigma of being 'poor man's meat'. The pulses grown in these European Mediterranean countries include broad beans, peas and legumes characteristic of the area but in addition the introduced American *Phaseolus vulgaris* is much more important than in North Africa. Pulses decline in importance from the Mediterranean northwards in Europe until in Scandinavia they are of only very minor significance. In these areas animal proteins are relatively more abundant, and population pressures on land have been less heavy than in the Mediterranean, the cradle of classical European civilisation. It remains to be seen whether pulses in Europe as a whole and in the world at large will come to enjoy a new vogue if, as might be anticipated, animal protein continues to become scarce and more expensive while the industrial processing of pulses enables attractive meat substitutes to be produced. The current vogue for pulses as 'health foods' and burgeoning anti-meat sentiments could provide stimulus for production of pulses. The Mediterranean area could, for example, produce more soyabeans than at present if a stable market were established. Other pulses might come to be processed and more use might be made of blended cereal–pulse protein mixtures. It would not be realistic to expect pulses alone to save the day as regards future protein supplies for the human race. The co-ordinated and integrated use of pulses with cereals and oilseed meals (cotton and sesame for example) in an industrial context could amply meet the need for balanced plant protein for some long time to come.

2.3. Utilisation of edible grain legumes

Pods and seeds of edible species can be utilised in a variety of ways and the actual nutritional value may be influenced very markedly by the way in which they are in fact used. There are six general ways in which they can be consumed.

These are:

1 mature dry seed;
2 green mature and immature seed;
3 immature pods;
4 germinated seed;
5 fermented products;
6 extracted seed proteins.

Mature dry seed

Although reliable figures are not available, it seems highly probable that the bulk of pulses are consumed in this form. The dried seed is storable and if it can be effectively protected against the common storage pests it is an efficient way of ensuring continuity of food supplies from one harvest to the next. Nutritional values does in fact decline gradually in storage, and tenderness ('cookability') of the seed also declines. Viability of the seed in good storage conditions is maintained well except for the leguminous oilseeds (soyabean and groundnut) which retain viability in ambient temperature storage for only a single or relatively few seasons.

The seed may be used whole, as are haricot beans, or split and used as a dhal, as the pigeonpea and garden pea commonly are.

Green mature and immature seed

Pulses, which can be used fully mature and dry, can also be used when mature but before normal drying out occurs. The soyabean can also be used easily at this stage, although when fully mature it cannot be used efficiently as a pulse. This is one of the main uses of the garden pea, which may be used either mature or immature. The latter use is becoming popular in peas grown for freezing as these tend to be sweet, tender and with the best flavour. Lima beans can also be used at the green mature stage, often mixed with sweet corn in succotash. Groundnuts and groundbeans are both consumed as green mature seeds, boiled in the pod.

Immature pods

Pulses which produce pods that remain fleshy for two or three weeks after setting can be picked green and used as a vegetable. This is one of the common ways in which *Phaseolus vulgaris* and *P. coccineus* are used in western Europe and elsewhere. It is possible to exploit some species (for example of *Canavalia*), whose seeds are not used as pulses, in this way also.

Nutritionally immature fruits have quite a different value from that of the mature seed. The protein content is lower but they are relatively richer in vitamins and soluble carbohydrates.

Germinated seed

The practice of using germinated seed and young seedlings of pulses as a fresh vegetable is widespread in the Orient. The storage of dried seed, and their sprouting as required, enables a continuous supply of a fresh vegetable to be produced at times when they cannot be produced in the field. The vitamin C content rises from negligible levels in the seed to 12 mg per 100 g after 48 h germination. Riboflavin and niacin contents increase significantly and insoluble carbohydrates are also mobilised. Germinated seed and young seedlings have been used in sailing ships and more recently in prisoner of war camps as a source of vitamin C for the prevention of scurvy. Although green gram is the most widely used species in producing bean sprouts, the soyabean, broad bean, chickpea and other species have also been used in this way. The mobilisation of cotyledonary reserves during germination improves all-round digestibility, conspicuously so in the soyabean, whose sprouts may be used as a salad or a vegetable.

Fermented products

In south-east Asia, as Stanton (1969) observed, an indigenous food technology of remarkable sophistication has developed. A range of micro-organisms are commonly employed in the fermentation of both plant and animal materials. The fermentation of soyabeans is of particular interest because both digestibility and palatability are improved. Although fermentation in Indonesia is carried on as a cottage industry, it has been developed on an industrial scale in Japan (Nicholls *et al.*, 1961). Typical products are soy sauce, soyabean paste, tempe and natto. The fermenting organisms include *Aspergillus oryzae*, *Rhizopus oryzae* and *Bacillus subtilis*. These have the very useful capacity of breaking down the toxic materials (lectins) and anti-metabolites (protease inhibitors) which make direct utilisation of the soyabean so very difficult.

Extracted pulse proteins

A separation of readily digestible protein from that which is less readily assimilated is a common practice in the Far East. The two products most commonly produced are soyabean curd ('tofu') and soyabean milk. Soyabean protein extracts can be used as dairy product substitutes and are useful alternatives to animal milk for those allergic to cow's milk, for example. Soyabean milk can also be fermented to produce a cheese-like product.

Useful proteins can also be extracted from soyabean meal residues, after oil extraction, to produce textured vegetable protein (TVP) and to enrich cereal products. They are also used as meat extenders and in the preparation of low-fat cream substitutes.

A combination of fermentation and protein extraction technology could perhaps be usefully extended to other species of legume which produce potentially valuable protein but the use of which is inhibited by the presence of toxic materials, alkaloids and essential amino acid analogues, for example.

1.4. The future for legumes as protein

Since the publication of the joint FAO–WHO report (1973) in which estimates of minimum protein requirements for adequate nutrition were drastically revised downwards, there has been a tendency to consider that it is sufficient to ensure adequacy of carbohydrate food supplies and that proteins will take care of themselves. Although this may generally be true there are undoubtedly exceptions. However, as a result, protein content and even quality has become much less of a burning issue than it was prior to 1973 (Payne, 1978). None the less the quality of life can in part be seen in terms of the protein content of the diet. Proteins contribute greatly to taste and texture of food and the organoleptic aspect cannot be ignored. It would be a poor existence if the carbohydrate staple foods were not seasoned with something of more gastronomic interest, to make eating something more than just a biological necessity.

3 The groundnut, *Arachis hypogaea* L.

The major stimulus to detailed studies of the biosystematics of the genus *Arachis*, to which the groundnut belongs, came directly from practical interest in the crop. This work, initiated in the United States and Argentina, is associated with the names of Gregory, Krapovickas and their co-workers. Although it has not yet reached a definitive stage, sufficient has been published to provide a sound and effective biosystematic framework within which to consider the origin, evolution and germplasm resources of the crop. Indeed, few legumes have been more effectively investigated by what is a surprisingly small number of individual workers. Data and observations from a variety of sources have been integrated to produce a workable taxonomic scheme for the groundnut and its relatives. Morphological evidence is of course paramount, but this has been very effectively supplemented by studies of experimental hybridisation, comparative cytology, cytogenetics and biochemistry. Arguably this has been more effective for the groundnut and its allies than for any other grain legume; as a result, a very satisfactory taxonomic synthesis is emerging.

3.1. Biosystematics of *Arachis*

The present state of *Arachis* taxonomy

The genus *Arachis* is morphologically well defined and clearly delimited from its closest relatives by the development of a peg and its geocarpy. *Arachis* is placed, with its relatives *Stylosanthes*, *Chapmannia*, *Arthrocarpum* and *Pachecoa*, in the subtribe Stylosanthinae of the tribe Aeschynomenae, on the basis of the shared morphological characters of a staminal tube with anthers attached alternately basally and dorsally, with flowers in terminal or axillary spikes or small heads (sometimes raceme-like), leaves pinnate, leaflets few, without stipels (*vide* Taubert, 1894).

Although no recent monograph of the genus has been published, publications of Krapovickas and Gregory (Gregory *et al.*, 1973; Krapovickas, 1973; Gregory *et al.*, 1980) have outlined a taxonomic scheme which provides a useful basis for biosystematic discussion. The problems and difficulties in producing a satisfactory classification of the genus have been fully discussed by Gregory *et al.* (1973) and the following is a brief summary of their views and conclusions. Prior to Bentham's (1841) description of five wild species, *A. glabrata*, *A. pusilla*, *A. villosa*, *A. prostrata* and *A. tuberosa*, the only member of the genus known to science was *A. hypogaea*, described by Linnaeus (1753). Although 23 species of the genus have been described and diagnoses published, it

Fig. 3.1. *Arachis hypogaea*, the groundnut.

seems probable that at least an equal number remains to be described. Recognised species are: *A. hypogaea* L. (1753), *A. villosa* Benth. (1841), *A. tuberosa* Benth. (1841), *A. glabrata* Benth. (1841), *A. prostrata* Benth. (1841), *A. pusilla* Benth. (1841), *A. marginata* Gard. (1842), *A. hagenbeckii* Harms (1898), *A. paraguariensis* Chod. *et* Hassl. (1904), *A. guaranitica* Chod. *et* Hassl. (1904), *A. diogoi* Hoehne (1919), *A. nambyquarae* Hoehne (1922), *A. angustifolia* (Chod. *et* Hassl.) Killip (1940), *A. villosulicarpa* Hoehne (1944), *A. lutescens* Krap. *et* Rig. (1957), *A. helodes* (Martius) Krap. *et* Rig. (1957) (material of this species was collected by Martius in 1839), *A monticola* Krap. *et* Rig. (1957), *A. burkartii* Handro (1958), *A. martii* Handro (1958), *A. repens* Handro (1958), *A. rigonii* Krap. *et* Greg. (1960), *A. batizocoi* Krap. *et* Greg. (in Gregory *et al.*, 1980).

Although the status of most of the validly described species is unquestioned, it is doubtful whether *A. nambyquarae* should be regarded as anything other than a form of *A. hypogaea*. The status of *A. monticola* as a distinct species can also be questioned; if it is in reality a wild form conspecific with *A. hypogaea*, as breeding experiments suggest (Hammons, 1970), then it could be more correctly regarded as a sub-species of *A. hypogaea*.

Chevalier (1933 *et seq.*), Hoehne (1940), and Hermann (1954) have all published monographs of the genus which Gregory *et al.* (1980) considered to be unsatisfactory, largely because of deficiencies in herbarium material which had been collected prior to about 1950. It was not until fresh material, of entire plants of a wide range of species, was collected from type localities and other areas of South America that it became possible for Krapovickas and Gregory to propose taxonomic subdivisions of the genus. An outline of their scheme is reproduced below (Table 3.1). However, the scheme has not been validly published according to the International Code of Botanical Nomenclature and therefore all sub-generic epithets are *nomina nuda* (Resslar, 1980). These considerations notwithstanding, the scheme is workable and of considerable practical value.

The unsatisfactory nature of taxonomic schemes advanced prior to the publication of the views of Krapovickas and Gregory is illustrated by the treatment accorded to the genus by successive monographers. Chevalier (1933 *et seq.*) listed eight species, although descriptions of eleven were validly published at the time. Hoehne (1940) recognised eleven species, while Hermann (1954) reduced the number to nine, although thirteen valid descriptions had been published of which only one (*A. nambyquarae* Hoehne) would be questioned now. The present tally of good species, validly described, is 23 (as already noted); satisfactorily distinct but undescribed forms comprise another eleven species. It is a matter of conjecture as to how many more species will be decribed from the mass of materials collected and listed by Gregory *et al.* (1973) and that which has been collected subsequently (Gregory *et al.*, 1980).

The morphological species concept in *Arachis*

As has been noted, the unsatisfactory quality of much *Arachis* material deposited in the major herbaria of the world has been a stumbling block in developing a sound morphological basis for species recognition in *Arachis*. A problem of interpretation, arising from the sparse and incomplete herbarium material, derives from the fact that strong morphological convergence has occurred in the aerial vegetative parts of taxa which are not very closely related. For example there is strong morphological resemblance between *A. hagenbeckii*, *A. chacoense* and some erectoid

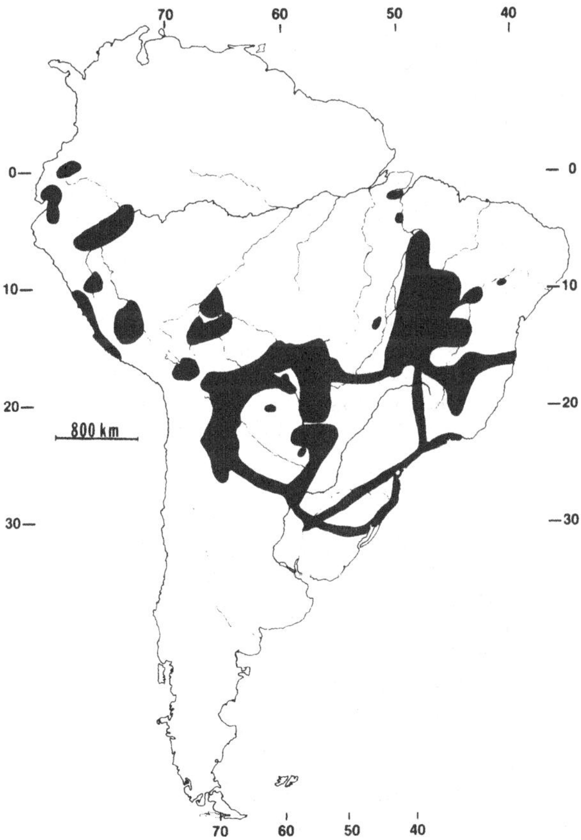

Fig. 3.2. Areas of *Arachis* germplasm collection, 1976–1983 (from Simpson, 1984).

Table 3.1. *Taxonomic sub-division of the genus* Arachis *(after Gregory* et al., *1980, and P. M. Resslar, 1980)*

Section *Arachis nom. nud.*

Plant tap-rooted with vertical pegs, flowers without red veins on back of standard.

Series *Annuae* Krap. *et* Greg. *nom. nud.* flowers medium to small, standard 14 mm wide × 12 mm high; short-lived, usually annual; $2n = 2x = 20$.

1. *A. batizocoi* Krap. *et* Greg.
2. *A. duranensis* Krap. *et* Greg. *nom. nud.*
3. *A. spegazzinii* Greg. *et* Greg. *nom. nud.*
4. *A. stenosperma* Greg. *et* Greg. *nom. nud.*
5. *A. ipaensis* Greg. *et* Greg. *nom. nud.*

Series *Perennes* Krap. *et* Greg. *nom. nud.* flowers medium to large, standard 14 mm wide × 12 mm high; perennial; $2n = 2x = 20$.

6. *A. helodes* Martius *ex* Krap. *et* Rig.

7a. *A. villosa* Benth. var. *villosa*

7b. *A. villosa* var. *correntina* Burk.
(*A. correntina* (Burk.) Krap. *et* Greg. *nom. nud.*)

8. *A. diogoi* Hoehne
9. *A. cardenasii* Krap. *et* Greg. *nom. nud.*
10. *A. chacoense* Krap. *et* Greg. *nom. nud.*

Series *Amphiploides* Krap. *et* Greg. *nom. nud.* flowers small to large, standard 10–21 mm wide × 8–14 mm high; short-lived; $2n = 4x = 40$.

11. *A. hypogaea* L.
12. *A monticola* Krap. *et* Rig.
13. *A.* × *batizogaea* Krap. *et* Fern. (of experimental hybrid origin).

Section *Erectoides* Krap. *et* Greg. *nom. nud.*

Plants tap-rooted or with thickened lomentiform roots or with tuberiform hypocotyl; plants erect or prostrate; pegs horizontal or nearly so, flowers medium to large, 16–24 mm × 12–20 mm; $2n = 2x = 20$.

Series *Trifoliolatae* Krap. *et* Greg. *nom. nud.* hypocotyl tuberiform; leaves trifoliolate.

14. *A. guaranitica* Chod. *et* Hassl.
15. *A. tuberosa* Benth.

Series *Tetrafoliolatae* Krap. *et* Greg. *nom. nud.* plants erect or prostrate; hypocotyls not tuberiform; leaves tetrafoliolate; standard orange.

16. *A. benthamii* Handro
17. *A. martii* Handro
18. *A. paraguariensis* Chod. *et* Hassl.
19. *A. oteroi* Krap. *et* Greg. *nom. nud.*

Series *Procumbensae* Krap. *et* Greg. *nom. nud.* plants prostrate; standard yellow.

20. *A. rigonii* Krap. *et* Greg.
21. *A. lignosa* (Chod. et Hassl.) Krap. *et* Greg. *nom. nud.*

Section *Caulorrhizae* Krap. *et* Greg. *nom. nud.*

Plants with hollow stems, rooting at nodes; pegs vertical, standard yellow; $2n = 2x = 20$.

Table 3.1 (*cont.*)

22. *A. repens* Handro
23. *A. pintoi* Krap. *et* Greg. *nom. nud.*

Section *Rhizomatosae* Krap. *et* Greg. *nom. nud.*
Plants rhizomatous, solid stems; flowers large.
Series *Prorhizomatosae* Krap. *et* Greg. *nom. nud.* plants delicate; flowers large, red veins on both faces of standard; $2n = 2x = 20$.
24. *A. burkartii* Handro

Series *Eurhizomatosae* Krap. *et* Greg. *nom. nud.* plants usually robust; flowers large, without red veins on back of standard; $2n = 4x = 40$.
25. *A. glabrata* Benth.
26. *A. hagenbeckii* Harms

Section *Extranervosae* Krap. *et* Greg. *nom. nud.*
Plants with thickened lomentiform tuberoid roots; pegs vertical, sometimes producing adventitious roots; flowers small to medium, with red veins on back; $2n = 2x = 20$.
27. *A. marginata* Gard.
28. *A. lutescens* Krap. *et* Rig.
29. *A. villosulicarpa* Hoehne
30. *A. macedoi* Krap. *et* Greg. *nom. nud.*
31. *A. prostrata* Benth.

Section *Ambinervosae* Krap. *et* Greg. *nom. nud.*
Plants tap-rooted; pegs vertical; flowers very small, 8 mm to 6 mm, standard with red veins on front and back; $2n = 2x = 20$.
No species names, valid or invalid, have been given to forms in this section.

Section *Triseminalae* Krap. *et* Greg. *nom. nud.*
Plants tap-rooted; pegs horizontal; flowers small, 10–12 mm × 8–10 mm, purple mark inside orange standard; fruits often three-segmented; $2n = 2x = 20$.
32. *A. pusilla* Benth.

forms, erroneously identified as *A. diogoi*. Similar close resemblances are apparent between *A. pusilla* and *A. duranensis*; *A. rigonii* and *A. cardenasii*; *A. lignosa* and *A. helodes*. It is only when morphological studies are made of reproductive parts and subterranean vegetative parts that a sensible basis for distinctions emerges, and confusion can be avoided between some members of different sections.

The morphological characters with the greatest diagnostic value can be enumerated briefly. An important distinction is possible on the basis of the root system. The major types are tap-rooted (axonomorphic) and tuberous-rooted; the latter can be sub-divided further into those in which both the primary root and hypocotyl become tuberous and those in which only lateral roots are so affected. The production of rhizomes and the

(a) (b) (c) (d)

Fig. 3.3. Germplasm collection in the field. (*a*) *Arachis glabrata* – growing wild, Porto Estralia, Mato Grosso, Brazil. (*b*) *Arachis* wild species habitat: open country near Caceras, Mato Grosso, Brazil. (*c*) *Arachis* wild species habitat: woodland, 'cerradon' near Caceras, Mato Grosso, Brazil. (*d*) Collecting *Arachis* wild species germplasm near Cuiaba, Mato Grosso, Brazil (left to right, Messrs Valls, Krapovickas and Silva) (C. E. Simpson).

ability to produce (spontaneously) adventitious roots at the nodes are characters of high diagnostic value. Behaviour of the peg during its growth phase, whether this is vertical or mainly horizontal, delimits important taxa within the genus. In floral morphology, actual size of flowers, pigmentation, presence and location of red venation on the standard, although these might generally be regarded as trivial characters, are of considerable importance in *Arachis* taxonomy.

This morphological scheme of classification had developed sufficiently by 1964 to have been made use of by Smartt (1965) in his study of inter-specific hybridization between the cultigen and wild species. Subsequently it has been developed and expanded until the broad lines of the classification have now been confirmed by experimental studies (Gregory and Gregory, 1979).

The biological species concept in *Arachis*

From the plant breeder's point of view this is the species concept of greatest significance, since the biological species comprises all the populations which actually or potentially can interbreed freely. Hard and fast demarcation between biological species may not always exist, in which case genetic introgression can be induced, which is of great practical value in improving the cultivated species.

This particular taxonomic approach (i.e. the establishment of biological species) is concerned with the evolution of isolating mechanisms. Where genetic isolation is complete we have no difficulty in distinguishing taxa at the species level or above. In the absence of complete isolation, species delimitation is perhaps more subjective. The evolution of isolating mechanisms cannot be considered apart from the evolution of the genus. Other things being equal, the more ancient evolutionary lineages tend to be more isolated genetically from each other than are those of relatively recent origin. This is likely to be true in a genus such as *Arachis*, which is predominantly self-pollinated (although cross-pollination can and does occur) and where selection pressures tending to establish isolating mechanisms, by suppressing effective inter-specific cross-pollination, would not be expected to be high. In these circumstances genetic isolation might be expected to evolve rather slowly by gradual and progressive accumulation of genetic differences. Therefore, where genetic isolation is incomplete between taxa, there is a strong probability that evolutionary divergence is of comparatively recent origin.

Two significant publications by Gregory *et al.* (1980) and Gregory and Gregory (1979) review evolutionary trends in the genus and present probably a definitive treatment of inter-specific relationships within the genus as determined by actual or attempted inter-specific hybrid production. The treatment of evolution in *Arachis* by Gregory *et al.* (1980) is the first published attempt to bring together geographical, geomorphological and ecological evidence to produce a reasoned synthesis and establish a credible evolutionary hypothesis. Their conclusions are well worthy of an extended summary.

In South America the genus ranges geographically from the equator near the mouth of the Amazon to 34° S on the northern bank of the Rio de la Plata in Uruguay. From the Atlantic coast it ranges westward to the Paraná and the eastern foothills of the Andes. The northern boundary is marked by the southern extent of the Amazonian rainforest. In this area, a great diversity of climatic and ecological conditions (e.g. soil type) occur, which are frequently quite extreme. The geocarpic habit of groundnuts is probably advantageous from the standpoint of survival in harsh environments but imposes considerable restrictions on distribution. The geocarpic fruit of *Arachis* can only be distributed effectively over long distances by agents which can physically move soil and fruits together, and therefore the only plausible natural agent is water. The effectiveness of moving water in distributing *Arachis* is apparently supported by distributions of taxa which are closely associated with specific drainage basins of both recent and ancient times.

From these considerations Gregory *et al.* (1980) infer that the centre from which the present distribution has been achieved is the 'planaltine ellipse', demarcated by plotting distributions of all *Arachis* collections from above 550 m on the Brazilian shield. Geomorphological changes have produced changes in drainage pattern and have isolated, in currently distinct drainage basins, forms which have evolved unique patterns of variation and genetic isolation from other isolated forms. This has been a major factor in the differentiation of the major sub-generic taxa.

Studies of inter-specific hybridisation in *Arachis*

Initial studies of inter-specific hybridisation in *Arachis* involved the use of *A. hypogaea* as seed parent. In subsequent crosses wild species were used as both pollen and seed parents, an approach which Gregory and Gregory (1967, 1979) developed extensively to determine taxonomic relationships between species.

The first recorded attempt at inter-specific hybridisation was reported by Hull and Carver (1938) between *A. hypogaea* and *A. glabrata*, but no hybrid seed were recovered. A similar attempt by Gregory (1946) was also unsuccessful, as were the crosses *A. hypogaea* × *A. villosulicarpa* and *A. hypogaea* × *A. 'diogoi'*. The first reported viable inter-specific hybrid was produced by Krapovickas and Rigoni (1951) between *A. hypogaea* and *A. villosa* var. *correntina*. Subsequently the same cross was produced by Kumar *et al.* (1957) and Raman (1959*b*). Johansen and Smith (1956) made a study of embryo development in the unsuccessful crosses *A. hypogaea* × *A. 'diogoi'* (this material was apparently not authentic *A. diogoi* Hoehne, *vide* Gregory and Gregory, 1979) and of *A. hypogaea* × *A. glabrata*. Fertilisation apparently occurred but growth of embryo and endosperm was retarded; hypertrophy of the testa was noted. Hybrid embryos died without differentiation occurring. Johansen and Smith (1956) found that, at maturity, pods arising from inter-specific hybridisation attempts were empty except for the shrivelled remains of aborted embryos and testas, as had been observed previously by Gregory (1946), by workers at the East African Agricultural and Forestry Research Organization (1954–6) and subsequently by Tuchlenski (1958), Smartt (1964) and Bharathi and Murty (1984). The first attempt to study systematically the cross-compatibility relationships between *A. hypogaea* and a broad cross-section of wild species was reported by Smartt (1965)

Fig. 3.4. River systems of South America. Water is an important agent of distribution in the genus *Arachis* (from Simpson, 1984).

and Smartt and Gregory (1967). Seven viable inter-specific hybrids were reported between *A. hypogaea* and wild species including *A. villosa*, *A. villosa* var. *correntina*, *A. duranensis*, *A. cardenasii*, *A. chacoense*, *A. helodes* and *A.* sp. 9901 GKP (see Gregory *et al.*, 1973). The cross *A. spegazzinii* × *A. hypogaea* succeeded only with the wild species as seed parent. Additional crosses between the cultigen and wild species *A. batizocoi* and *A. stenosperma* have been obtained by Gregory and Gregory (1979) and with *A. diogoi* by Pompeu (1977, 1979, 1983). On morphological grounds all species which cross successfully with

Fig. 3.5. (*a*) Distribution of botanical groups of *Arachis* above 550 m on the Planalto (black areas): *Erectoides* and *Eurhizomatosae* to the south-west and *Extranervosae* to the north-east. When inscribed in common areas, these describe the 'planaltine ellipse'. (*b*) Dissemination from planaltine ellipse. Letters signify sections of the genus: A, *Arachis*; C, *Caulorrhizae*; E, *Erectoides*; R, *Rhizo-*

A. hypogaea, although clearly distinct from it, appear to be close relatives and are included with it in the same section. The wild tetraploid *A. monticola* is so freely cross compatible with the tetraploid cultigen *A. hypogaea*, and progeny produced are so fertile and vigorous through F1 and beyond, that they can be regarded with some justification as conspecific (Krapovickas and Rigoni, 1954; Hammons, 1970; Kirti *et al.*, 1982).

Gopinathan Nair *et al.* (1964) reported production of a viable hybrid *A. hypogaea* × *A. glabrata* var. *hagenbeckii*; Raman and co-workers

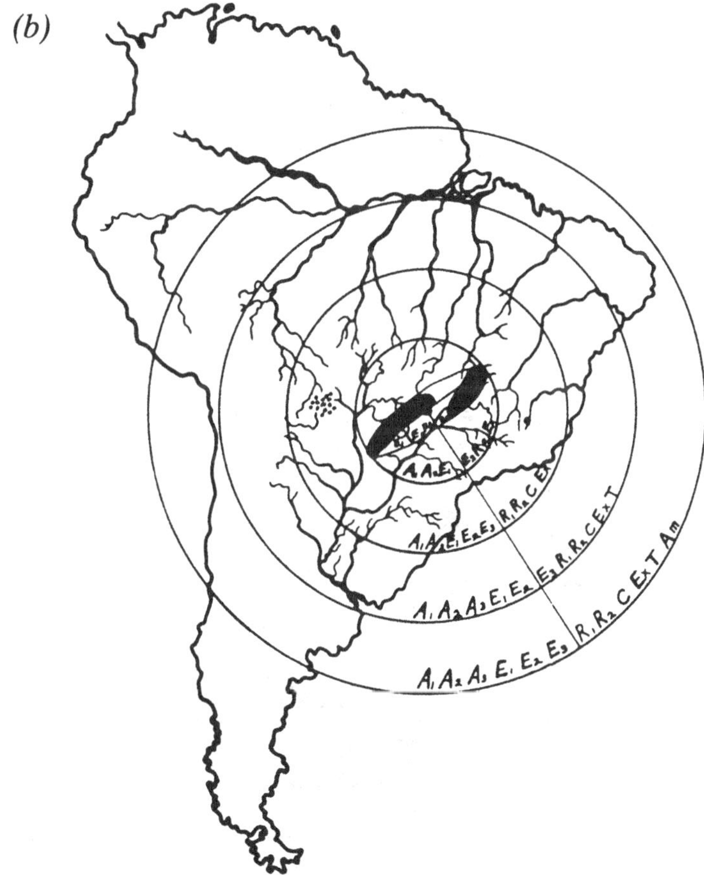

matosae; T, *Triseminalae*; Am, *Ambinervosae*; and Ex, *Extranervosae*; numbers indicate series within sections. (From Gregory *et al.*, (1980). British Crown Copyright.) Reproduced with permission of the Controller, Her (Britannic) Majesty's Stationery Office and the Trustees, Royal Botanic Gardens, Kew, © 1980.)

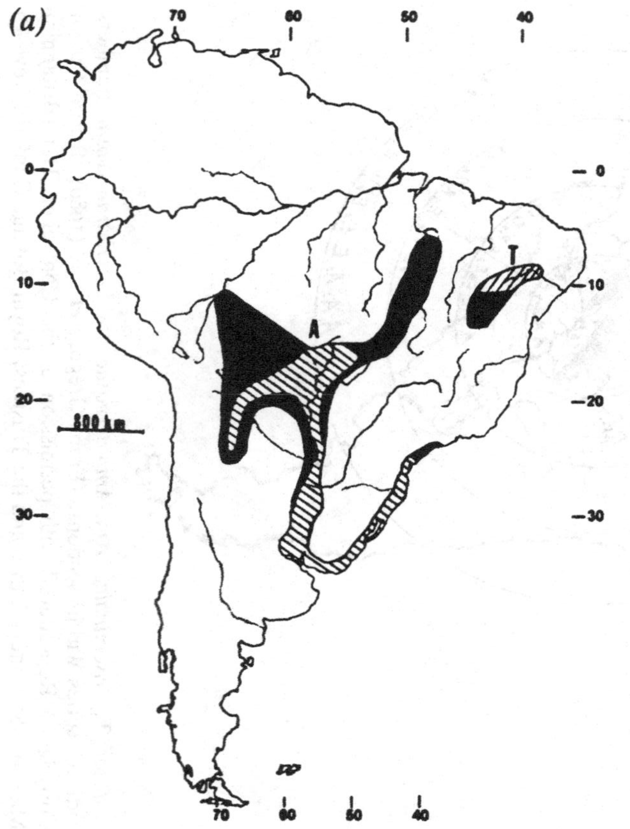
(a)
70
60
50
40
0
10
20
30
800 km
A
T

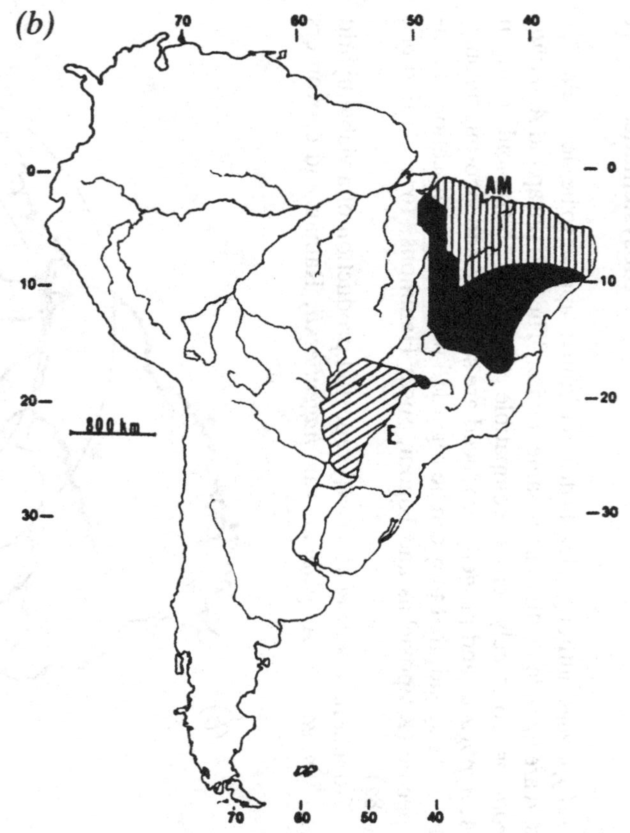
(b)
70
60
50
40
0
10
20
30
800 km
AM
E

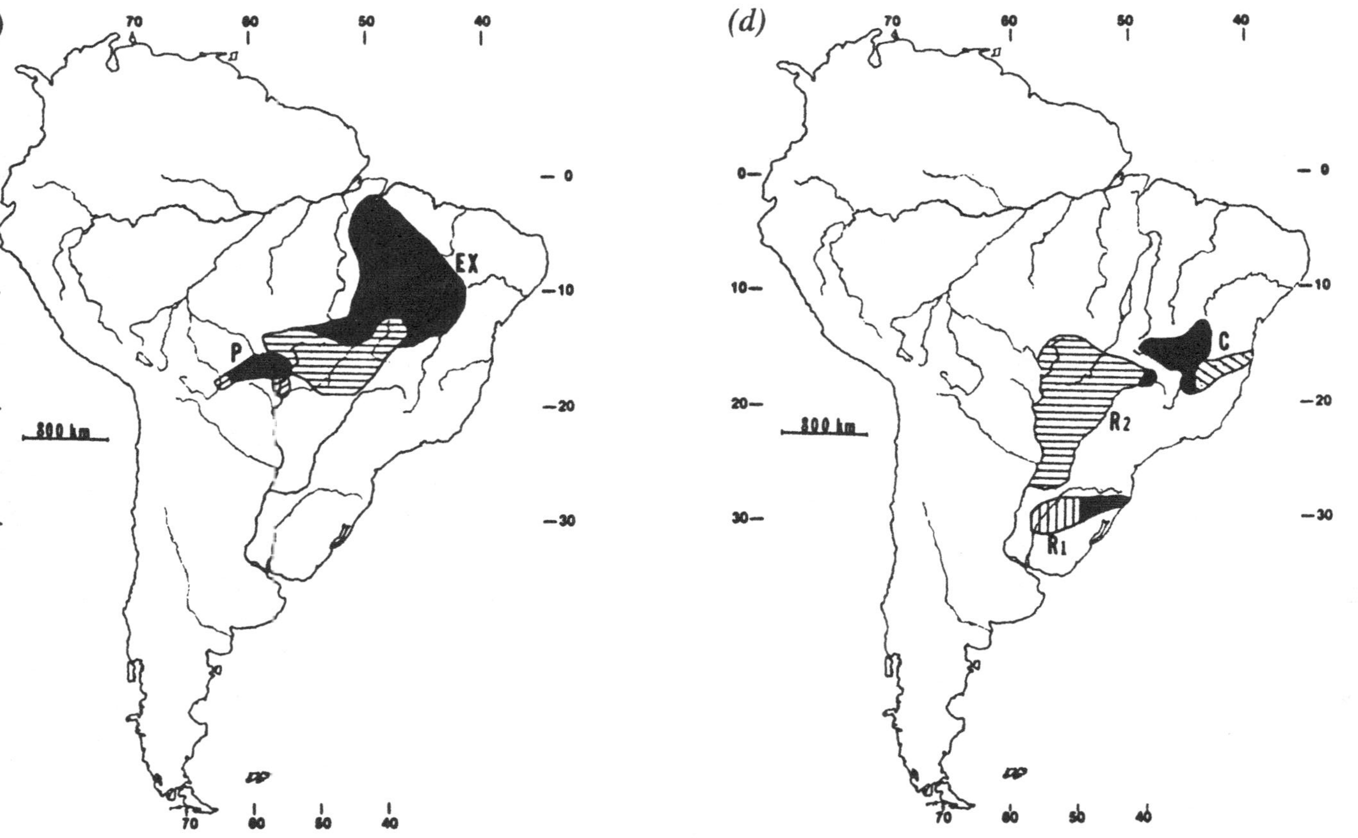

Fig. 3.6. Distribution of sections in the genus *Arachis*. (*a*) Sections *Arachis* and *Triseminalae*; (*b*) *Ambinervosae* and *Erectoides*; (*c*) *Extranervosae* and *Procumbense* (*Erectoides*); (*d*) *Caulorrhizae* and *Rhizomatosae*. (From Simpson, 1984.)

(1957 *et seq.*) and Varisai Muhammad (1973*a–d*) have reported viable hybrids between *A. hypogaea* as seed parent with *A. 'diogoi'* (see Johansen and Smith, 1956), *A. glabrata* and *A. villosulicarpa*, also between *A. monticola* and the species *A. 'diogoi'* and *A. marginata* as well as *A. villosa* × *A. hagenbeckii* and *A. duranensis* × *A. villosulicarpa*.

Fig. 3.7. Inter-specific hybridisation in *Arachis*. (*a*) *A. hypogaea*; (*b*) *A. villosa* (note light-coloured main veins of leaflets); (*c*) *A. hypogaea* × *A. villosa* (note expression of light-coloured leaflet vein of pollen parent); (*d*) leaf-forms of parents and hybrid of *A. hypogaea* × *A. villosa* cross.

Pompeu (1977) has been unable to repeat any of this work using materials from the same sources. Gregory and Gregory (1979), who have examined material of putative hybrid origin (*A. hypogaea* × *A. glabrata*) are of the opinion that it is pure *A. hypogaea*. It is possible that this material could have arisen by some undetected lapse in crossing technique or perhaps by sporadic apomixis (Smartt, 1979). Gregory and Gregory (1979) remain convinced that all successful inter-specific crosses involving *A. hypogaea* are with closely related species only, i.e. within sect. *Arachis*. The use of sophisticated embryo rescue techniques has produced small plantlets from some incompatible crosses but not, as yet, mature plants (Mallikarjuna and Sastri, 1985).

Crosses between wild species are of particular interest because they might well shed light on the origin from diploid progenitors of the tetraploid *A. hypogaea*. The first inter-specific hybrid between wild species reported in the literature was produced by Raman and Kesavan (1962). Gibbons and Turley (1967) produced hybrids *A. batizocoi* × *A. duranensis*, × *A. villosa*, × *A. villosa* var. *correntina*; *A. spegazzinii* × *A. duranensis*, × *A. batizocoi*; *A. villosa* × *A. villosa* var. *correntina*. The most interesting feature of these crosses is that F1 progeny were highly fertile except where *A. batizocoi* was one of the parents. Resslar and Gregory (1979) and Stalker and Wynne (1979) have reported additional hybrids between species of sect. *Arachis* in which only those involving *A. batizocoi* were completely pollen sterile. Gregory and Gregory (1979) published a comprehensive listing of viable inter-specific hybrids. Extensive further reports of the production and behaviour of inter-specific hybrids are to be found in the ICRISAT *Arachis* cytogenetics report (1985).

Chemotaxonomy

Three different groups of chemical compounds have been studied chemotaxonomically in *Arachis*. These are seed proteins, nucleic acids and flavonoids.

Proteins

Seed proteins have been studied using the techniques of both immunoelectrophoresis and disc electrophoresis. Daussant *et al.* (1969*a*, *b*) produced the first immunoelectrophoretic characterisation of *A. hypogaea* seed proteins. The use of this technique was applied to other species of *Arachis* by Neucere and Cherry (1975). Their immunoelectrophoretic analyses suggested inter-specific relationships which were consistent with the taxonomic scheme of Krapovickas and Gregory (Gregory *et al.*, 1980). A similar conclusion was reached by Cherry (1975), using disc electrophoresis. Tombs and Lowe (1967) have noted that one of the major seed storage proteins, arachin, is polymorphic and have identified three forms of it.

More sophisticated chemotaxonomic protein studies of *Arachis* have been carried out by Klozová *et al.* (1983*a*, *b*). These included immunochemical and polyacrylamide gel analyses. Sixteen species were studied using the former technique and twelve using the latter. Seed proteins were extracted and the albumin and globulin fractions separated. They were then tested with antisera against the albumin and globulin fractions

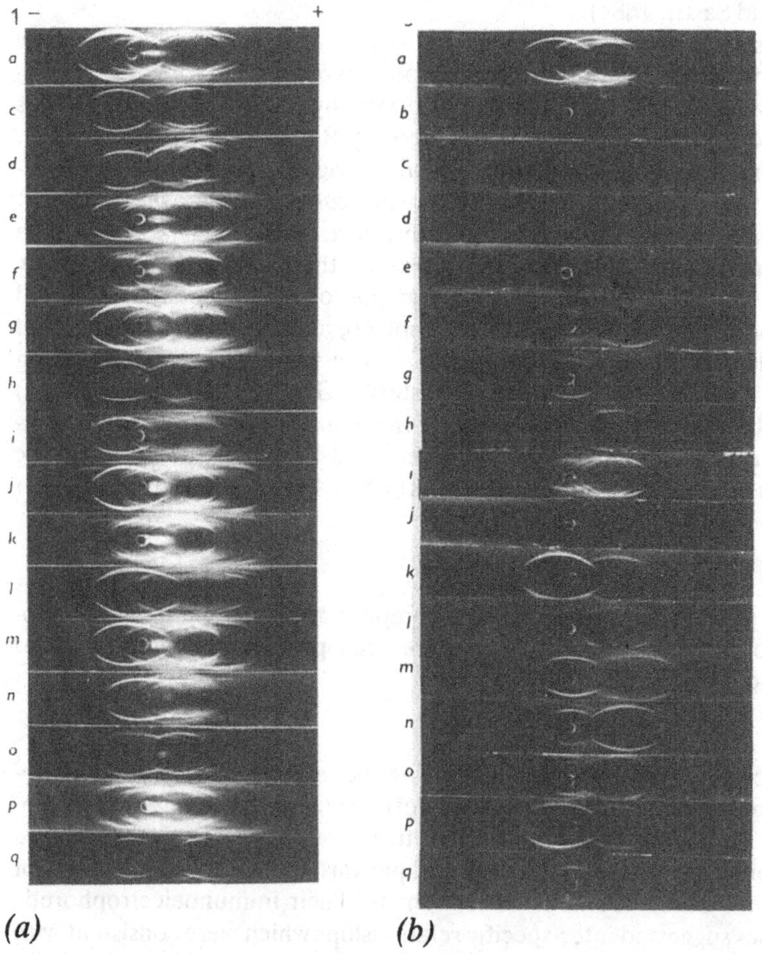

Fig. 3.8. Immunoelectrophoretograms of *Arachis* species. (*a*) Reactions of unabsorbed *Arachis* species seed antigens with *A. hypogaea* albumen antiserum. (*b*) Corresponding reactions of absorbed antigens. (*c*) Reactions of unabsorbed

of *A. hypogaea*. Strong cross-reactions were obtained with species in all sections of the genus, a situation which contrasts with that in *Phaseolus* where cross-reactions are weaker with less closely related species (Kloz and Klozová, 1968). This had led Klozová *et al.* to suggest that protein characters in *Arachis* have been relatively conservative in evolution. Similarities between proteins of *A. hypogaea* and *A. monticola* are particularly close. There are especially strong similarities between albumins of *A. batizocoi* and *A. villosa* and those of *A. hypogaea*. Results given by

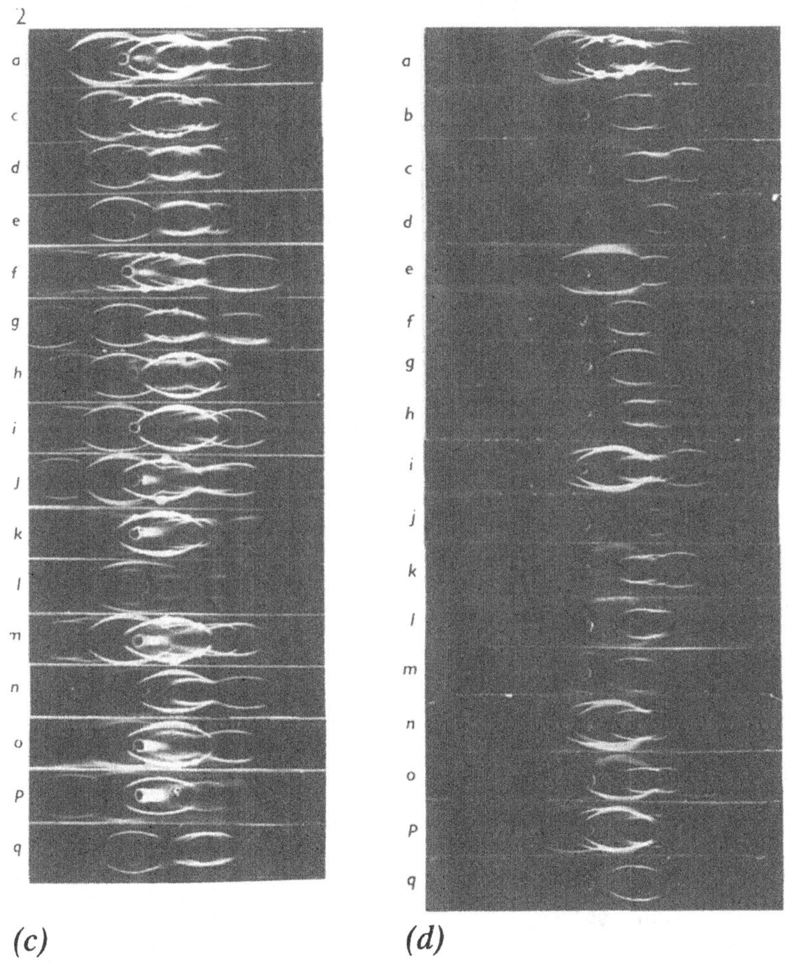

seed antigens with *A. hypogaea* globulin antiserum. (*d*) Corresponding reactions of absorbed antigen. (From Klozová *et al.*, 1983).

polyacrylamide gel analyses are much more difficult to interpret. This technique seems to lack the fine discriminating power of the immunochemical techniques.

Cytophotometric studies of cell DNA contents

Resslar (1979) determined 2*C* amounts of DNA for 12 taxa in sect. *Arachis*. He found these to range from 4.92 to 5.98 pg DNA per cell in diploid species and 10.36–11.35 pg DNA in the tetraploids. Annual diploids (series *Annuae*) averaged 1 pg less per cell than the diploid

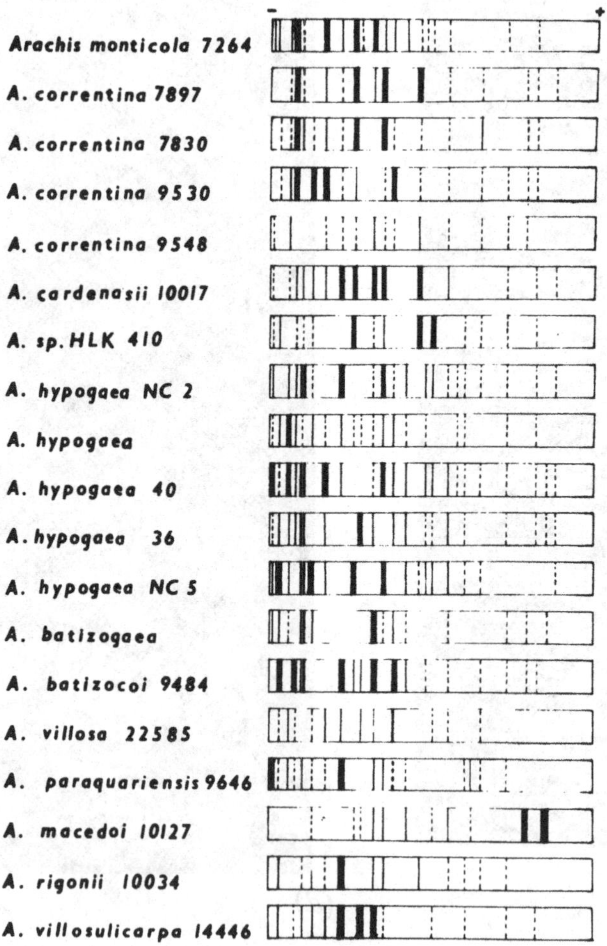

Fig. 3.9. PAGE separation of *Arachis* species seed proteins (Klozová *et al.*, 1983).

perennials (series *Perennes*). Variation was found in the tetraploids (series *Amphiploides*) between *A. monticola* and *A. hypogaea* in the first place and also between the sub-species *hypogaea* and *fastigiata* Waldron of *A. hypogaea*.

Flavonoids

Flavonoid chromatography of leaf extracts has been undertaken by Seeligmann and Krapovickas (Krapovickas, 1973; Krapovickas *et al.*, 1974). More than twenty compounds have been detected in the genus as a whole with no more than twelve of these, and usually fewer, found in any one taxon. The data obtained are not easy to interpret and considerable variation obviously exists within the species *A. hypogaea*. Additive inheritance of flavonoids has been shown in an inter-specific hybrid derivative, *A.* × *batizogaea* Krap. *et* Fern. (Krapovickas *et al.*, 1974). In general Krapovickas (1973) has found that the centres of variation for chemical and morphological characters coincide reasonably well.

The role of studies on chemical variation

Published work certainly indicates that interesting and potentially useful variation exists for chemical characters in the genus. The data are not as yet so extensive as to supplement greatly the volume of taxonomically useful information. Flavonoids derived from leaf tissue could potentially be of value in resolving the problems of classifying largely clonal material in sect. *Rhizomatosae*. Such studies might also be useful in establishing affinities between incomplete herbarium specimens and material from living collections.

At present the preferred source of material for protein chemotaxonomic studies is the seed. Rhizomatous forms produce seed very sparingly, and alternative sources of proteins such as leaves could be investigated with possible taxonomic advantage.

Studies of nucleic acids are clearly in a very preliminary phase. The differences in nuclear DNA contents observed between the series of sect. *Arachis* by Resslar (1979) suggest that a comprehensive study of the whole genus would be worth while.

The protein studies of Klozová and co-workers (1983*a*, *b*) are of particular interest. Unfortunately it has not as yet been possible to take these studies to a definitive conclusion but the indications are that the immunochemical studies in particular are worth pursuing further. Elucidation of immunochemical affinities between the cultigen and other species within sect. *Arachis* would be especially valuable. There is obviously considerable scope for molecular and biochemical techniques in characterisation of *Arachis* species and in genome analysis (Atreya *et al.*, 1985).

3.2. Cytology and cytogenetics of *Arachis*

The somatic chromosome complement

Chromosome number

The earliest comprehensive reports on chromosome number, morphology and behaviour were those of Husted(1933, 1936) on *A. hypogaea*. Kawakami (1930) had earlier reported a somatic complement $2n = 40$ and a gametic number $n = 20$ and Husted (1931) had confirmed the somatic complements of *A. nambyquarae* and six cultivars of *A. hypogaea* to be $2n = 40$. These reports contradicted the finding of Badami (1928) of complements $2n = 20$, $n = 10$ in some lines of cultivated groundnut. Stalker and Dalmacio (1986) suggested that the infra-specific taxa within *A. hypogaea* have developed distinctive cytological characters.

The first chromosome count reported for a wild species was $2n = 40$ for *A. glabrata* (Gregory (1946) in Johansen and Smith, 1956). This count was confirmed by Conagin (1962) and Smartt and Gregory (1967). Mendes (1947) published counts of $2n = 20$ chromosomes for *A. diogoi*, *A. marginata*, *A. prostrata* and *A. villosulicarpa*. This gave the first indication of the existence of two series of chromosome numbers in the genus, $2n = 20$ and $2n = 40$. Although the nomenclature of some of Mendes' material can be questioned (Gregory *et al.*, 1973, 1980) it does appear that at least five clearly distinct wild species were studied. Krapovickas and Rigoni (1949, 1950, 1951, 1953) reported $2n = 20$ for two forms of *A. villosa* ('*typica*' and var. *correntina*) and $2n = 40$ for *A. pusilla* (correctly *A. monticola*). Subsequently the same authors (Krapovickas and Rigoni, 1957) published counts of $2n = 40$ for *A. hagenbeckii* and *A. monticola* and $2n = 20$ for *A. pusilla* (correctly *A. duranensis*). Krapovickas and Gregory (1960) found $2n = 20$ in the species *A. rigonii*. Conagin (1962, 1963, 1964) recorded counts of $2n = 20$ for both *A. lutescens* and *A. repens* and listed counts of $2n = 40$ for *A. glabrata*, *A. hagenbeckii* and *A. monticola* together with counts of $2n = 20$ for '*A. prostrata*' and *A. villosa* var. *correntina*. Smartt (1965) and Gregory (1967) published counts in substantial agreement with those previously published and in addition counts of $2n = 20$ for *A. helodes*, *A. macedoi*, *A. benthamii*, *A. paraguariensis*, *A. cardenasii*, *A. chacoense*, *A. lignosa*, *A. batizocoi*, *A. oteroi* and *A. spegazzinii*.

From these data it became clear that two series of chromosome numbers occur in the genus, $2n = 2x = 20$ and $2n = 4x = 40$. Polyploidy has apparently arisen independently at least twice in the genus, in the immediate ancestor of the cultivated groundnut itself and in sect. *Rhizomatosae*. Primitive rhizomatous forms are diploid; the more abundant and more robust forms are tetraploid (Gregory *et al.*, 1973). These

Table 3.2. *Reported chromosome numbers of named Arachis species in chronological order*

Species	2n	Reference
A. hypogaea	40	Kawakami, 1930
A. glabrata	40	Gregory, 1946
A. diogoi	20	Mendes, 1947
A. marginata	20	"
A. prostrata	20	"
A. villosulicarpa	20	"
A. villosa ('typica' and var. *correntina*)	20	Krapovickas and Rigoni, 1949
A. pusilla (correctly *A. monticola*)	40	"
A. hagenbeckii	40	Krapovickas and Rigoni, 1957
A. monticola	40	"
A. pusilla (correctly *A. duranensis*)	20	"
A. rigonii	20	Krapovickas and Gregory, 1960
A. lutescens	20	Conagin, 1963
A. repens	20	"
A. helodes	20	Smartt and Gregory, 1967
A. macedoi	20	"
A. benthamii	20	"
A. paraguariensis (*A.* sp. 9646, 10585)	20	"
A. cardenasii (*A.* sp. 10017)	20	"
A. chacoense (*A.* sp. 10602)	20	"
A. lignosa (*A.* sp. 10598)	20	"
A. batizocoi	20	"
A. oteroi (*A.* sp. 10541)	20	"
A. spegazzinii (*A.* sp. 10038)	20	"
A. ipaensis	20	Gregory and Gregory, 1979
A. stenosperma	20	"

authors also report chromosome complements of $2n = 20$ for species of their sections *Ambinervosae* (*Pseudouxonomorphae*) and *Triseminalae*, this latter including the true *A. pusilla*. A listing of chromosome counts is given in Table 3.2.

Aneuploidy

Aneuploid complements have been reported in *A. hypogaea* sporadically since Husted (1936) first reported a plant showing $2n = 41$ chromosomes plus a fragment. The most extensive reports of aneuploidy in the genus have arisen as a result of inter-specific hybridisation. Kumar and D'Cruz (1957) obtained a plant with $2n = 41$ from the backcross (*A. hypogaea* × *A. villosa*) × *A. hypogaea*. This behaved as trisomic. Smartt (1965) and Smartt and Gregory (1967) reported material with aneuploid complements ranging between $2n = 38$ and $2n = 60$ arising from *A. hypogaea*

× section *Arachis* diploid species hybrids. Jahnavi and Murty (1985) reported similar complements. Interestingly enough Davis and Simpson(1976) report aneuploid chromosome complements in the ranges 32–43 and 32–48 in the F7 generation of allohexaploids derived from F1 hybrids *A. hypogaea* × *A. cardenasii* produced by Smartt (1965). The origin of these aneuploids is unclear; they could have arisen through crosses with the cultivated form producing pentaploids, the meiosis of which would tend to produce aneuploids at the sub-pentaploid level. Alternatively they could have arisen through erosion of the hexaploid complement by multivalent formation and unequal chromosome segregation in meiosis. It is interesting to note that all selections made by Stalker *et al.* (1979) for good agronomic characters from material of the same origin as that of Davis and Simpson (1976) had chromosome complements of $2n = 40$. Aneuploidy can also arise from the effects of ionising radiation on cells in division (Patil and Bora, 1961; Patil, 1968; Madhava Menon *et al.*, 1970).

Chromosome morphology and karyotype evolution

Ghimpu (1930) in his study of *A. hypogaea* chromosomes notes in addition to the complement being $2n = \pm 40$ that centromeres were median and that chromosomes of bunch and runner types were essentially similar. Husted (1933, 1936) identified two distinctive chromosome pairs: one he termed 'A' chromosomes, which were decidedly smaller than any other pair; the other, termed the 'B' chromosomes, showed a marked secondary constriction. These observations were confirmed by Babu (1955) and D'Cruz and Tankasale (1961). Raman (1959*b*) observed the presence of one pair of 'A' chromosomes in *A. villosa* var. *correntina* and suggested a relationship between this genome and one of the presumably distinct genomes of *A. hypogaea*.

Smartt (1965) confirmed Raman's observation on the occurrence of 'A' chromosomes in *A. villosa* var. *correntina* and noted that in all species of section *Arachis* in which he had been able to examine karyotypes an 'A' chromosome pair was to be found. He noted also the apparent absence of this distinctive chromosome pair in the sect. *Erectoides* species *A. paraguariensis* (*A.* sp. 9646). The suggestion was made that the origin of the cultivated groundnut from diploid ancestors could have occurred by the hybridisation of a form with a karyotype like that of *A. villosa* and another with a karyotype like that of *A. paraguariensis*. This suggestion raised some difficulties in that inter-sectional species hybrids *Arachis* × *Erectoides* are difficult to produce experimentally and are probably not formed naturally, only two examples being confirmed (Gregory and Gregory, 1979).

The publication of the description, chromosome counts and photo-

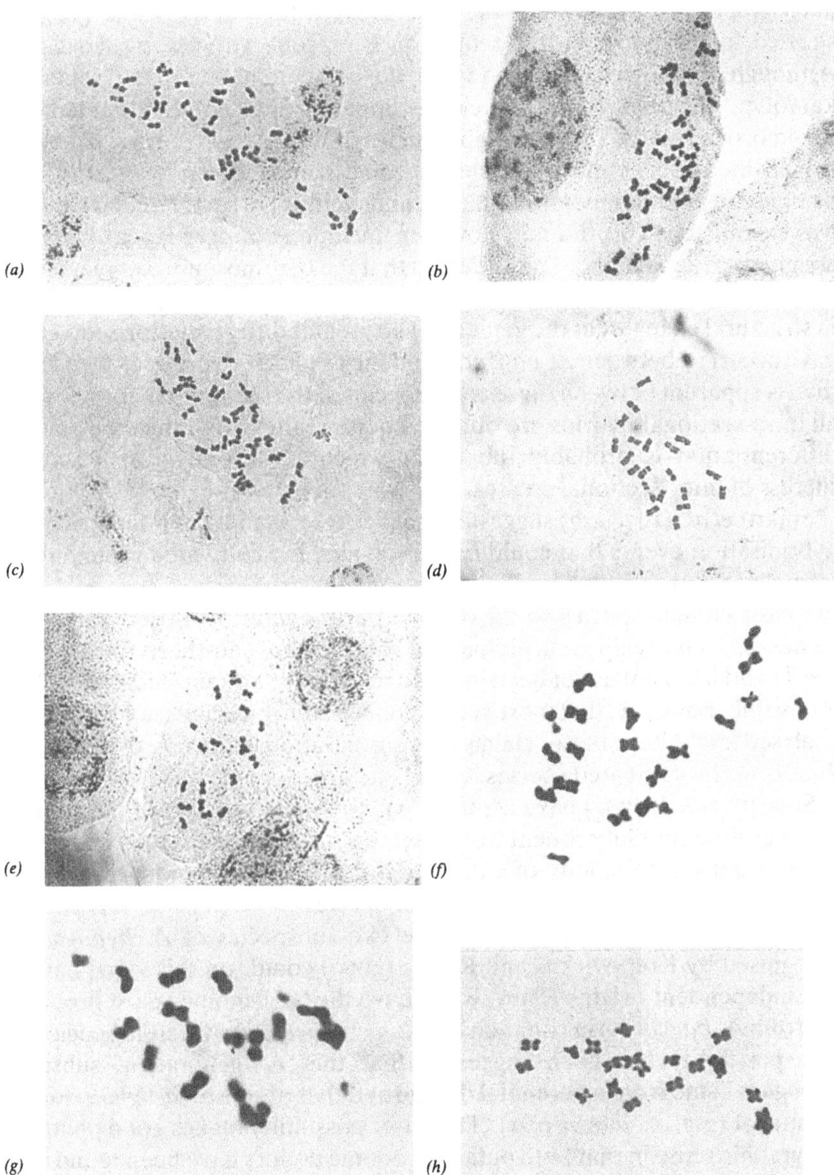

Fig. 3.10. Somatic chromosomes of *Arachis* species. (*a*) *A. hypogaea* (× 780); (*b*) *A. monticola* (× 1040); (*c*) *A. glabrata* (× 780); (*d*) *A. batizocoi* (× 1560); (*e*) *A. duranensis* (× 780); (*f*) *A. cardenasii* (× 1560); (*g*) *A. spegazzinii* (× 2080); (H) *A. paraguariensis* (× 1040). (Parts (*a*)–(*e*) courtesy of A. K. Singh.)

micrographs of chromosomes of *A. batizocoi* by Krapovickas *et al.* (1974) showed that cytological differentiation is present within sect. *Arachis*. Although Krapovickas *et al.* (1974) do not comment on the general karyotype of this species, it is clearly apparent that no identifiable 'A' chromosome pair is present. Subsequently Smartt *et al.* (1978*a*, *b*) confirmed the absence of an 'A' chromosome pair in *A. batizocoi* and its presence in all other material (mitotic and meiotic) of sect. *Arachis* which was examined. (This did not, however, include *A. diogoi* Hoehne, now assigned to the section.) They inferred that the chromosome complement of *A. batizocoi* differed largely from that of other species in sect. *Arachis* in structural changes; at the genic level no greater differentiation seems to have occurred between *A. batizocoi* and the other species of sect. *Arachis* than is apparent between these other species of the section. Furthermore, all intra-sectional hybrids are obtained quite readily. By inference genic differentiation is probably the factor which so severely restricts the success of inter-sectional crosses.

Smartt *et al.* (1978*a*, *b*) suggested that we have a model of inter-specific hybridisation events that could have produced the cultivated groundnut from diploid progenitors within sect. *Arachis*. These authors suggest that the most eligible species so far collected are *A. batizocoi* itself and *A. cardenasii*. The reciprocal diploid F1 hybrids between these forms are sterile and have not as yet been induced to produce any amphidiploids. It is possible, however, that more recent collections which have not yet been analysed might have better claims for consideration in the role of genome donors to the cultivated species.

Smartt *et al.* (1978*a*) have assumed a monophyletic origin of the cultivated groundnut. Subsequent workers, most notably Singh (1985 *et seq.*), have raised the possibility of a dual origin, with perhaps more than one source of the 'A' genome. The annual species *A. duranensis* has been suggested as one possible donor. The two subspecies of *A. hypogaea* recognised by Krapovickas and Rigoni (1960) could, on this view, have had independent origins. They would have the 'B' genome (most probably from *A. batizocoi*) in common but 'A' genomes from different species. The possibility has been suggested that the 'A' genome of subsp. *hypogaea* came from a perennial diploid and that of subsp. *fastigiata* from an annual (e.g. *A. duranensis*). The latter possibility makes good phytogeographic sense in that both putative genome donors have been found in north-western Argentina and adjacent parts of Bolivia. While morphological and phytogeographical evidence supports the possible role of *A. cardenasii* as an 'A' genome donor, this hypothesis is not well supported by chemotaxonomic evidence; *A. villosa* has been suggested by A. K. Singh (personal communication) as a stronger contender for the role of 'A' genome donor and *A. correntina* by Murty and Jahnavi (1986).

These are less plausible on phytogeographic grounds but credible morphologically, biochemically and cytogenetically.

If one accepts a monophyletic origin then the best identification of genome donors at present is *A. duranensis* (A) and *A. batizocoi* (B). If a dual origin is favoured then an 'A' genome donor closely related to *A. villosa* from the north-west Argentina–Bolivia area needs to be identified.

The above studies have made use of the presence or absence of one chromosome pair as markers of genomes. It is clear that other recognisable karyotype differences exist, for example in the morphology of nucleolar organiser chromosomes. Some attempts have been made to gain more karyotype information from preparations in which chromosomes are not strongly contracted (Stalker and Dalmacio, 1981). In such preparations the distinctness of the 'A' chromosomes can be reduced, for example in *A. cardenasii*. This has also been reported by Singh and Moss (1982), who assert that *A. cardenasii* lacks the short 'A' chromosome pair recorded in other species of the group (i.e. sect. *Arachis*). The preparative techniques used by Singh and Moss (1982) are different from those used by Smartt *et al.* (1978*a*, *b*) for somatic chromosomes. However, in meiotic preparations of *A. cardenasii* pollen mother cells, a distinctly small bivalent is readily apparent, corresponding to the 'A' chromosome pair. The use of different spindle inhibitors may well result in different degrees of contraction. It appears, therefore, that where maximum chromosome contraction is achieved 'A' chromosomes can be recognised in both mitotic and meiotic divisions of all sect. *Arachis* species so far examined except for *A. batizocoi*.

Meiotic chromosomes

Chromosome behaviour

The first detailed study of meiotic chromosome behaviour in *Arachis* was carried out by Husted (1936). The material studied was in fact all *A. hypogaea* (this included forms such as *A. rasteiro* and *A. nambyquarae*, now regarded as synonymous with *A. hypogaea*). In most metaphase I plates studied, pairing was 20II (ranging from 88.2% in 'White Spanish' to 97.1% in 'Pearl', another bunch form). The runner cultivar 'Improved Virginia' showed 94.0% normal bivalent pairing. Departures from this pattern in these cultivars included formation of trivalents and univalents in addition to bivalents (18II + 1III + 1I) and (19II + 2I). In other cultivars 18II + 1IV associations were also found. In 'Nhambiquaras' Husted (1936) reported 3IV + 2III + 11II; in hybrids 'Improved Virginia' × 'White Spanish' configurations observed were 20II for the most part but 18II + 1IV, 17II + 1IV + 2I, 14II + 1VI + 2III, 14II + 2VI, 17II +

2III and 17II + 1VI were also observed. Because of the low frequencies of such configurations it can be inferred that although the cultivated groundnut is tetraploid it is effectively well diploidised. Multivalent association could be due to homoeologous pairing (the formation of quadrivalents or a trivalent plus a univalent) between chromosomes of the two genomes. In the case of pairs of trivalents or hexavalents the probability of segmental interchanges having occurred in the differentiation of the genomes is high. The enhanced production of such associations in the 'Virginia' × 'Spanish' F1 hybrid discussed by Husted suggests that there may be chromosome structural differences between different forms (i.e. sub-species) of the cultigen, a suggestion made more recently by Gregory *et al.* (1980) on the basis of reduced fertility in hybrids between sequentially branching and alternately branching forms. Subsequent study by Raman (1976) also confirms Husted's conclusions. In these studies occasional aneuploidy was observed in addition to sporadic occurrence of chromatin fragments in meiotic cells. These could well originate, as has been suggested by the authors cited, as a result of departures from normal diploid pairing.

Wild species meiosis

Meiotic studies in wild species have been reported by Raman (1976) for both tetraploid and diploid wild species. The behaviour of *A. monticola* is comparable to that of *A. hypogaea* with normally 20II but occasionally with 18II + 1IV. Apparently meiosis is less regular in the tetraploid rhizomatous species which form up to four quadrivalents. All authorities who have studied meiosis in diploid wild species uniformly conclude that meiosis is regular with 10II forming (Smartt, 1965; Raman, 1976; Resslar and Gregory, 1979; Smartt *et al.*, 1978*a*; Stalker and Wynne, 1979).

Meiosis in inter-specific hybrids

The authenticity of some inter-specific hybrids claimed to have been produced by Raman (1976) and Varisai Muhammad (1973*a–d*) has been questioned (Gregory and Gregory, 1979; Smartt, 1979). For this reason only the meiotic behaviour of inter-specific hybrids of unquestioned authenticity will be reviewed. The first inter-specific hybrids obtained in *Arachis* were produced with *A. hypogaea* as seed parent. These were between the tetraploid cultigen and diploid species of sect. *Arachis* and as a result functionally sterile triploids were produced. Natural or artificially induced hexaploidy restored a relatively high level of fertility (Kumar *et al.*, 1957; D'Cruz and Chakravarty, 1961; Smartt and Gregory, 1967).

The first inter-specific hybrid reported between diploid *Arachis* species was produced by Raman and Kesavan (1962) between *A. duranensis* and

A. villosa var. *correntina*. These authors found meiosis to be almost completely regular, a conclusion which has been confirmed and amplified by Resslar and Gregory (1979) and Stalker and Wynne (1979). Regular meiotic pairing has been found in all inter-specific hybrids that have been made between species within section *Arachis* except for those involving *A. batizocoi* (Smartt *et al.*, 1978*a*, *b*; Stalker and Wynne, 1979). In F1 inter-specific hybrids involving the latter species meiosis is extremely irregular and sterility virtually complete (Gibbons and Turley, 1967; Smartt *et al.*, 1978*a*, *b*; Stalker and Wynne, 1979). Irregular meiosis in this instance appears to be due to extensive differences in arrangement of structural elements between *A. batizocoi* and other species in the section.

Stalker (1978, 1981) reported meiotic behaviour in a complex triploid hybrid (sect. *Erectoides* 4*x* × sect. *Arachis* 2*x*). The *Erectoides* 4*x* form is itself an amphidiploid derived from the F1 hybrid between *Arachis rigonii* and *A.* sp. FKP9841 (*vide* Gregory *et al.*, 1973). This was crossed successfully with the two accessions *A.* sp. HLK 410 and *A. duranensis* from sect. *Arachis*. The resulting hybrids were ± triploid and sterile (some were triploid, $2n = 30$, others aneuploid, $2n = 31, 32$). Trivalents were formed at low frequencies, suggesting that at least some homology exists between the chromosomes of the *Arachis and Erectoides* species involved.

Further meiotic studies of inter-sectional hybrids could yield valuable information on genomic homologies. The difficulty with which such hybrids are produced suggests that, within each section, the genome or genomes have evolved highly effective genetic isolation from those of other sections. The most numerous inter-sectional hybrids have arisen from combinations *Erectoides* × *Rhizomatosae* and *Arachis* × *Rhizomatosae*. Considerably fewer have arisen from other combinations such as *Erectoides* × *Arachis* and *Erectoides* × *Caulorrhizae* and none has been produced by the great majority of inter-sectional combinations (Gregory and Gregory, 1979).

It remains to be seen whether application of techniques such as protoplast fusion or culture *in vitro* of early F1 hybrid embryos could produce further inter-specific combinations. The pattern of inter-sectional cross-compatibility observed has led Gregory and Gregory (1979) to suggest that both sections *Arachis* and *Erectoides* have some affinity with the 4*x* *Rhizomatosae*. It is possible that, since *Arachis* and *Erectoides* are almost completely cross-incompatible, one of the two *Rhizomatosae* genomes confers compatibility with the *Erectoides* and the other with species of sect. *Arachis*. In section *Arachis* it is only members of the series *Annuae* which have demonstrated inter-sectional cross-compatibility. Neither the perennials nor the tetraploids of this section have produced inter-sectional hybrids. The presumed presence of a genome from a perennial species (Smartt *et al.*, 1978*a*) may explain that lack of cross-compatibility between *A. hypogaea* (and *A. monticola*) and any other section (Gregory and Gregory, 1979).

Technical and interpretative aspects

The chromosomes of *Arachis* species are far from ideal material for cytological study. The chromosomes are small, 1–4.5 μm (the actual lengths observed in preparations vary according to duration of pre-treatment), and are prone to stickiness in both mitotic and meiotic preparations. This latter problem can be overcome by taking precautions during preparation (Fernandez, 1973) and avoiding conditions of stress (H. T. Stalker, unpublished).

Somatic chromosomes

The karyotype of *Arachis* species is at first sight relatively featureless, but Smartt *et al.* (1978*a*) have shown that such distinctive features as are exhibited can be of value. It seems highly probable that different technical approaches to the preparation of chromosomes for examination could be of value in different ways. The simplest procedure is to reduce either pre-treatment times or concentration of the spindle inhibitor reagent, to minimise the degree of chromosome contraction while retaining effective spindle inhibition. This could be expected to maximise expression of differences in chromosome morphology, and ensure consistent expression of features such as secondary constrictions and satellites, which are frequently lost in preparations of strongly contracted chromosomes. The second and potentially much more valuable approach is that of chromosome banding. Resslar (1979) has obtained results which show that the technique has promise, but production of high-quality material in adequate quantity is a problem. Banding patterns could well be of value in characterisation of the genomes in different sections of the genus and the tracing of chromosome homologies between species.

Meiotic chromosomes

On the whole, production of high-quality meiotic preparations is not excessively difficult and satisfactory preparations for cytogenetic analysis can be obtained fairly readily in most species. Polyploidy is a feature of the genus in both sections *Arachis* and *Rhizomatosae* and as in other F1 inter-specific hybrids induction of amphidiploidy by colchicine treatment often, but not invariably, improves fertility. Interpretation of meiosis in polyploids is thus of considerable importance; it is, however, not without pitfalls.

The interpretation of pairing relationships in diploid inter-specific hybrids is quite simple and straightforward. In instances where meiotic pairing is high, fertility may also be high (Raman and Kesavan, 1962;

Resslar and Gregory, 1979; Stalker and Wynne, 1979); where it is reduced, fertility is also low (Smartt *et al.*, 1978*a*, *b*). Pairing relationships in triploids are much more difficult to interpret and very often there is a strong temptation to go farther in interpretation than the evidence warrants. In triploids and higher polyploids, the extent to which multivalents form is determined by chromosome length and chiasma frequency in addition to homology. In an autotriploid it is possible to envisage a situation in which pairing is entirely (II + I) due largely to low chiasma frequency combined with short arm length. Similarly in an autotetraploid, for the same reason, pairing could be exclusively (II + II). In allopolyploids the situation is more complex.

Smartt (1965) observed that in triploid F1 hybrids the frequency of trivalents varied widely according to the wild species used as pollen parent with *A. hypogaea*. In *A.hypogaea* × *A. villosa* var. *correntina* a mean of 0.95 trivalents (range 0–2) per cell was recorded; in *A. hypogaea* × *A. duranensis* this was 2.15 (0–5) trivalents per cell while in *A. hypogaea* × *A. helodes* 3.40 (range 0–6) trivalents per cell were recorded. The major variable in these three situations is the wild species genome. It is reasonable to assume that, within limits, the more homologous the wild species genome is with one of those already present in *A. hypogaea*, synapsis in the triploid meiosis will occur more rapidly and tend to exclude the second *A. hypogaea* genome. A lower level of homology could reduce the rate and extent of synapsis and permit more multivalent associations to arise. The true extent of homology between the genomes of *A. hypogaea*, as indicated by meiotic pairing relationships, would be best exemplified in a haploid *A. hypogaea*, but this has never been found. Another culture might eventually produce such haploids, which would be extremely valuable for cytological analysis.

Raman's (1959*b*) interpretation of genomic homology between *A. villosa* var. *correntina* and *A. hypogaea* is probably correct. However, he had no way of knowing that all incoming chromosomes of *A. villosa* var. *correntina* were pairing with one genome of *A. hypogaea* as he assumed. It is also possible that *A. villosa* var. *correntina* chromosomes could have been pairing with members of both *A. hypogaea* genomes, or that the two *A. hypogaea* genomes could have been pairing with each other to a greater or lesser extent.

Similar caution is advisable in the interpretation of chromosome pairing situations in artificially produced allotetraploids and allohexaploids as to the implications of both production and non-production of multivalent associations. An example from another leguminous amphidiploid is instructive. Smartt and Haq (1972) produced an amphidiploid from the F1 hybrid *Phaseolus vulgaris* L. × *Ph. coccineus* L. and observed in successive generations a reduced frequency of multivalent associations in meiosis. Propagation by seed imposes selection for a more regular and diploidised meiosis through selection for high levels of seed production. In amphidiploids genomic homologies would be indicated by meiotic associations of III and IV but these could also arise from interchange heterozygosity. Higher multivalent associations (III + III), (V + I) or (VI) etc. could also indicate genomic differentiation by segmental interchange. In allohexaploids, associations of (V + I) or (VI) could indicate some homology of all three genomes present. The formation of quadrivalents only could indicate that two of the genomes had sufficient homology to pair but would not definitely exclude the possibility of

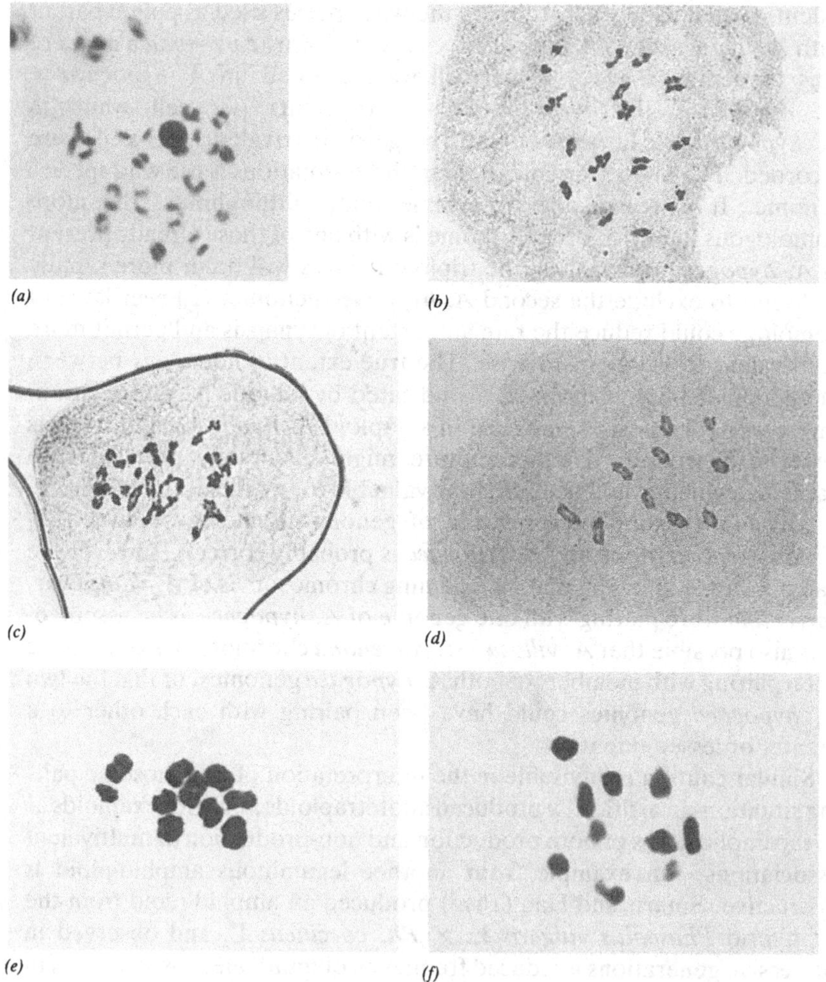

homology between all three genomes, more particularly if more than one quadrivalent per cell was observed. Conversely, normal diploid pairing patterns in allopolyploids do not necessarily indicate lack of the capability of homoeologous pairing between genomes.

Genomic divergence in *Arachis hypogaea*

Divergence between evolving genomes can come about through changes at individual genetic loci and also through rearrangement of chromosome segments. In the long-standing differentiation between genomes of different sections in the genus it is probable that the lack of inter-specific

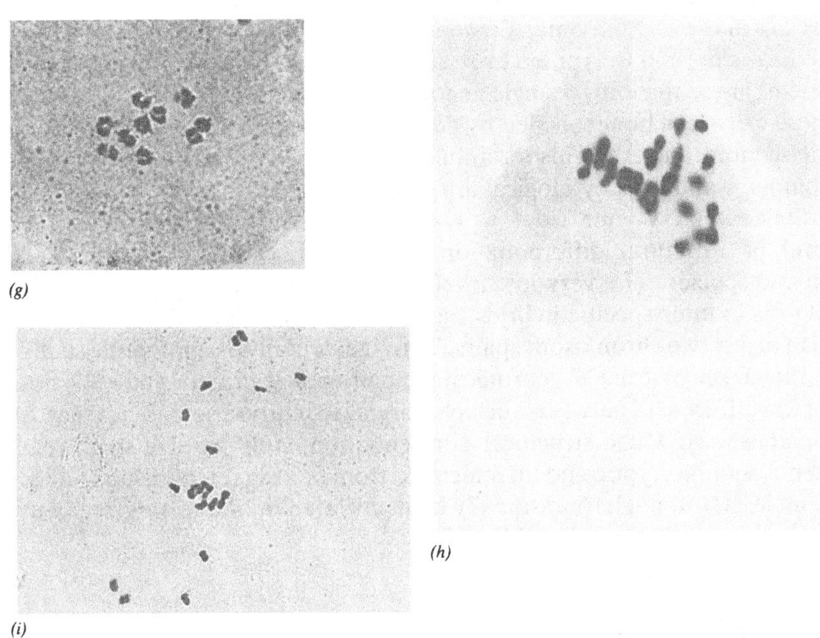

Fig. 3.11. P.M.C. meiotic chromosomes of *Arachis* species. (*a*) *A. hypogaea* (diakinesis) (× 810); (*b*) *A. hypogaea* MI (metaphase) (× 810); (c) *A. glabrata MI* (× 540); (*d*) *A. duranensis* MI (× 810); (*e*) *A. cardenasii* MIO (× 1620); (*f*) *A. helodes* MI (× 1620); (*g*) *A. batizocoi* MI (× 810); (*h*) *A. hypogaea* × *A. helodes* MI (× 1080); (*i*) *A. batizocoi* × *A. correntina* MI (× 540). ((*b*)–(*d*), (*g*) and (*i*), A. K. Singh.

cross-compatibility is due primarily to genetic divergence but also, perhaps, to plasmon differentiation (Ashri, 1976*a*, *b*). This may well be accompanied by extensive chromosome structural rearrangement, although we have no means at present of measuring its extent. Differentiation of genomes within a section can occur and has obviously progressed in section *Arachis*. However, *A. batizocoi* has not diverged genetically from other members of this section to the point where it can no longer hybridise with them, even though the hybrids produced are sterile. Smartt *et al.* (1978*a*) and Stalker and Wynne (1979) ascribe this to chromosome structural rather than genetic divergence. These authors take somewhat different stands on whether the genome of *A. batizocoi* can be regarded as constituting a different genome to the genome of other species in the section. Smartt *et al.* (idem) call them A (typical of the section *Arachis*) and B, typified by *A. batizocoi*, whereas Stalker and Wynne (idem) argue that only a single genome is found in sect. *Arachis*. Perhaps these views can be reconciled by designating the genomes A1 (typical of the section) and A2. This terminology would convey both their genetic homology and their cytological differentiation.

Stalker and Wynne (idem) probably under-estimate, however, the level of structural differentiation between *A. batizocoi* and related diploid species. The very low level of fertility and the highly disrupted meiosis in inter-specific hybrids suggests more structural differentiation than in just two chromosome pairs. This, incidentally, is apparent visually in the absence of the 'A' chromosome pair in *A. batizocoi* and morphological differences between nucleolar organiser chromosomes (J. Smartt, unpublished). Much structural differentiation could involve small segments, and be cryptic and undetectable from pairing relationship studies in meiosis, but might occasionally be manifested in bridge and fragment formation in anaphase I and II.

Evolutionary implications

The production of structural divergence in genomes within sect. *Arachis* provides an insight into a probable mode of evolution for the cultivated groundnut. Extensive chromosome structural changes, such as apparently have occurred in the divergence between *A. batizocoi* and the other diploid species of sect. *Arachis*, effectively reduce and perhaps inhibit gene exchange between diverging forms. The low order of genetic divergence apparent could still permit hybridisation. Extreme structural heterozygosity would render sterile any inter-specific hybrids carrying both structurally differentiated genomes. Doubling of the chromosome complement would provide structurally congruent chromosome pairs in meiosis, and fertility might well improve. Selection for fertility would then tend to reduce multivalent formation.

Since the genus *Arachis* is largely autogamous, a relatively high chiasma frequency is likely to be favoured by selection. This would create no problems in diploids, but in tetraploids this could increase multivalent formation unless it was suppressed or eliminated by chromosome structural reorganisation, or a genetic mechanism such as that in *Triticum aestivum* L. This would establish a diploid meiotic pairing pattern while still permitting high rates of recombination of linked genes following occasional hybridisation. The importance of this is considerable in the contexts of both evolution and practical plant breeding.

Genome evolution in different sections of the genus

From the studies by Gregory and Gregory (1979) of inter-specific cross-compatibility, it is possible to establish tentatively a series of genomes. Some of the taxa concerned represent only a single or a pair of species, i.e. series *Procumbense* (sect. *Erectoides*): *A. rigonii*, *A. lignosa*; series *Prorhizomatosae* (sect. Rhizomatosae): *A. burkartii*; sect. *Triseminale*: *A. pusilla*; and sect. *Caulorrhizae*: *A. repens* and *A. pintoi*. This narrow range of species provides a very restricted base for inference. However, the larger and more species-rich taxa provide reasonably satisfactory bases from which to draw conclusions.

On the basis of crossing relationships established by Gregory and Gregory (1979), it seems probable that the following distinct genomes have evolved.

1. Am: *Ambinervosae*
2. T: *Triseminale*
3. C: *Caulorrhizae*
4. Ex: *Extranervosae*
5. E: *Erectoides* (sub-genomes E1, E2, E3 ± corresponding to series within the genus).

The sections above are all diploid and raise few problems. Section *Erectoides* comprises three series and there may well be corresponding sub-genomes. The situation considered in sect. *Arachis* is rather different; here sub-genomes developed do not conform with the delimitation of series. The series *Annuae* embraces species possessing one or other sub-genomes (A1 or A2), the series *Amphiploides* species probably contain both (A1 + A2), while series *Perennes* species apparently all possess the same sub-genome (A1). The *Rhizomatosae* pose a particular set of problems. Compatibilities *Erectoides* × *Rhizomatosae* and *Arachis* × *Rhizomatosae* are high for inter-sectional crosses. This suggests that the tetraploid rhizomatous species have one genome with *Erectoides* affinities, the other perhaps closer to section *Arachis*. In terms of apparent evolutionary age *Rhizomatosae* is older than *Arachis*, but it is highly unlikely that sect. *Arachis* evolved from *Rhizomatosae*. The diploid rhizomatous *A. burkartii* is genetically isolated from all other *Arachis* species and its affinities remain uncertain. It is known that even within a species such as *A. hypogaea* some genotypes are extremely poor parents in both intra-specific and inter-specific crosses (Smartt, 1965). The failure of *A. burkartii* to cross successfully, therefore, may be a reflection of the genotypes used in the crossing programme rather than of fundamental cross-incompatibility.

The genomes in the two sections *Rhizomatosae* and *Arachis* could be designated R1 (*prorhizomatosae*), R2 and R3 (*Eurhizomatosae*) and A1, A2 for sect. *Arachis*. These suggestions are tentative and could be expected to be modified as plant exploration and experimental hybridisation studies proceed.

3.3. Germplasm resources

While the information which has been obtained by investigators of the systematics of the genus *Arachis*, and of *A. hypogaea*, is of considerable intrinsic scientific interest, it is of even greater importance to those seeking to improve the cultivated groundnut. A soundly based taxonomy of the genus, using morphological characters in the first place, establishes the affinities of the cultigen and indicates the taxa within the genus most likely to be accessible to the breeder. Attempts to delimit biological species in the genus by the study of hybridisation patterns and hybrid behaviour are also of the highest importance. It is perhaps fortunate that the sect. *Arachis*, to which the cultivated groundnut belongs, is probably one of the more recently evolved and most rapidly evolving taxa within the genus at the present time. As a result, barriers to inter-specific gene flow are less than they appear to be in the apparently more ancient sections such as the *Extranervosae* and *Erectoides*.

The general position can be summed up by a definition of ordered gene pools which are available for groundnut improvement. We can consider a first order gene pool which consists of all cultivated varieties (highly selected and landraces), together with all breeding lines derived from them. A second order gene pool would be constituted by *A. monticola* and any other wild tetraploid forms with a similarly high level of cross-compatibility with *A. hypogaea*. The wild diploid species of sect. *Arachis* would comprise a third order gene pool, which should be reasonably accessible to breeders. A fourth order gene pool of low or indeterminate accessibility is constituted by remaining sections of the genus. Some exploitation of this large resource may be possible through the use of bridging inter-sectional crosses, for example sect. *Arachis* × sect. *Erectoides* (Banks, 1974).

It can be anticipated that gene pools of the fourth order will be exploited only in rather exceptional circumstances and that such efforts will be expensive and time-consuming, with little chance of ultimate success. A desirable gene from a species in sect. *Erectoides*, for example, will not necessarily be equally attractive when transferred to a species (e.g. *A. hypoogaea*) in sect. *Arachis*. The most accessible genetic resources will obviously be the most heavily exploited and their actual breeding value is likely to be more predictable.

The characters of wild species which have the most immediate attraction to groundnut breeders concern immunity, resistance and tolerance to pests and diseases. A considerable effort has been devoted to the evaluation of the pest and disease resistance of wild *Arachis* species and, most notably, leafspot resistance has been identified in three species within sect. *Arachis* (Abdou, 1966; Gibbons and Bailey, 1967; Abdou *et al.*, 1974; Seetharam *et al.*, 1974; Nevill, 1978; Foster, 1979; Foster *et al.*, 1979; Company *et al.*, 1982; Gardner and Stalker, 1983; Stalker, 1984). Resistance to nematodes, lesser cornstalk borer, spider mites, rosette virus, stunt virus, peanut rust, tobacco thrips, web blotch and tolerance of southern blight have been reported (Leuck and Hammons, 1968; Kousalya *et al.*, 1972; Kamal, 1976; Simpson, 1976; Banks, 1976; Johnson *et al.*, 1977; Hassan and Beute, 1977; Moss, 1980; Amin, 1985; Subrahmanyan *et al.*, 1985*a*, *b*; Rathnaswamy *et al.*, 1986).

Another area which has attracted some attention is the possible use of wild germplasm to improve the chemical composition of the groundnut in terms of protein and oil content and composition (Cherry, 1977). Groundnut seed protein is unusually low in lysine for a legume seed and, more typically, it is low in the content of sulphur amino acids and tryptophan. Amaya *et al.* (1977) were able to demonstrate a range in protein content of 21.35–33.35% in wild species. Tryptophan content in *A. villosulicarpa* varied between 1.444 and 1.661% per 100 mg protein, somewhat in excess of the best *A. hypogaea* line tested (1.407%). It is quite apparent that further detailed study of protein content and composition is required both in the cultigen and related wild species in order to determine the nature and extent of any protein polymorphisms that may exist and the scope these give for selection. In addition some cost–benefit analysis would be necessary to determine whether breeding effort in this direction would be justifiable economically.

Some physiological features such as drought tolerance might well be transferred from wild species with advantage. Improved general vigour and growth rate, photosynthetic efficiency and more effective *Rhizobium* symbiosis are additional characters which might possibly be improved by introgression. Furthermore, structural and anatomical changes in vegetative and reproductive parts, for example the pod and pegs, might effect useful improvement. The possibilities are very extensive and only just beginning to be appreciated. A full realisation of the potential breeding value of wild species will not come to pass until the wild species and their hybrids among themselves and the cultivated forms, and the progenies of these hybrids, are subjected to intensive study. Their biochemical and physiological behaviour is not well understood, nor is the range of feasible phenotypic manipulation known. Suffice it to say that present efforts have merely scratched the surface; a fuller and more comprehensive evaluation of the breeding value of *Arachis* germplasm resources is an urgent necessity. Conservation of resources is futile without their exploitation and utilisation. Germplasm resources are perhaps unique among our human resources in that their utilisation and exploitation does not necessarily exhaust them, and ideally should never do so.

In order to reap the benefit of our germplasm resources in the improvement of the groundnut crop, it is essential that efficient and effective breeding strategies are developed. It is here that cytogenetic studies fulfil a very important role. As Smartt *et al.* (1978*a*, *b*) have pointed out, since *A. hypogaea* is an allotetraploid which is quite effectively diploidised, the existence of two more or less distinct genomes can be accepted. One genome, held in common with most diploid species of sect. *Arachis*, is much more easily subjected to introgression than the other. This means that genetic improvement of characters controlled by duplicated loci in both genomes is complicated if, as suggested, the two genomes differ substantially as a result of chromosome structural change. If, as further suggested by Smartt *et al.* (idem), species similar in chromosome structure to *A. duranensis* and *A. batizocoi* (but not necessarily these species themselves) are involved in the ancestry of *A. hypogaea*, genetic recombination between the genomes is unlikely to occur very extensively. It may therefore be necessary to induce segmental interchanges and other chromosome structural changes in order to effect specific gene transfers. It might also be possible to produce chromosome addition or substitution lines involving more remote germplasm. Some suggestions have been made on how best to exploit wild species germplasm and these can now be considered.

Breeding strategies for the exploitation of wild species germplasm

In devising breeding strategies for the incorporation of exotic germplasm in the cultivated groundnut the following considerations must be borne in mind. The probable presence of two structurally differentiated genomes in *A. hypogaea* has dual implications. Firstly, the arrangement of chromosome segments in the two genomes will determine the ease with which the necessary introgression can be achieved. Secondly, the high level of genetic homology which probably still exists between the two genomes implies that many qualitative characters may be under the control of duplicate loci.

Transferring a desirable dominant character may present few problems; if it were recessive and duplicate inheritance occurred, producing homozygosity at homologous loci in both genomes would be very difficult. A less serious problem could be encountered where the genes in question had an additive effect; however, maximum expression could not be achieved unless both genomes were introgressed.

Ploidy level manipulation

It is fortunate that differences in the ploidy level of *Arachis* species are not in themselves barriers to hybridisation and may not be insuperable barriers to gene flow. There is, therefore, the possibility of operating effectively at different ploidy levels despite the problem of reduced fertility. This is important when breeding materials can range from diploid to hexaploid as is the case in *Arachis*.

The question of breeding for improved leafspot resistance provides a good illustration of the nature of the problems involved. Leafspot is caused by two species of fungi, *Cercosporidium personatum* (Berk. & Curt.) Deighton and *Cercospora arachidicola* Hori (late and early leafspot respectively). *A. cardenasii* has been reported as immune to *C. personatum* and *A. chacoense* as resistant to *C. arachidicola* (Abdou, 1966). In most groundnut-growing areas, resistance to both pathogens is desirable. It would therefore be pertinent to consider whether each resistance should be bred into the cultigen separately or whether, as Smartt *et al.* (1978*b*) suggest, it would be more efficient to combine both resistances at the diploid level and then cross a doubly resistant segregant to the cultigen. Such a cross would be triploid and more or less sterile. It is frequently possible to produce hexaploids from such triploids artificially and/or spontaneously (D'Cruz and Chakravarty, 1961; D'Cruz and Upadhyaya, 1962; Smartt and Gregory, 1967; Spielman and Moss, 1976) and backcross these to the cultigen to produce pentaploid progeny. Pentaploids in *Arachis* have variable fertility but those capable of reproduction would probably lose chromosomes in meiosis and tend to produce progeny whose chromosome number would stabilise at the tetraploid level.

Selection for both resistances could be practised and a doubly resistant tetraploid breeding line produced. Moss (1980) suggests an alternative strategy of crossing one diploid species to *A. hypogaea*, doubling the chromosome complement of this F1 hybrid and crossing the resulting hexaploid to a second diploid species to produce a tetraploid. However, this tetraploid could have (3 A1 + 1 A2) genomes and might be of reduced fertility as Smartt *et al.* (1978*b*) suggested. Both these alternative strategies, and the modification of inducing polyploidy before hybridisation with *A. hypogaea* suggested by Stalker and Wynne (1979), are probably worthy of trial. In addition the possibility of obtaining useful recombination between genomes (A1 and A2) in raw amphidiploids should not be overlooked.

Sharief *et al.* (1978) conclude that leafspot resistances are controlled multi-factorially. It would appear that some improvement in the level of leafspot resistance in the cultivated groundnut might therefore be achieved by introgression of the more accessible A1 genome but that this would not be maximised until both genomes were effectively introgressed.

The basic strategy suggested here could be employed with diploid species within sect. *Arachis* for a range of possible improvements. Results obtained to date suggest that this approach could be productive. Bridging the inter-sectional gaps is a very different problem and one likely to prove difficult. It would probably involve further development of techniques for anther, embryo and tissue culture as well as investigation of the physiology of differentiation in cultured cells and tissues (Bajaj *et al.*, 1982; Sastri and Moss, 1982). Where conventional hybridisation fails, protoplast fusion may yet succeed (Rugman and Cocking, 1985). However, it must be borne in mind that there is a distinct possibility that the genomes from different sections may be developmentally antagonistic, and preclude both normal reproductive processes and normal growth and development. Similar considerations may also apply to single chromosome pairs if these are substituted for homoeologues in a genome or added to it. Obviously wide crosses in *Arachis* from the standpoint of groundnut improvement are a last resort.

In conclusion it can be stated that we are indeed fortunate that the cultivated groundnut is a member of a recently evolved section of the genus, unlike *A. villosulicarpa*, the only other cultigen of long standing in the genus. Within section *Arachis* most if not all of the genetic resources should be accessible to the breeders. It is possible that more remote genetic resources than these might be mobilised, but the difficulties can be expected to be greater and the results less certain. Nevertheless, all the genetic resources within the genus should be properly evaluated.

3.4. The evolutionary synthesis

In considering the evolution of any cultigen the questions needing to be addressed are threefold: what happened? where did it happen? and, more problematically, how and when did it happen? Only partial answers can be given, but these enable us to break down broad and perhaps ill-defined questions into ones that are regrettably more numerous but more specific and better defined. The fact that we can now ask very much more specific and penetrating questions than we could a decade ago can be taken as a measure of our progress.

What happened?

A tetraploid cultigen evolved in a section of the genus *Arachis* in which the majority of the taxa known and described are diploid. The only described wild tetraploid in the relevant section (*Arachis*) of the genus is obviously very closely related to the cultigen and differs principally in characters which, on the one hand, are essential for survival in the wild but on the other are highly disadvantageous in cultivation. This form, *A. monticola*, is obviously close to the progenitor type. In considering polyploid crop species the nature of the polyploidy itself (allopolyploidy versus autopolyploidy) and the source or sources of the genome or genomes both require investigation.

Where did it happen?

It is clearly apparent (Gregory *et al.*, 1980) that the centre of distribution of the genus is in Brazil in the area of Mato Grosso and its neighbouring states. For a time it was assumed that this was the probable centre of origin of the cultigen also. The distribution of sect. *Arachis*, to which the groundnut belongs, is in fact peripheral, for the most part to the south and east of the main area of distribution. The distribution area of *A. monticola* as far as is known is very restricted, in the eastern foothills of the Andes in north-west Argentina and south Bolivia. This region is a key area in the search for diploid putative ancestral forms (Krapovickas and Rigoni, 1956).

How did it happen?

The development of polyploidy in *Arachis* is by no means confined to sect. *Arachis*; in fact it is even more strongly developed in sect. *Rhizomatosae*, in which most taxa are tetraploids. The origin of polyploidy in these two groups has been quite independent. Significantly, there are some parallels in the consequences of polyploidy in the two groups. In the *Rhizomatosae*, the tetraploids are much more vigorous and competitive than the only known diploid, while the tetraploid cultigen *A. hypogaea* has proved to have been more effective in establishing itself as a crop than the diploid domesticate *A. villosulicarpa*. Under domestication *A. hypogaea* has produced a great diversity of forms both in South America and in post-Columbian times elsewhere.

The difference in the course of events in the *Rhizomatosae* and sect. *Arachis* may be due to the possibly different course of events in the production of the tetraploids. The species in sect. *Arachis* are thought to be of comparatively recent evolutionary origin and this is supported by the fact that, although morphological divergences have evolved, isolating mechanisms are not very effective. But *A. hypogaea* is a true amphidiploid, a sexually reproducing derivative of sexually reproducing diploid species. In the *Rhizomatosae*, only one diploid taxon has, as yet, been collected and described. The probable course of events in the polyploidisation is obscure but the outcome is clearly apparent.

Although much rhizomatous material has been collected, only one diploid species, *A. burkartii*, has been identified and only two tetraploids, *A. glabrata* and *A. hagenbeckii*, have been described. There is no lack of morphological diversity in collected rhizomatous material but it has proved to be difficult to delimit species satisfactorily, on the basis of morphological discontinuities. Seed production has almost been suppressed in the tetraploids and reproduction is largely vegetative. This has possibly resulted in a proliferation essentially of somatic mutants and made the section a critical taxonomic group. Whereas polyploidy in sect. *Arachis* is clearly allopolyploidy, it is possible that in the *Rhizomatosae* it is autopolyploidy. This could produce meiotic difficulties and have given vegetative reproduction a substantial selective advantage.

When did it happen?

Beyond the conclusion that the production of the amphidiploid ancestor of the cultivated groundnut must have been a relatively recent event in the evolutionary history of the genus, it is not possible to go. Since speciation in this section is itself probably a recent event, polyploidisation must of necessity have been even more recent. At the present time it is not possible to suggest a time scale for these events. A similar problem also exists in setting a time scale for the domestication and evolution of the cultigen. Where physical remains can be found, radiocarbon dating may be possible. The status of preserved ancient materials, as wild collections or domesticates, can be extremely difficult to determine. Although evidence for the practice of agriculture only goes back 10000 years, this is a very conservative estimate indeed. The duration of the pre-agricultural phase of plant domestication might have been quite protracted.

With regard to archaeological evidence, it is a matter of blind chance and coincidence whether materials capable of preservation are actually deposited in conditions favourable for this to occur. Archaeological datings of groundnuts go back less than four thousand years in Peru: 1200 to 1500 BC. Findings from other areas such as Mexico and Argentina are of much more recent date, the earliest of which have been dated in the first millennium AD from the Tehuacan valley in Mexico (AD 100) and Costa de Royes and Catamarca, Argentina (AD 600) (Hammons, 1973*a*, 1982; A. Krapovickas, unpublished, 1975). Since the very oldest archaeological materials are from an area favourable for the preservation of plant materials but outside the crucial geographical area, it is reasonable to conclude that the history of the groundnut under domestication pre-dates 1500 BC but by how much it is impossible to estimate.

3.5. Evolution of the domesticated form

It is a curious fact that evolutionary studies of food legumes have tended to concentrate on pin-pointing the geographical area of domestication and identifying the ancestral or progenitor type. Much less attention has been paid to genetical analysis of the changes which have been established under domestication *vis-à-vis* the ancestral condition. In order that such analyses can be carried out effectively, the probable evolutionary sequences need to be worked out so that testable hypotheses can be generated.

Morphological changes

In the cultigen there is a range of growth form showing greater or lesser departure from that of the progenitor type, as represented by *A. monticola*. The growth habit of all wild species within sect. *Arachis* is uniform. Essentially, a relatively short vertical main axis is produced, and four major horizontal branches arise in the axils of the cotyledons and the first pair of foliage leaves. Further branching customarily occurs on these four major branches, and this branching is commonly of two types, alternate and sequential. In alternate branching pairs of axillary vegetative branches are produced in an alternating sequence with axillary inflorescences, while in sequential branching there may be either a basal sequence of axillary vegetative branches followed by a distal sequence of inflorescences, or an unbroken sequence of axillary inflorescences. Both branching systems have been found in *A. monticola* (Gibbons, 1966).

Growth habit

The range of variation found under domestication includes forms essentially similar in growth form to wild species. These are the most prostrate runner genotypes in which the main axis remains short and the lateral branches grow horizontally in close contact with the soil. The lateral branches can grow to 1 m in length or more. Other runner genotypes are known in which there is a tendency towards ascending growth of the branches, but the length of branches themselves dictates horizontal growth although with ascending tips. A tendency to the production of shortened internodes has produced more compact growth habits with both alternate and sequential branching patterns. Genetic analysis of growth form has been attempted and the subject has been reviewed by

Hammons (1973*b*) and Wynne and Coffelt (1982) and is complex; it may involve both nuclear and cytoplasmic factors and their interaction. In some schemes as many as five loci have been suggested. This complexity is not surprising in view of the allopolyploid nature of the groundnut; a complementary gene situation in the ancestral diploids would become more complex in the polyploids and could involve additive gene action in addition to epistatic effects. This is borne out by the observation that novel growth habits can arise in segregants from crosses between parents differing in growth habit.

Plant type

Among the most widespread cultivated groundnuts, the plant types commonly found can be designated Spanish, Valencia and Virginia. It is

(*a*)

Fig. 3.12. *Arachis monticola*, the presumed wild prototype of *A. hypogaea*. (*a*) A young plant showing initial development of the prostrate growth habit. (*b*) Collection localities of diploid species of sect. *Arachis* in relation to the approximate area of *A. monticola* distribution and its probable area of origin (circled). ((*b*) from Simpson, 1984.)

reported (Hull, 1937) that Valencia type plants arise from crosses Spanish × Valencia in F2. The genetic constitutions postulated for the three types are Valencia va1 va1, va2 va2, Spanish Va1 Va1, va2 va2 and Virginia va1 va1 Va2 Va2.

Branching patterns

There are two aspects of branching to be considered, the pattern of branching alternate or sequential on the primary branches in the first place, and secondly the extent of production of secondary and tertiary branches, low in Valencia and Spanish and high in Virginia. The alternate pattern and abundant branching both behave as dominant traits but no entirely satisfactory genetic analysis has been produced since deviant branching patterns are also known (IBPGR, 1985).

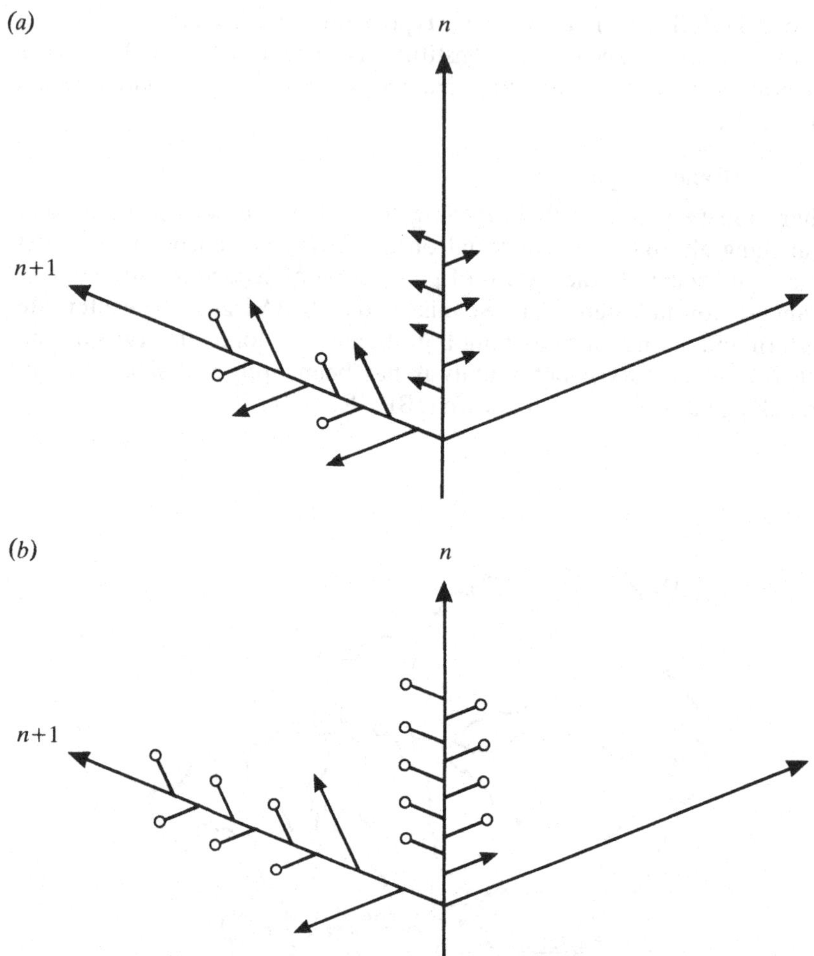

Fig. 3.13. Branching patterns of *Arachis hypogaea*. (*a*) Alternate alternating pairs of vegetative and reproductive branches on laterals. (*b*) Sequential: sequences of reproductive branches on laterals.

Inflorescence location

A consistent morphological difference between the Virginia and the Spanish–Valencia groundnuts is that no axillary inflorescences are produced on the main axes of Virginia cultivars, in contrast to their occurrence in leaf axils on those of Spanish–Valencia types. The character difference has been analysed by Hammons (1971); he postulated two pairs of interacting loci J1, J2 and K1, K2. The interaction of J1 or J2 with K1 or K2 produces vegetative branches in main-stem leaf axils.

Leaf colour

Characteristically, leaves of Virginia groundnuts are darker green than those of Spanish or Valencia. Two different explanatory hypotheses have been advanced, monogenic with either complete dominance (Badami, 1923) or incomplete dominance (Balaiah *et al.*, 1977). The leaf colours seen in the cultigen have parallels in the diploid species, with annuals tending to be lighter in colour than the perennials. Simple genetic differences, arising from mutation in the ancestral amphidiploid, could explain the establishment of the two leaf colours in the cultigen.

Flower colour

The range of flower colour found in wild *Arachis* species is not great. Flowers are either orange-yellow or yellow without any suggestion of red pigments. In the cultigen the orange-yellow colour predominates, but brick-red flowers are found in a number of geographical races, most notably 'Nhambyquarae'. White flowers are also known. Genetic control is relatively simple; Hammons (1973*b*) quotes reports of simple, monofactorial inheritance with incomplete dominance, or with a pair of duplicate genes and complete dominance.

Seed size

There is a considerable range of seed size in wild *Arachis* species. Among close relatives of the cultigen, that of *A. cardenasii* is very small, while that of *A. correntina* is relatively large, approaching that of the smallest-seeded genotypes of *A. hypogaea*. The largest-seeded forms of *A. hypogaea* are without doubt the 'Nhambyquarae' lines, with individual seed weights of ± 1 g. The smallest-seeded *Arachis* species have seed weights in the range of 20–50 mg.

Seed dormancy

Seed dormancy relationships have not been well studied in the genus. In the cultigen two distinct types of dormancy behaviour occur, a complete absence of dormancy in the Spanish–Valencia cultivar groups and a short-term dormancy of a few months in the Virginia group (Bhapkar *et al.*, 1986). In wild species it may be difficult at times to induce germination. Inhibition of germination is almost certainly due to physiological mechanism since the paper-thin testas are not impermeable to water. It may even be that viability of seed in some species may be short-lived; oil-rich seeds can produce effectively lethal concentrations of free fatty acids and their derivatives if oxidation and/or hydrolysis of the stored oil occurs. Stability of stored oil is important under domestication, both as it affects viability of the seed and its organoleptic qualities.

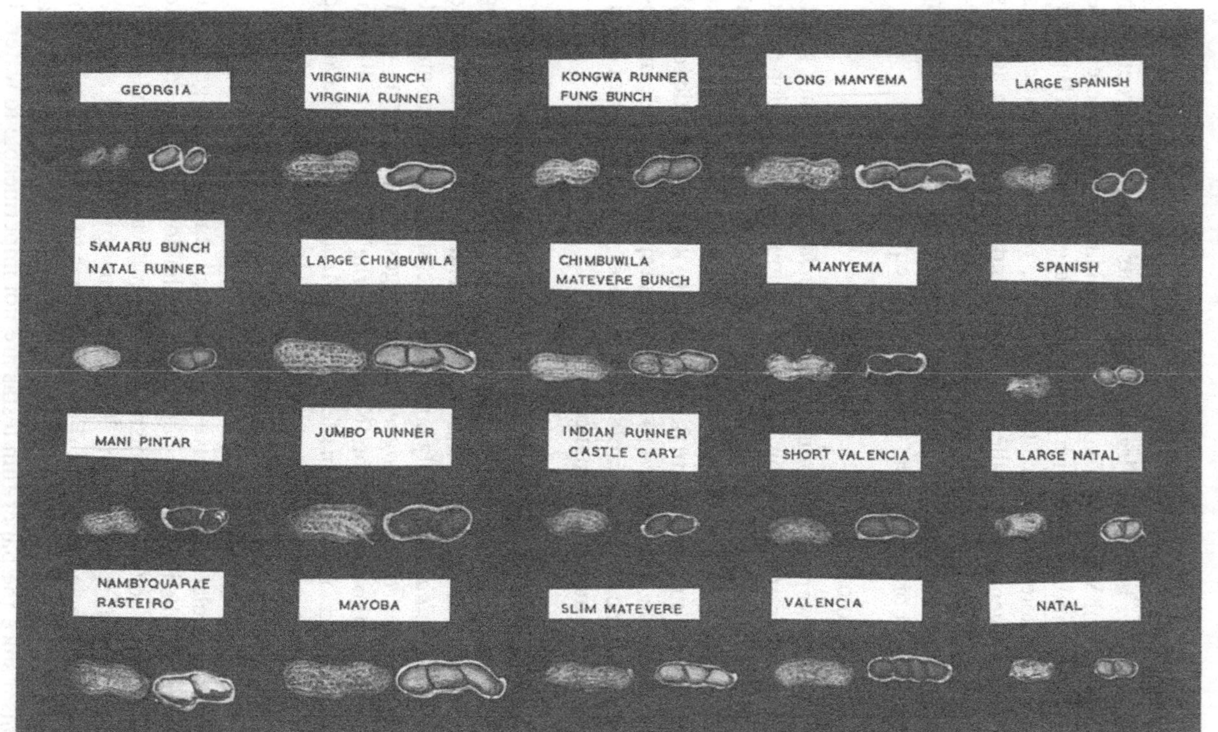

Fig. 3.14. Variation of pod morphology within *A. hypogaea*.

Pod and seed morphology and fruiting habit

Wild *Arachis* species share a common pod morphology. Pods are extremely lomentiform. Not only is the pod strongly constricted between seeds, but the constricted portion of the pod is very extended. In the development of the pod, meristems arise at the base of the ovary and in the pod between the seeds. The whole pod structure can become extremely elongated and even exceed 1 m in length (Gregory *et al.*, 1980). The wild counterpart of *A. hypogaea* has essentially a typical lomentiform pod. Under domestication it would appear that a reversion to a more basic form of legume pod has occurred. This has come about by suppression of meristematic activity in portions of the pod between seeds. In addition the length of peg between the inflorescence axis and the proximal portion of the pod has been reduced. These changes concentrate pods and seeds in a smaller soil volume, in much the same way that the tubers of cultivated potatoes are concentrated by production of shorter stolons than those of wild species. These effects are accentuated by change of growth habit from procumbent to erect, which serves to concentrate fruiting close to the crown of the plant.

Pod morphology has very important commercial consequences. For many purposes, especially in the confectionery trade, uniform (or near uniform) size and shape are extremely important. In practice the ideal pod type is one with characteristically two seeds and a moderately constricted pod. The least desirable pod type is without constrictions, in which seed size is variable and flat contact faces develop on the seeds. The end seeds will each have one such face and the intermediates two. Variability in seed size is greater between seeds in unconstricted pods than in those with constrictions. Extremely constricted pods, in which the pod walls closely invest the seed, are also disadvantageous since they are difficult to shell, either by hand or by machinery, and the proportion of damaged seed is often very high. Another determinant of seed shape is pod length. In the highest quality, large-seeded (Jumbo) groundnuts, the seed is relatively elongated, whereas in the Bolivian maní pintado landraces the proportions are different, with rather shorter seeds related to the short pods characteristic of these genotypes.

Seed size, shape and uniformity are characters of importance to processors who are reluctant to accept substantial changes from the standards to which they are accustomed. The confectionery trade has exacting standards in this regard while the oilseed trade is more accommodating. This is especially so as more oil expression is tending to occur in producing countries. A character which seems trivial, but of by no means negligible economic importance, is testa colour. The preferred colour is brown, ranging from light tan to russet brown. Pink testas which become brown with age are acceptable but darker colours, red or purple, can be discriminated against. White or yellow testas would probably be ideal were it not for their susceptibility to staining and discoloration. The curious thing is that for confectionery purposes the seed is blanched and the testa removed and it is only for the 'roasting in the shell' trade that the testa would remain. The remains of pigmented testas do not pose any serious problems in groundnut seed cake after oil expression. However capricious some commercial standards appear to be, neither the producer nor the breeder can afford to disregard them.

3.6. Classification and nomenclature of the cultigen and its variant forms

Certain difficulties arise in the taxonomic treatment of biological species which include both wild and domesticated forms. The situation is well examplified by the groundnut. The biological species to which the groundnut belongs obviously includes *A. monticola* (Hammons, 1970) and logically the nomenclature should reflect this. *A. monticola* could be recognised as a sub-species (ssp. *monticola*) of *A. hypogaea*; the cultigen could be referred to as sub-species *hypogaea*. However, Krapovickas (1968), after a detailed study of the variability of the cultigen and the geographical range and distribution of the variants, proposed the following taxonomic breakdown of the cultivated forms.

Arachis hypogaea L.
- ssp. *hypogaea*: alternately branched forms.
  - var. *hypogaea* (Virginia group): short main axis, Bolivia, Amazonian region.
  - var. *hirsuta* Kohler: long main axis, Peru (*vide* Banks, 1986).
- ssp. *fastigiata* Waldron: sequentially branched forms.
  - var. *fastigiata* (Valencia): simple inflorescences, Guarani region, Goias, Minas Gerais, north-east Brazil and Peru.
  - var. *vulgaris* Harz (Spanish): compound inflorescences, Guarani region, Goias, Minas Gerais and north-east Brazil.

Gibbons *et al.* (1972) produced a scheme of cultivar classification in which each taxon is broken down into a series of cultivar groups on the basis of differences in vegetative morphology and that of pods and seeds. Their scheme was based largely on African material and some recent introductions of South American provenance. This type of breakdown has not yet been attempted on the bulk of South American germplasm collections but it could be very instructive.

The inclusion of *A. monticola* in a comprehensive taxonomy of the biological species *A. hypogaea* could be accomplished by designating the sub-species of the Krapovickas scheme as botanical varieties and the botanical varieties as *formae*. It is a moot point whether this would be acceptable. Nevertheless there is a very definite patterning of the morphological variability in *A. hypogaea* which lends itself to a hierarchical treatment; the nomenclature of these levels does present some difficulty. Recognition and subordination of the levels of the hierarchy within the cultigen presents few problems, in African material at least. Probably the most serious problem in varietal classification of the groundnut arises from the fact that irregular forms of branching are to be found in some South American landraces, which are neither strictly alternate nor sequential. It is possible that allocation to one or other of the two major divisions might be made on correlated characters, e.g. presence or absence of reproductive branches on the main axis (IBPGR, 1985).

3.7. Recent evolutionary history in relation to germplasm resources

Tracing the ancestry of a polyploid cultigen is only a part of the evolutionary studies that have been undertaken. Arguably the most practically useful germplasm resources are in the main those which have evolved subsequent to domestication. The use of these will entail fewer problems than those provided by wild species. The post-domestication history of the groundnut has been reviewed by Krapovickas (1968, 1969) and Hammons (1973*a*, 1982). Possibly the most significant point to emerge from Krapovickas' study of the distribution of variability in American groundnut landraces is the very wide distribution of the crop in South and Middle America, which had probably been achieved in pre-Columbian times. It was probably being exploited agriculturally in suitable habitats from what is now Argentina in the south to Mexico in the north. Expansion of cultivation into most areas of North American production can be regarded as essentially post-Columbian. Although this is commonly thought to have arisen from African introductions there is the possibility, as Hammons (1973*a*, 1982) points out, that introduction to North America could have been from the Caribbean.

The actual centre of origin of the groundnut is thought to be in the general area of north-west Argentina – south Bolivia (Krapovickas and Rigoni, 1956); there is general agreement on this point (Gregory and Gregory, 1976; Smartt *et al.*, 1978*a*). Earlier it was thought that it probably originated closer to the centre of origin or diversity of the genus in the Mato Grosso (Krapovickas, 1973). The botanical, phytogeographical, cytogenetic and biochemical evidence all support a peripheral rather than a central origin of the cultigen. From this probable point of origin (i.e. north-west Argentina – south Bolivia (Krapovickas and Rigoni, 1956)) dispersal of the domesticate has occurred and in the course of time centres of diversity have developed in South America. Krapovickas (1968, 1969, 1973) recognises five such centres while Gregory and Gregory (1976) postulate the existence of six, adding north-east Brazil to the five centres of Krapovickas, namely (1) the Guarani region, (2) Goias and Minas Gerais (Brazil), (3) Rondônia and north-west Mato Grosso (Brazil), (4) the eastern Andean foothills of Bolivia, and (5) Peru. Krapovickas (1968, 1969) has also pointed out that exploration of areas to the north of the identified centres is incomplete; the most important is the Caribbean area. Here, however, there may be very considerable difficulties in identifying local landraces. In much of the Caribbean area the indigenous population has virtually disappeared and been replaced by immigrants from Europe, Africa and Asia. The disappearance of indigenous people from the Caribbean islands could have initiated genetic erosion and extinction of local landraces. Hammons' (1973) suggestion that introduction of the groundnut to North America could have been from the Caribbean and not from Africa cannot be ignored. However, if such Caribbean materials were themselves introduced from Africa rather than indigenous, then North American germplasm could still substantially be of African origin. Be this as it may, typical African, North American and Caribbean germplasm of subsp. *hypogaea* is of Virginia type. It is also true to say that it is in the African continent that the greatest variation on the Virginia theme is produced and this variation has an identifiable geographical basis. The spectrum of variants produced in Central Africa is characteristic of this region and quite distinct from that of West Africa. If one follows Gregory and Gregory (1976) and regards Krapovickas' genocentres as secondary centres of diversity, then the African centre could be regarded as a tertiary centre. It is rather perplexing that, apparently, no other centres of diversity have evolved elsewhere in the Old World where the groundnut has been introduced and cultivated for as long as in Africa, that is in Asia and Europe. This may not be unrelated to the African cultivation of another geocarpic legume, the groundbean or Bambara groundnut (*Vigna subterranea*). The obvious superiority of the groundnut may have led to its enthusiastic adoption as a partial substitute for an indigenous crop rather than as a complete novelty. Acceptance may therefore have been earlier and exploitation more intensive than elsewhere in the Old World.

As Hammons (1973*a*, 1982) has pointed out there is virtually no documentation of the early post-Columbian movement of the crop. Reliable documentary evidence dates from the seventeenth century and subsequently; however, the evolution of distinctive geographic forms has enabled the paths of migration to be inferred. The Peruvian type (*A.h.*

hypogaea hirsuta) is a very distinctive groundnut and this has been noted in Mexico, indicating a pre-Columbian dispersal, and in China and elsewhere in Asia as a result of post-Columbian movement. The probable paths of movement have been traced from Acapulco in Mexico to Manila in the Philippines. The Spanish encountered the groundnut in Hispaniola when they established themselves there in the early sixteenth century. It is therefore possible that movement of the crop to both Europe and Africa could have occurred soon after the establishment of the Spanish and Portuguese in the New World. The Portuguese, with large African colonies as well as Brazil, probably transported groundnuts (along with maize) to Africa and broad affinities between African and Brazilian landraces of New World crops are often apparent.

Once in Africa, New World crops which have been accepted have apparently generated variability, which is perhaps largely recombinational in nature, but with some uniquely African characters emerging, such as the strongly striated type of pod found in West Africa (C. Harkness, personal communication). The genetic variability which has evolved in Africa is clearly patterned and has been discussed and described by Gibbons *et al.* (1972). However, no comparable study of the greater and more complex variability found in the New World has yet been attempted. There are indications that many of the distinctions easily made on African material, e.g. into alternately and sequentially branching forms, are more difficult to make on South American landrace material (IBPGR, 1985). A monographic analysis of the variability found in the South American landraces is long overdue. This would greatly facilitate its effective utilisation. At the present time this has not apparently been done on a systematic basis for older collections, which contain much invaluable germplasm.

3.8. The present state of germplasm resources for groundnut improvement

Harlan and de Wet (1971) introduced the concept of a series of gene pools of varying accessibility for the improvement of crop plants. This concept can usefully be applied to legumes in general and particularly effectively to the groundnut. The primary gene pool (GP1) is co-extensive with the biological species, that is all those forms which can inter-breed freely. In the groundnut this includes all forms of the cultigen plus the wild *A. monticola*. Harlan and de Wet distinguish two components of the primary gene pool: A, that part represented by the genetic variability of the cultigen, and B, that of wild races. In practical terms, these represent first and second choices for plant breeders in exploiting genetic resources. In the groundnut all the enormous variability found in the cultigen constitutes the A component, while the B component is represented by the comparatively minuscule genetic variability of *A. monticola*. An exploitable secondary gene pool (GP2) exists for the groundnut, comprising the whole of sect. *Arachis*. Curiously enough this is at the present time being exploited as a source of resistance to the early and late leafspot fungi in preference to that which exists in some genotypes of the cultigen itself. The reason for this is simple: resistance has been studied more intensively in the wild species and has been identified and characterised better than in the germplasm collection of the cultigen. It is questionable perhaps if this strategy is justifiable in the long term. The present size of germplasm collections and the lack of characterisation can be a very considerable disincentive to their exploitation. While recognising that collection and conservation are prime first steps in countering genetic erosion, the question of characterisation must be addressed urgently. We would otherwise be failing in our duty to plant breeders, who would be forced by our dilatoriness to attempt effecting difficult inter-specific gene transfers, when comparatively simple intra-specific crosses might well give as good or better results for less effort.

The question of input of effort also applies to the exploitation of the very extensive tertiary gene pool (GP3) constituted by sections of the genus other than sect. *Arachis*. The exploitation of this genetic resource is likely to be extremely difficult and the results, if successful, could be disappointing. The real question that must be asked is, whether we are devoting our efforts to what is really necessary and likely to be most effective in improving the groundnut crop, or to the more appealing

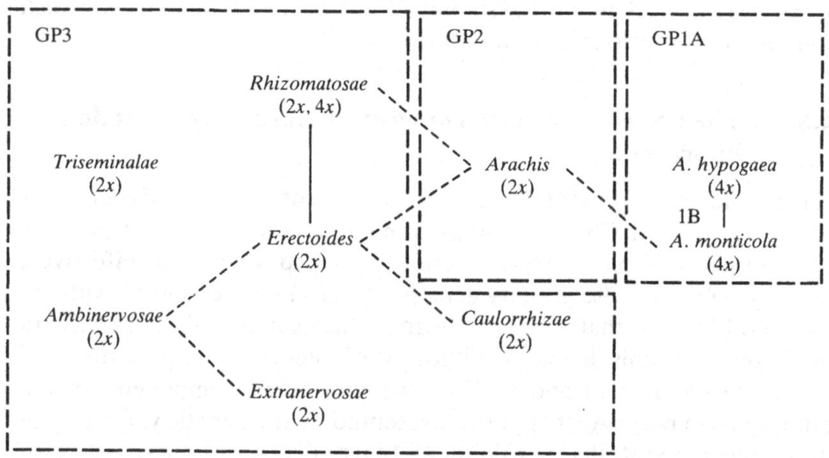

Fig. 3.15. Accessibility of genetic resources of *Arachis* for use in groundnut breeding.

Fig. 3.16. An *Arachis* wild species germplasm nursery (C. E. Simpson).

research problems. Germplasm evaluation and documentation could degenerate into a tedious and routine exercise if handled unimaginatively. With flair and enthusiasm such an endeavour could substantially extend and deepen our understanding of the evolutionary processes which have produced our crop plants.

4 The New World pulses: *Phaseolus* species

It is probably true to state that the evolution of cultigens within the genus *Phaseolus* is as well understood as that of any grain legume species. This is because a great diversity of pertinent evidence has come to light from a range of disciplines in addition to the biological, principally from archaeology and chemistry. At the present time *Phaseolus* beans are widespread in use both as pulses and green vegetables, particularly the common bean *P. vulgaris*, and there is considerable interest in their improvement emanating from a wide range of interests. They are of very considerable importance as a subsistence crop in Central and South America as well as in parts of Africa. They are also of commercial interest to the canning and frozen food industries in addition to pulse merchants. Quite obviously the requirements of all these markets are very different. There is therefore the broadest possible interest in the range and extent of the germplasm resources existing in the genus, which can be mobilised to meet the enormous range of actual and potential breeding objectives. The incentives for extensive collection, efficient conservation and evaluation are therefore very considerable and there is ample economic justification for investment in these activities.

One of the most remarkable features of the cultigens of this genus is that they combine very similar basic patterns of morphological divergence from ancestral forms with very marked agro-ecological differentiation. This has resulted in the evolution of cultigens within the genus which can be produced successfully in a climatic range from the cool temperate climates of northern Europe and America to humid tropical environments in America, Asia and Africa. This situation is in marked contrast to that of the Asiatic *Vigna* species, which cover a much narrower climatic and agro-ecological range. Over the whole range of agricultural environments *Phaseolus* species can be regarded as being complementary to each other rather than in competition.

The economic significance of *Phaseolus* beans, although considerable world wide, varies markedly in different geographical areas. The consumption of pulses shows an enormous range; in parts of Asia and in most

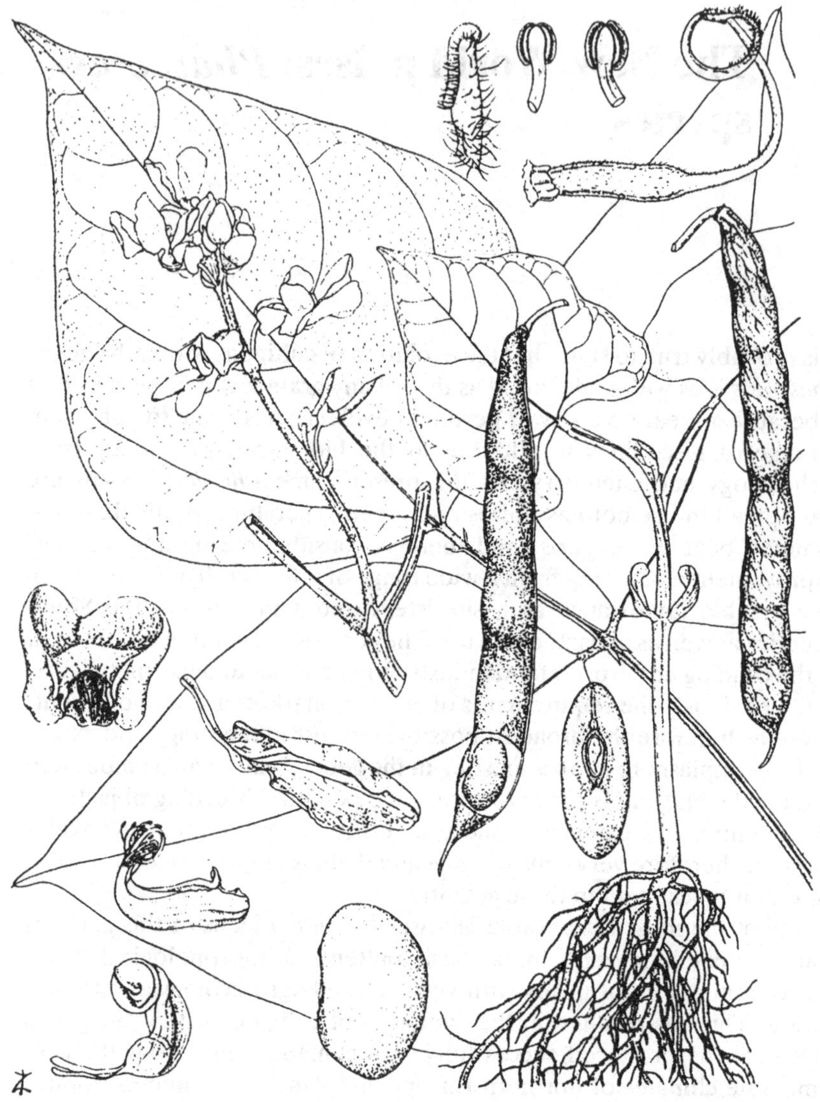

Fig. 4.1. Common bean, *Phaseolus vulgaris* cv. Purple Bedelle (from Westphal, 1974).

Fig. 4.2. Runner bean, *Phaseolus coccineus* cv. Jima Giant (from Westphal, 1974).

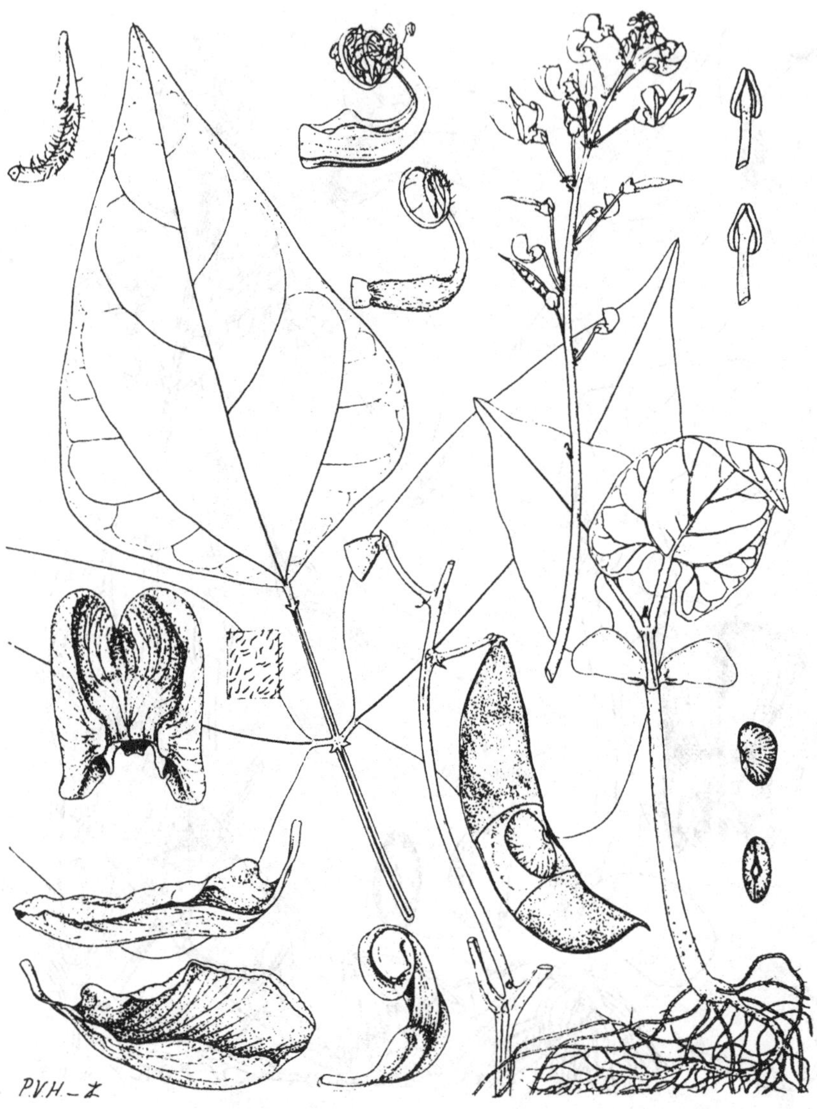

Fig. 4.3. Lima bean, *Phaseolus lunatus* cv. Majo Moon (from Westphal, 1974).

of Latin America it is considerable, while in northern Europe consumption is low, although this is to some extent compensated for by significant consumption of green beans as a vegetable. The consumption of pulses in the USA is relatively high for a developed economy (Smartt, 1976*a*) but on the whole there is a 'North–South' divide in which pulses make a relatively greater dietary contribution at present in developing than in developed economies. The situation is, however, anything but static. In the developing world where 'Green Revolution' agriculture has become established, pulse production has suffered in competition with that of the new 'miracle' cereals. This has exposed one of the major shortcomings of the Green Revolution strategy, based as it was on the promotion of a single staple crop (be it rice, wheat or perhaps maize) rather than balanced farming systems. A corollary of this has been the undermining of the local agricultural economy, rendering at one and the same time the production of subordinate crops unattractive and that of staple foods uneconomic to all but the most efficient producers. Research on legume grain crops is a very practical step to retrieve this situation and restore hope in the broad agricultural community in the developing world. While the Green Revolution itself clearly demonstrated that an agro-economic revolution was possible, it also pinpointed and underscored the very delicate economic, political and sociological balances involved.

4.1. Classification and biosystematics

Taxonomy of *Phaseolus*

The genus *Phaseolus* is now as well understood in biosystematic terms as most other economically important genera in the Leguminosae. This has arisen from the convergent activities of plant breeders, geneticists and taxonomists interested in exploring the biosystematic relationships of beans for their own sake or for the exploitation of this knowledge in the improvement of bean crops. The culmination of this endeavour was the production of a monograph of *Phaseolus* and related genera by Maréchal *et al.* (1978). This revision has been generally welcomed, its broad findings readily accepted and subjected to criticism only on matters of detail (Smartt, 1981*b*). The genus *Phaseolus* as defined is a very distinctive and natural group with a clearly defined American geographic base, probably of comparatively recent evolutionary origin as compared with the arguably more ancient but related genus *Vigna*.

The scheme of classification shown in Table 4.1 can be criticised on the following relatively minor grounds. It is unfortunate that the infraspecific taxa in *P. vulgaris*, *P. coccineus*, *P. acutifolius* and *P. lunatus* are not recognised consistently at the sub-species level since they are

Table 4.1. *Taxonomy of* Phaseolus *L.*

Genus		*Phaseolus* L.
Section		*Phaseolus*
	1	*Phaseolus vulgaris* L.
		var. *vulgaris*
		var. *aborigineus* (Burk.) Baudet
	2	*Phaseolus coccineus* L.
		ssp. *coccineus*
		ssp. *obvallatus* (Schlecht.) Maréchal, Mascherpa & Stainier
		ssp. *formosus* (H.B.K.) Maréchal, Mascherpa & Stainier
		ssp. *polyanthus* (Greenman) Maréchal, Mascherpa & Stainier
	3	*Phaseolus glabellus* Piper
	4	*Phaseolus augustii* Harms
	5	*Phaseolus acutifolius* A. Gray
		var. *acutifolius*
		var. *latifolius Freeman*
	6	*Phaseolus filiformis* Bentham
	7	*Phaseolus angustissimus* A. Gray
	8	*Phaseolus wrightii* A. Gray
	9	*Phaseolus grayanus* Woot. & Standley
	10	*Phaseolus polystachyus* (L.) Britt, Sterns & Pogg
		var. *polystachyus*
		var. *sinuatus* (Nutt.) Maréchal, Mascherpa & Stainier
	11	*Phaseolus pedicellatus* Bentham
	12	*Phaseolus oaxacanus* Rose
	13	*Phaseolus polymorphus* S. Watson
	14	*Phaseolus microcarpus* Mart.
	15	*Phaseolus anisotrichus* Schlecht./subsp *incisus* Piper
	16	*Phaseolus ritensis* Jones
	17	*Phaseolus lunatus* L.
		var. *lunatus*
		var. *silvester* Baudet
	18	*Phaseolus tuerckheimii* Donn. Smith
	19	*Phaseolus brevicalyx* Micheli
	20	*Phaseolus pachyrrhizoides* Harms
	21	*Phaseolus sonorensis* Standley
	22	*Phaseolus xanthotrichus* Piper
	23	*Phaseolus micranthus* Hooker & Arn.
Section		*Alepidocalyx* (Piper) Maréchal, Mascherpa & Stainier
	24	*Phaseolus parvulus* Greene
	25	*Phaseolus amblyosepalus* (Piper) Morton
Section		*Minkelersia* (Mart. & Galeotti) Maréchal, Mascherpa & Stainier
	26	*Phaseolus galactoides* (Mart. & Galeotti) Maréchal, Mascherpa & Stainier
	27	*Phaseolus nelsonii* Maréchal, Mascherpa & Stainier
	28	*Phaseolus pluriflorus* Maréchal, Mascherpa & Stainier
	29	*Phaseolus vulcanicus* (Piper) Maréchal, Mascherpa & Stainier
	30	*Phaseolus chacoensis* Hassler

After Maréchal, Mascherpa and Stainier (1978).

obviously comparable in their respective levels of morphological differentiation. A more fundamental criticism can be made of the infra-specific breakdown of *P. coccineus*. On the one hand the recognition of ssp. *obvallatus* and *formosus* seems to be based on relatively trivial characters, whereas the substantial differences between ssp. *polyanthus* and ssp. *coccineus*, which in the eyes of some (Smartt, 1973) probably merit specific rank, are made light of. It is significant that in the hierarchy produced ssp. *polyanthus* is closer to *P. vulgaris* than to *P. coccineus*. It would appear to be sensible to follow Greenman and recognise *polyanthus* as a species, albeit closely related to both *P. vulgaris* and *P. coccineus*, but not completely isolated genetically from either. With the discovery of wild *polyanthus* material from Guatemala (R. Maréchal, personal communication), there is a growing trend to accept Greenman's perception of *P. polyanthus* (Debouck, 1986).

Speciation patterns

Any assessment of the germplasm resources of a cultigen must take into account phylogenetic relationships within its genus and its own species. These can be and have been elucidated in three major ways: (i) by inter- and intra-specific hybridisation; (ii) by cytogenetic studies; and (iii) by sero- and chemo-taxonomic investigation. These can all be treated in a summary fashion.

Inter-specific and intra-specific hybridisation

The first recorded inter-specific hybridisation in the genus *Phaseolus* was that of Mendel (1866), who crossed *P. vulgaris* × *P. coccineus*. This cross has been repeated on several occasions; the most detailed studies have been those of Lamprecht (1966). The reciprocal cross *P. coccineus* × *P. vulgaris* is much more difficult to make and the first reciprocal hybrid was not reported until the success of Tschermak-Seysenegg (1942). Occasional success has been reported subsequently (Smartt, 1979; Hucl and Scoles, 1985). This inter-specific combination is of particular interest in that the F1 hybrids show some fertility; paradoxically, it is the rather rare *P. coccineus* × *P. vulgaris* hybrid which tends to show higher levels of fertility than the reciprocal hybrid, which is the more easy to produce (Hoover *et al.*, 1985*a*, *b*). Introgression between these two species is thus possible and can easily be brought about. The position is complicated somewhat by a third form, *P. polyanthus* Greenm. (syn. *P. dumosus* Macf., *P. flavescens* Piper) which has, as already noted, presented some taxonomic difficulty. The most satisfactory course to follow would appear to be to allow Greenman's diagnosis of *P. polyanthus* to stand and regard the trio *P. vulgaris*, *P. coccineus* and *P. polyanthus* as a syngameon between which gene exchange is possible, since all three species give at least partially fertile hybrids. Smartt and Haq (1972) succeeded in pro-

(a)

(b)

(c)

(d)

(e)

Fig. 4.4. Inter-specific hybridisation *Phaseolus vulgaris* × *P. coccineus*. (*a*) F1 hybrids: necrotic dwarf and normal; (*b*) F1 hybrids: stunted with leaf-curl and normal; (*c*) F2 segregation for vigour: hybrid breakdown; (*d*) F2 segregation for inflorescence type; (*e*) F2 segregant with *P. coccineus* type inflorescence.

ducing amphidiploids from F1 inter-specific hybrids *P. vulgaris* × *P. coccineus* but had no success with the reciprocal cross.

Other inter-specific hybrids have been reported but these, though viable, are sterile. The first was the hybrid *P. lunatus* × *P. polystachyus* (a wild species) reported by Lorz (1962). Le Marchand *et al.* (1976) also produced this hybrid, which was sterile, as previously reported by Dhaliwal *et al.* (1962). From this hybrid Fozdar (1963) produced a fairly fertile amphidiploid. Honma and Heeckt (1958, 1959) claimed to have produced viable hybrids between *P. vulgaris* and *P. lunatus*; the authenticity of these hybrids has been questioned by Smartt (1979) who doubts whether such hybrids can be produced by conventional crossing techniques. Mok *et al.* (1978) claim to have produced hybrids *P. vulgaris* × *P. lunatus* by embryo culture, but the young plants produced *in vitro* failed to mature.

A number of hybrids have been produced involving *P. acutifolius*. The first claim to have produced a hybrid *P. vulgaris* × *P. acutifolius* was advanced by Honma (1956). However, Smartt (1979) again has questioned the authenticity of this hybrid on the grounds that Homna's putative F1 inter-specific hybrid was fertile, but that obtained by Smartt (1970) was sterile. It is interesting to note that further inter-specific F1 hybrids (*P. vulgaris* × *P. acutifolius*) produced have also been noted as sterile (Hwang, 1979; Mok *et al.*, 1978; Alvarez *et al.*, 1981; Prendota *et al.*, 1982; Pratt and Bressan, 1983; Thomas and Waines, 1984; Pratt *et al.*, 1985; Gray *et al.*, 1985). The reciprocal cross *P. acutifolius* × *P. vulgaris* obtained by Smartt (1970) was inviable past the flowering stage, indicating plasmon differentiation between the species. Mok *et al.* (1978) have also produced reciprocal hybrids but these apparently ceased vegetative growth two weeks after transference to a soil medium.

Although Smartt (1970) had no success in producing amphidiploids from the sterile F1 inter-specific hybrid *P. vulgaris* × *P. acutifolius*, Prendota *et al.* (1982) and Thomas and Waines (1984) have been successful and produced fertile amphidiploids.

An interesting and apparently authentic cross between *P. acutifolius* and *P. coccineus* has been obtained by Coyne (1964). The F1 hybrid showed the red-flowered character of the pollen parent but was sterile. Honma and Heeckt (1958) claimed to have produced a hybrid *P. coccineus* × *P. lunatus* which showed hypogeal seed germination and high pollen fertility. It might well be expected that such a hybrid would be produced only with difficulty and would be sterile. No confirmation of such hybridisations has been obtained subsequently.

One inter-specific hybridisation of a cultigen × a wild species has already been mentioned, namely that by Lorz (1952) of *P. lunatus* × *P. polystachyus*. Others which have been reported include *P. vulgaris* ×

(a)

(b)

Fig. 4.5. Inter-specific hybridisation *Phaseolus vulgaris* × *P. acutifolius*. (*a*) Upper (l to r) *P. vulgaris* cv. Masterpiece; F1 hybrid; wild *P. acutifolius* PI 13199446. (*b*) Lower (l to r) *P. vulgaris* cv. Masterpiece (L16); F1 hybrid; *P. acutifolius* f. *acutifolius* (L320). (From Thomas and Waines, 1984.)

P. ritensis (Braak and Kooistra, 1975); *P. lunatus* × *P. ritensis* (Le Marchand and Maréchal, 1977); *P. lunatus* × *P. metcalfei* (Baudoin, 1981), *P. vulgaris* × *P. angustissimus* (Belivanis and Doré, 1986) and *P. vulgaris* × *P. filiformis* (Maréchal and Baudoin, 1978).

In contrast to hybrids between members of the *P. vulgaris* syngameon (*P. vulgaris*, *P. coccineus* and *P. polyanthus*) these F1 inter-specific hybrids are sterile, but as already mentioned amphidiploids have in some cases been produced. Additional artificial polyploids are produced from time to time, such as *P. vulgaris* × *P. filiformis* (de Tau *et al.*, 1986). Simple introgression at the diploid level does not appear to be possible; it is possible that transfer of genetic material might be feasible by means of bridging species as between *P. vulgaris* and *P. lunatus* via *P. ritensis*. This would involve and depend on successful circumvention of ploidy-level barriers, which could be troublesome. Such manipulation could produce chromosome substitutions and additions which might usefully be developed in breeding lines. An additional or alternative approach could be the production of translocations between genomes in which only relatively small chromosome segments were exchanged. This could well produce a cleaner genetic transfer of a desirable character.

In order to differentiate adequately between inter- and intra-specific hybridisation it is necessary to adopt as clear-cut a species concept as possible. For the plant breeder (if not always the taxonomist!) the biological species concept is ideal. Its use certainly reduces confusion which has arisen from past inconsistencies in the naming of taxa within the genus *Phaseolus*. Smartt (1979) has urged consistency in this regard. It is obvious from his experimental crosses (Smartt, 1970) that the four *Phaseolus* cultigens, *P. vulgaris*, *P. coccineus*, *P. lunatus* and *P. acutifolius*, all have wild counterparts which can readily be crossed with the corresponding cultigens to give fully fertile and viable F1 hybrids which in turn produce normal F2 progeny. R. Maréchal (personal communication) has also identified a wild counterpart of *P. polyanthus*; experimental crosses between this and the corresponding cultigen should provide interesting information and could help to resolve the biosystematic status of the species.

Harlan and de Wet's (1971) gene pool concept has been applied (Smartt, 1984) to the grain legumes generally and it appears to have very useful application in this group of crop plants. Each domesticated *Phaseolus* species has a primary gene pool (GP1) consisting of domesticated (GP1A) and wild (GP1B) components. The species *P. vulgaris*, *P. coccineus* and *P. polyanthus* also have secondary gene pools (GP2); that of *P. vulgaris* consisting of the other two species, similarly with *P. coccineus* and *P. polyanthus*. Both *P. lunatus* and *P. acutifolius* apparently lack a secondary gene pool; the tertiary gene pool (GP3) for these species comprises approximately all other species of the genus. For the members of the *P. vulgaris* syngameon it consists basically of those *Phaseolus* species outside the syngameon.

Genetic resources

The great merit of Harlan and de Wet's (1971) gene pool concept is that, apart from providing a good biological basis for a sound nomenclatural hierarchy, it provides a useful system for classifying genetic resources. The ease of utilisation and accessibility to the cultigen of genetic resources is clearly indicated. The GP1A gene pool presents the least problem for exploitation but varies greatly in extent from species to species; it is very large in *P. vulgaris* and small in *P. acutifolius*. There is similar variation in the extent of the wild component of the primary gene pool; in the two species mentioned it is again larger in *P. vulgaris* than *P. acutifolius*. This situation augurs well for the future of *P. vulgaris* but bodes ill for that of *P. acutifolius*, which has only a small total primary gene pool. The situation in the *P. vulgaris* syngameon is of interest in that the options open are much broader. There may be a choice, for example, of exploiting either the GP1B or the GP2. The advantage is not necessarily with the GP1B, particularly if the part of the GP2 to be exploited consists of domesticated forms of another species. For example, it is very easy indeed to introduce the *P. coccineus* inflorescence type into *P. vulgaris* from cultivars of *P. coccineus*; a very clean genetic transfer is possible. A similar type of gene transfer from wild *P. vulgaris* might be less easy to achieve cleanly, because more potentially undesirable wild type traits might be introduced.

The range of breeding strategies possible within the *P. vulgaris* syngameon increases the tactical range of possible procedures. In general morphological characters are more easily transferred than the less readily identified and quantified ecological factors. Smartt (1969) noted that the ecological ranges of the four cultigens (*P. vulgaris*, *P. coccineus*, *P. lunatus* and *P. acutifolius*) are complementary; permitting, as already noted, cultivation of *Phaseolus* species from cool temperate climates to the humid tropics. The climatic optimum for *P. vulgaris* can be designated as warm temperate and that of *P. coccineus* as cool temperate. The range of genetic variability in *P. vulgaris* is vastly greater than that of *P. coccineus*. If it were thought desirable to produce, in a cool temperate climate, beans with the market quality of *P. vulgaris*, it would be possible to combine the genetic resources of the two species in two ways: either introduce the ecological tolerance of *P. coccineus* into essentially a *P. vulgaris* genetical background, or introduce the desired commercial and agronomic characters of *P. vulgaris* into the background of *P. coccineus*. On *a priori* grounds it is not easy to predict the relative merits of these alternative strategies. It might well prove to be easier to fix the commercially important characters of *P. vulgaris* in a *P. coccineus* background than the less tangible and perhaps genetically more complex adaptive factors of *P. coccineus* in that of *P. vulgaris*. Much would of course depend on the extent to which the actual production environments differed from the optima for the two species.

A rather different situation exists in the exploitation of the genetic resources of *P. vulgaris* in a more xeric environment. In North America there is some interest in exploiting more arid environments for production of *Phaseolus* beans. The element of ecological adaptation would be supplied by *P. acutifolius* and commercial quality by *P. vulgaris*. It appears from the studies of Nabhan (1979) that water utilisation by *P. acutifolius* is the more efficient. Under a low rainfall régime, or one with supplementary irrigation, efficient utilisation of water resources is of the utmost importance to mature crops reliably. Where generous irrigation is used to minimise salinisation problems, a less thirsty crop might usefully help to reduce salt accumulation. Although viable inter-specific (*P. vulgaris* × *P. acutifolius*) F1 hybrids can be produced, these are sterile, and although amphidiploids can also be produced, the commercial quality of these is not likely to be competitive with *P. vulgaris*. Introduction of commercial quality from *P. vulgaris* into *P. acutifolius* or transferring ecological adaptation of *P. acutifolius* to *P. vulgaris* is unlikely to be achieved without complex genetic manipulation and exploitation of highly sophisticated biotechnological procedures. These would be necessary to achieve what would take place naturally in *P. vulgaris* × *P. coccineus* hybrids by normal processes of meiotic recombination. Whether the effort necessary to achieve appropriate genotypes ultimately derived from *P. vulgaris* × *P. acutifolius* could be justified is a very moot point indeed at the present time in North America. Such an approach might actually be of more practical value in drought-prone areas of the Third World than in North America.

Cytogenetic studies

The most comprehensive karyotype study of the genus was carried out by Maréchal (1969). The chromosome number in the genus is uniform, $2n = 2x = 22$; no naturally occurring polyploids have been reported. The general appearance of karyotypes in different species, insofar as they have been studied, is very similar. Some difference can be seen between those of *P. vulgaris* and *P. coccineus*, which are reflected in the pairing relationships between chromosomes in F1 inter-specific hybrids. These have been considered by Maréchal *et al.* (1978) in relation to palynological characters. It is of interest to note that, in meiosis of F1 inter-specific hybrids involving *P. ritensis*, 2 univalents were recorded in reciprocal hybrids (*P. lunatus* × *P. ritensis*) while 14 were observed in that between *P. vulgaris* and *P. ritensis*. From this it can reasonably be inferred that the

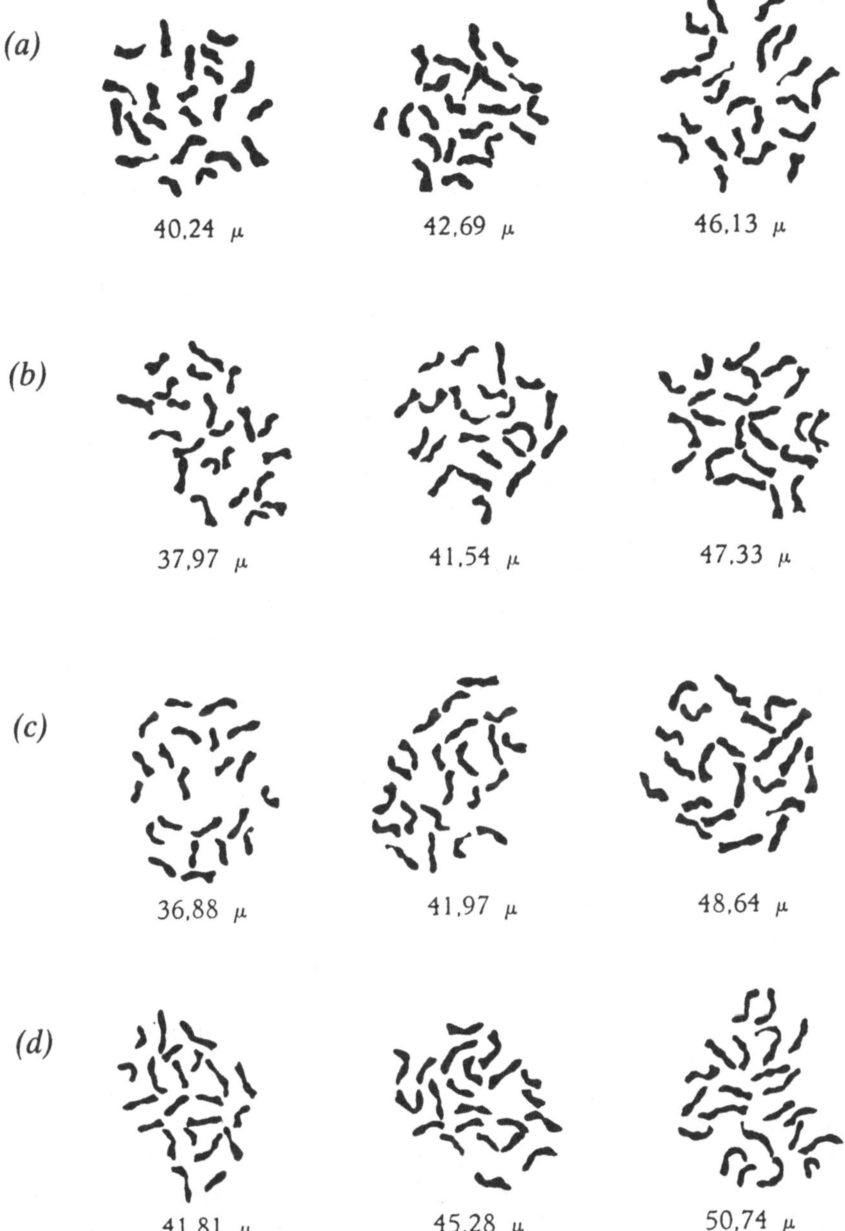

Fig. 4.6. Chromosome complements of *Phaseolus* species (× 2500). (*a*) *P. vulgaris*; (*b*) *P. coccineus*; (*c*) *P. lunatus*; (*d*) *P. acutifolius* (from Maréchal, 1969).

genetic differentiation between *P. vulgaris* and *P. lunatus* probably involved a large measure of chromosome structural change.

It is also of interest to note in the *P. vulgaris* syngameon (Smartt, 1980*c*) that, in terms of hybridisation, *P. vulgaris* appears to be more differentiated from *P. polyanthus* and *P. coccineus* than they are from each other. However, data have not been obtained from reciprocal hybrids in all cases, and neither has an adequate sampling of parental species genotypes been involved. Since considerable differences in pollen fertility of reciprocal hybrids (*P. vulgaris* × *P. coccineus*) have been observed, it is possible that differences in meiotic chromosome pairing occur in reciprocal hybrids. These have not been adequately studied. Again, cryptic structural differentiation may occur between different geographical races of the same cultigen as well as major structural changes such as interchanges which are occasionally observed (J. Smartt, unpublished). This point can be illustrated from the data reviewed by Maréchal *et al.* (1978). F1 hybrids *P. coccineus* ssp. *obvallatus* × *P. coccineus* ssp. *formosus* and *P. coccineus* × *P. coccineus* ssp. *obvallatus* both show perfectly regular meiotic pairing. However, whereas *P. vulgaris* × *P. coccineus* produced 0.32 bivalents per cell, *P. vulgaris* × *P. coccineus* ssp. *formusus* produced 1.64 bivalents. It would thus appear that the importance of differences of this magnitude based on relatively small samples of parental genotypes should not be exaggerated. It is nevertheless reasonable to conclude that chromosome structural differentiation has occurred within this group of species. Another example, that of *P. lunatus* × *P. polystachyus*, is also worthy of comment; the mean of 8 univalents per cell here obviously indicates substantial chromosome structural differentiation (Dhaliwal *et al.*, 1962).

Chemosystematic studies

The biochemical studies which have been undertaken on *Phaseolus* species have on the whole not had a biosystematic or evolutionary bias. They have usually been concerned with nutritional and related problems. The exceptions have been the work of Kloz, Klozová and co-workers (Kloz, 1971) and that of Gepts (1985), Gepts *et al.* (1986) and Gepts and Bliss (1985, 1986).

There is no doubt that biochemical studies could have very considerable impact on biosystematic understanding of the genus as a whole and the cultigens it contains. In order that this be achieved some attempt must be made to understand the broad biochemical context of the study. If, for example, isoenzyme studies are to shed light on, say, phylogenetic relationships within the genus, the distribution patterns of enzymes within the individual species and the genus as a whole will have to be known and understood. Chemosystematic studies are thus no easy shortcut to biosystematic understanding.

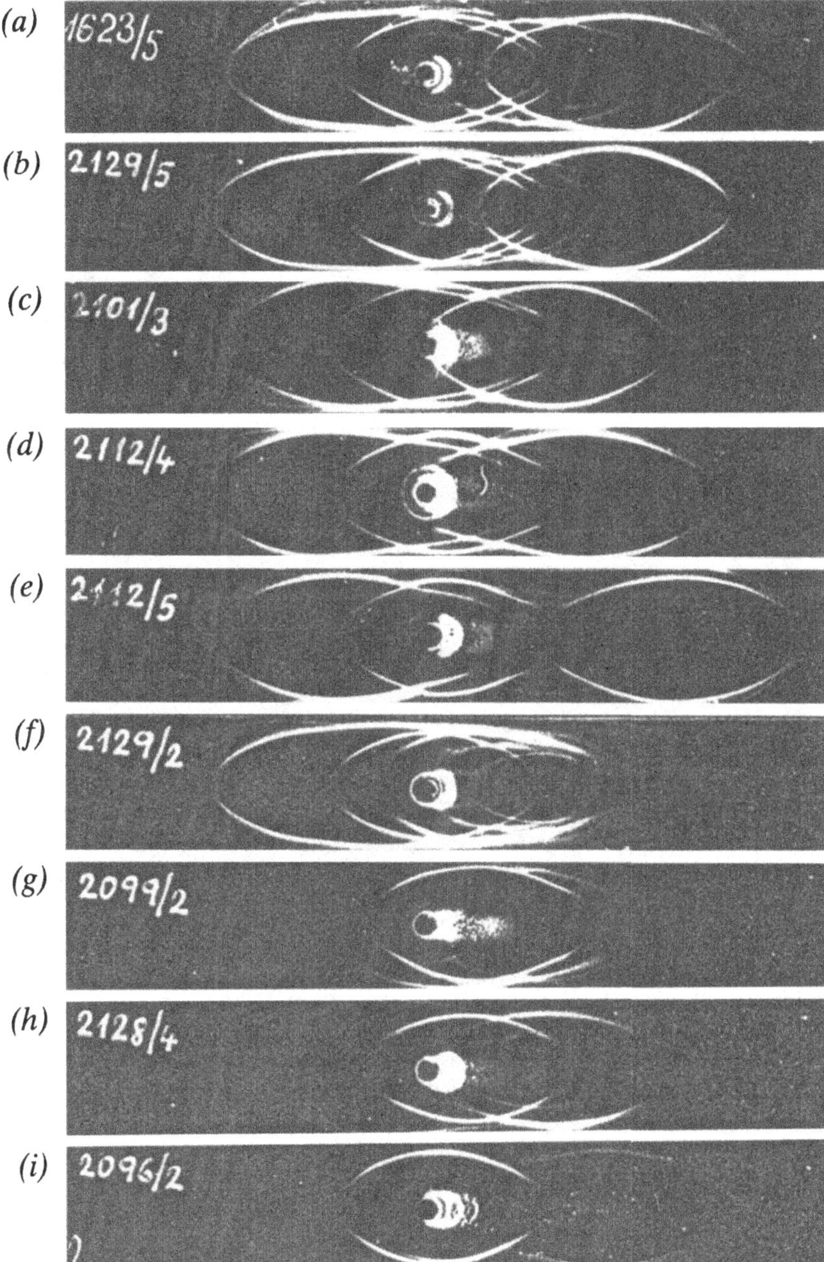

Fig. 4.7. Immunoelectrophoretograms of albumin seed proteins, detected with antiserum against the *Phaseolus vulgaris* seed albumin fraction. (From Kloz and Klozová, 1968.)

The serotaxonomic studies of Kloz and co-workers (Kloz, 1971) clearly indicate the pattern of investigation which needs to be followed. The starting point for their investigation was *P. vulgaris* itself. They prepared antisera against the seed proteins of *P. vulgaris*, by using rabbits, and showed first of all that the most informative cross-reactions were given by the albumin fraction of the seed protein as antigens with the appropriate

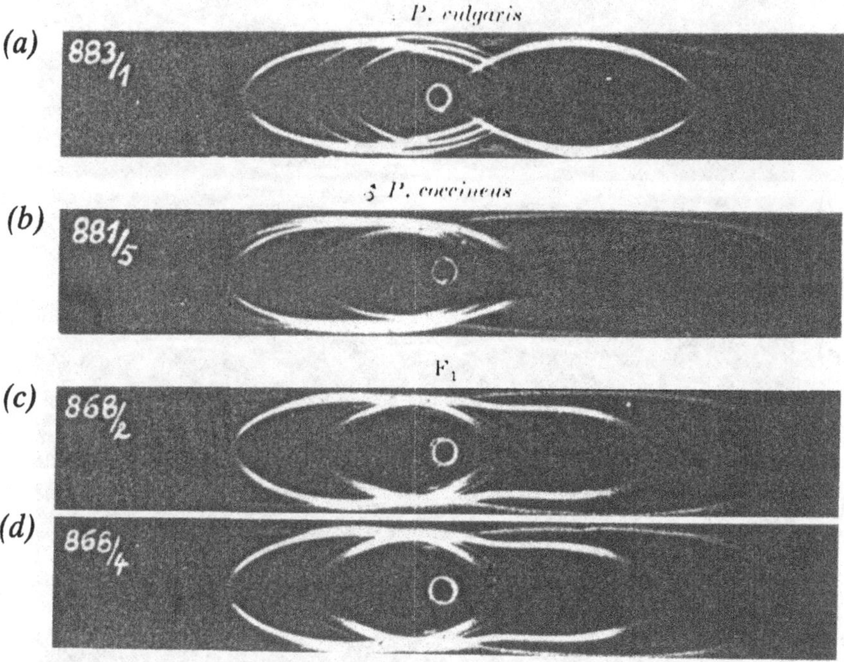

Fig. 4.8. Immunoelectrophoretograms of albumin seed proteins of parents, F1 and F2 hybrids of *Phaseolus vulgaris* × *P. coccineus* detected with antiserum against the *P. vulgaris* seed albumin fraction. (i) Parents (*a*) *P. vulgaris* cv. Veltruska Saxa (*b*) *P. coccineus* cv. Weisse Riesen; (ii) F1 hybrid *P. vulgaris* × *P. coccineus* (individuals *c* and *d*); (iii) F2 segregants from cross *P. vulgaris* × *P. coccineus* (individuals *e–l*). (From Kloz and Klozová, 1968.)

antibodies. The immunoelectrophoretograms produced give a readily appreciated visualisation of the pattern of production of the albumin antigens. They established a remarkably consistent pattern of antigen production in a very broad sampling of the species, including some wild material as well as several hundred cultivars.

They identified two major antigens which they termed proteins I and II

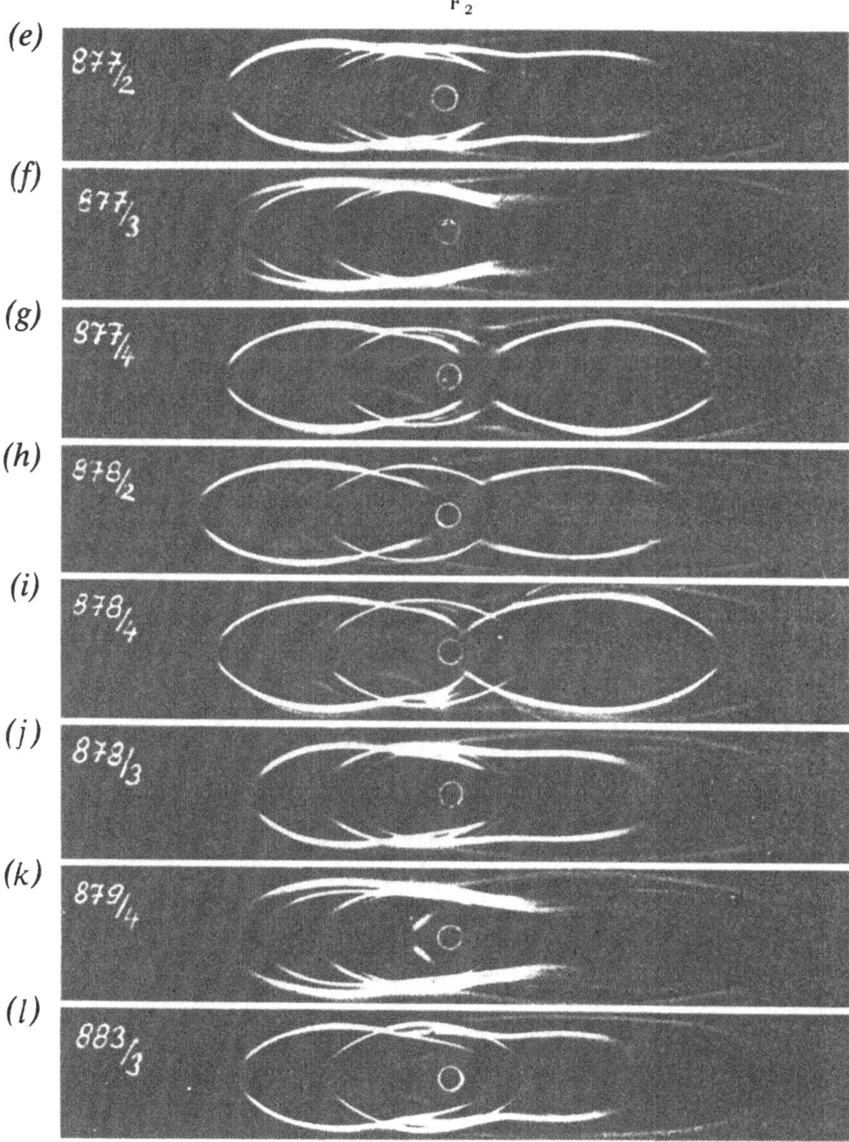

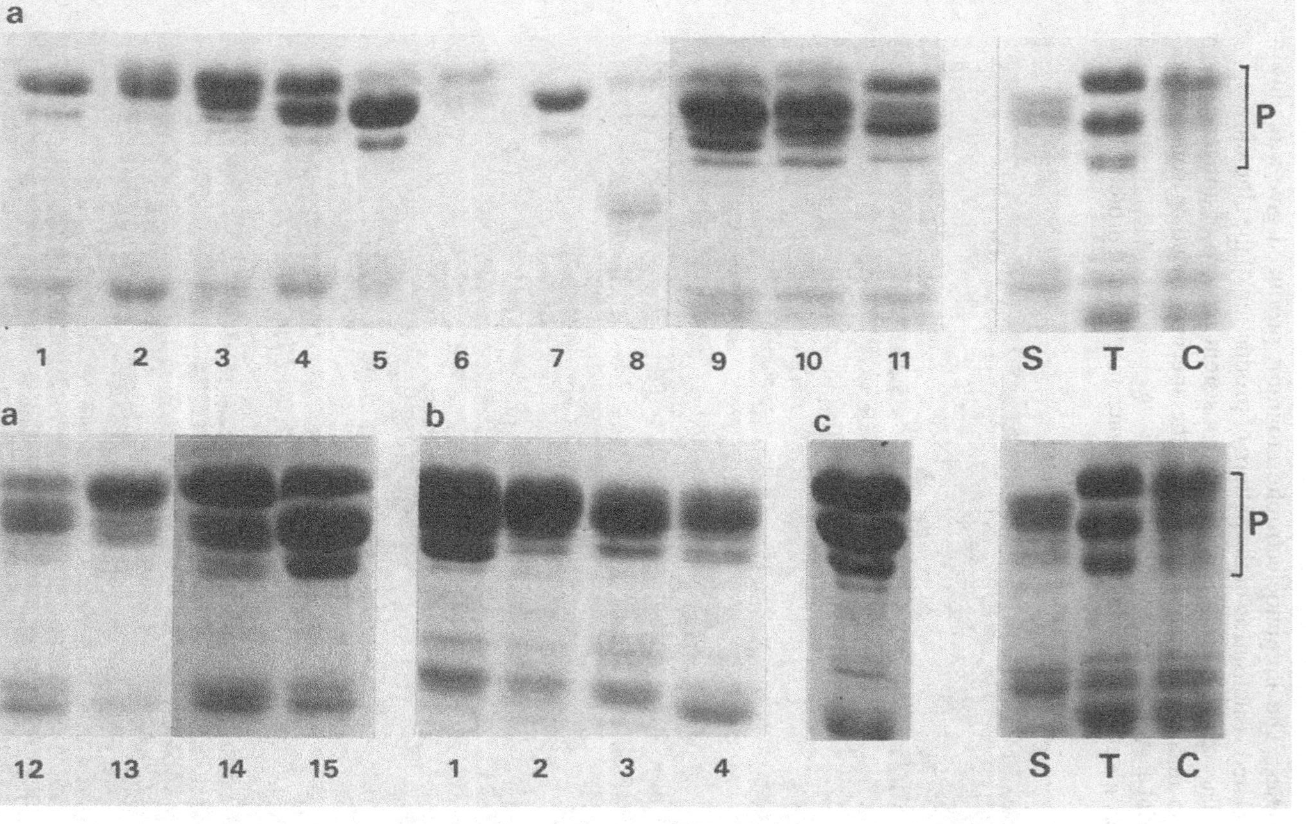

Fig. 4.9. One-dimensional SDS–PAGE of phaseolin (P) from wild *Phaseolus vulgaris* of Middle (a, b) and Andean South American (c) origins. S, T and C are reference phaseolins of cultivars Sanilac, Tendergreen and Contender respectively. (From Gepts *et al.*, 1986.)

and showed that a few cultivars were lacking one or both of these (Klozová and Turková, 1978). However, they were able to demonstrate the presence in these cultivars of proteins of similar mobility but with different antigenic action (Klozová *et al.*, 1976). These particular antigens are important because they are lectins and undesirable from the nutritional point of view. Lectin-free cultivars are thus of considerable interest (Bliss, 1986).

Cross-reactions of antibodies against *P. vulgaris* seed proteins were studied with seed protein extracts from other cultivated *Phaseolus* species. There were strong cross-reactions with those of *P. coccineus* and *P. polyanthus*, a weak cross-reaction with those of *P. acutifolius* and weakest of all with those of *P. lunatus*. These studies also revealed that the electrophoretic mobilities and serological specificities of the *P. coccineus* seed proteins were much more variable than those of *P. vulgaris*. This undoubtedly is a reflection of the high levels of heterozygosity in *P. coccineus*, arising from its effectively random system of pollination, but why this species should be intrinsically so much more polymorphic is not clear. An interesting study carried out by J. Kloz (personal communication) showed that protein patterns of inter-specific F1 hybrids (*P. vulgaris* × *P. coccineus*) obtained by immunoelectrophoresis were the same as those produced by a mechanical mixture of the parental species seed proteins.

This work could readily be expanded by extending the range of antisera used and the number of species studied. The intensity and pattern of cross-reaction can be exceedingly informative.

The work of Gepts (Gepts *et al.*, 1986; Gepts and Bliss, 1986) has followed a rather different line, that of characterising variant forms of phaseolin, the major seed storage protein in *P. vulgaris*, and correlating these in landraces with their geographic origin. This work has both established a satisfactory context for itself and produced evidence which can be interpreted in support of the view that there have been at least two major centres of *P. vulgaris* domestication in the Americas with at least one other minor centre. Equally valuable information could be provided without doubt by similarly conceived and executed studies of DNA sequence studies and re-association of DNA segments from different species.

4.2. Origins and domestication

The archaeological record of domestication

The archaeological record of *Phaseolus* cultigens has cast more light on the evolutionary history of these beans than is the case with most other

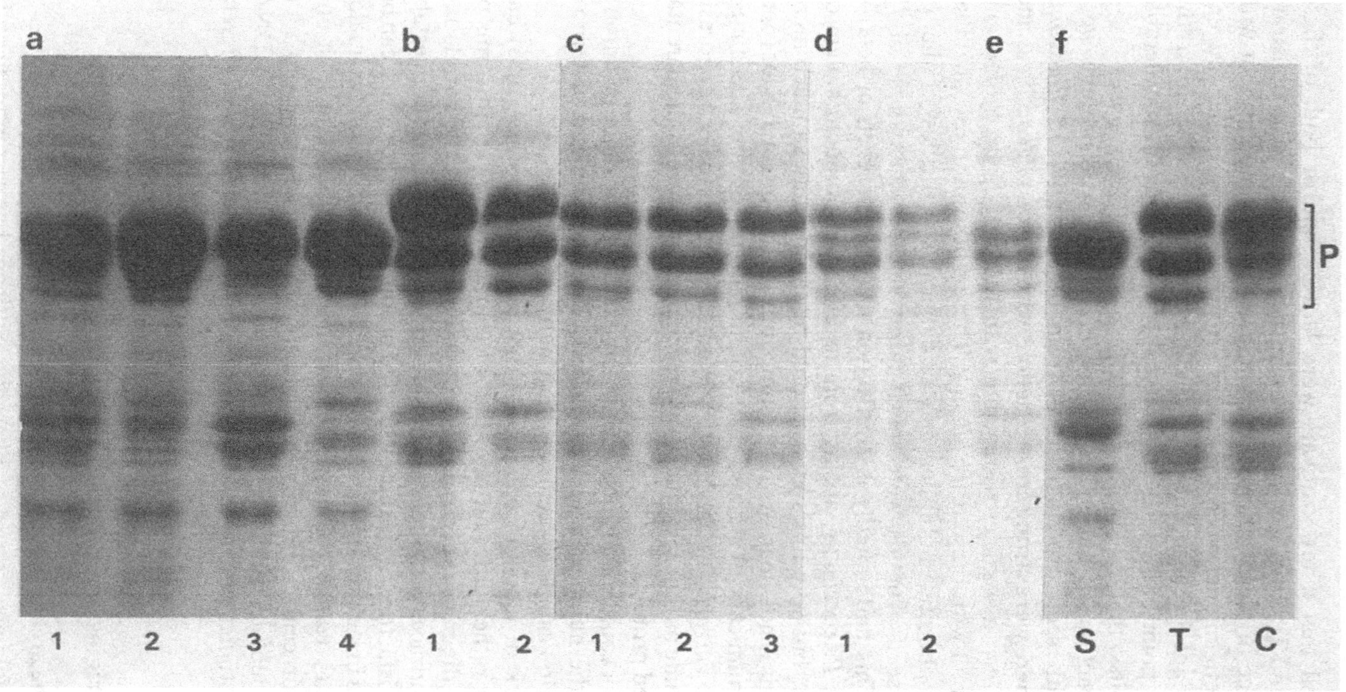

Fig. 4.10. One-dimensional SDS–PAGE of phaseolin (P) of landraces from Middle American (a, b) and Andean South American (c) origins. S, T and C are reference phaseolins of cultivars Sanilac, Tendergreen and Contender respectively. (From Gepts *et al.*, 1986.)

grain legumes. Archaeological study of *Phaseolus* has been linked with that of maize, especially in Mexico, a clear indication of the importance of the combined role of these two crops in the life of ancient populations. It is reasonable to infer that this was not significantly different from that which they perform at the present time (Kaplan, 1965).

The remains which have been found consist largely of seeds (whole and fragments) and pod material. Vegetative plant remains are not commonly found but some have been encountered and are of considerable interest. In Mexico (Tehuacan) the oldest remains of *P. vulgaris*, consisting of a pod valve, have been dated at 3500–5000 BC; in Peru, seed material has been dated at 5681 BC (Kaplan *et al.*, 1973; Kaplan, 1981). A rather puzzling feature of archaeological finds is that the seed material of domesticated *Phaseolus* (and *P. vulgaris* in particular) is essentially of modern type. Some wild *Phaseolus* material has been found in a Mexican site, Guila Naquitz, but this was clearly not *P. vulgaris* since it exhibited hypogeal germination; it could possibly be *P. coccineus*. Collection of seed from wild *Phaseolus* species for food was still being practised in recent times (Burkart and Brücher, 1953) in some regions; seed may still be so collected.

Kaplan (1981) was greatly exercised by the fact that there is no archaeological evidence for the collection of wild *P. vulgaris* by ancient populations in the vicinity of sites where preserved remains have been found, since wild *P. vulgaris* is found in close proximity at the present time. There could be at least two explanations for this: that the local people collected such material but their pattern of use was such that it was never exposed to conditions conducive to preservation; or that domestication of beans did not occur close to the preservation sites but elsewhere. Subsequently domesticated material could have been introduced to areas with pre-existing, wild material. In the regions of South America where archaeological beans are found Kaplan notes that wild *Phaseolus* species are not found in the vicinity; in such areas it is highly probable that fully domesticated material was introduced. We have no means of knowing the time scale for the full domestication of *Phaseolus*, but it could well be of the order of many thousands of years and not necessarily have been a rapid process. Kaplan (1981) also notes that the preservation of the older materials from Tehuacan is not particularly good. However, he considers that the evidence indicates the presence of fully domesticated beans in the vicinity of Tehuacan from 5500–7000 years ago. Evidence of earlier cultivation has already been noted in the form of seeds dated 5681 BC from Peru.

Remains of the other cultivated species have been obtained from archaeological sites in Mexico and Peru, in the case of *P. lunatus*, and from Mexican sites only in those of *P. coccineus* and *P. acutifolius*.

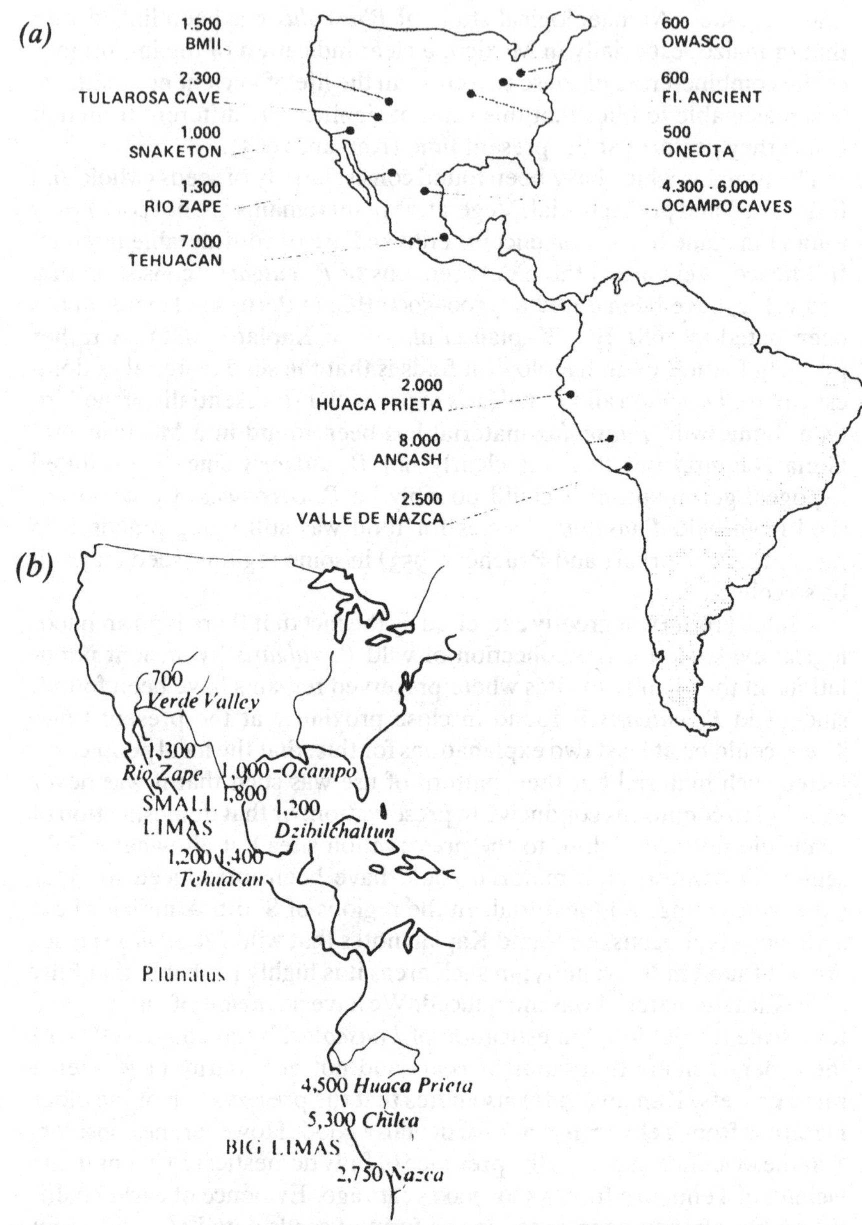

Fig. 4.11. Archaeological sites with remains of *Phaseolus* species. (*a*) Sources of *P. vulgaris* material; (*b*) Sources of *P. lunatus* material; (*c*) Sources of *P. coccineus* material; (*d*) Sources of *P. acutifolius* material; (*e*) *Phaseolus acutifolius* areas of present cultivation and known archaeological distribution (from Kaplan, 1965).

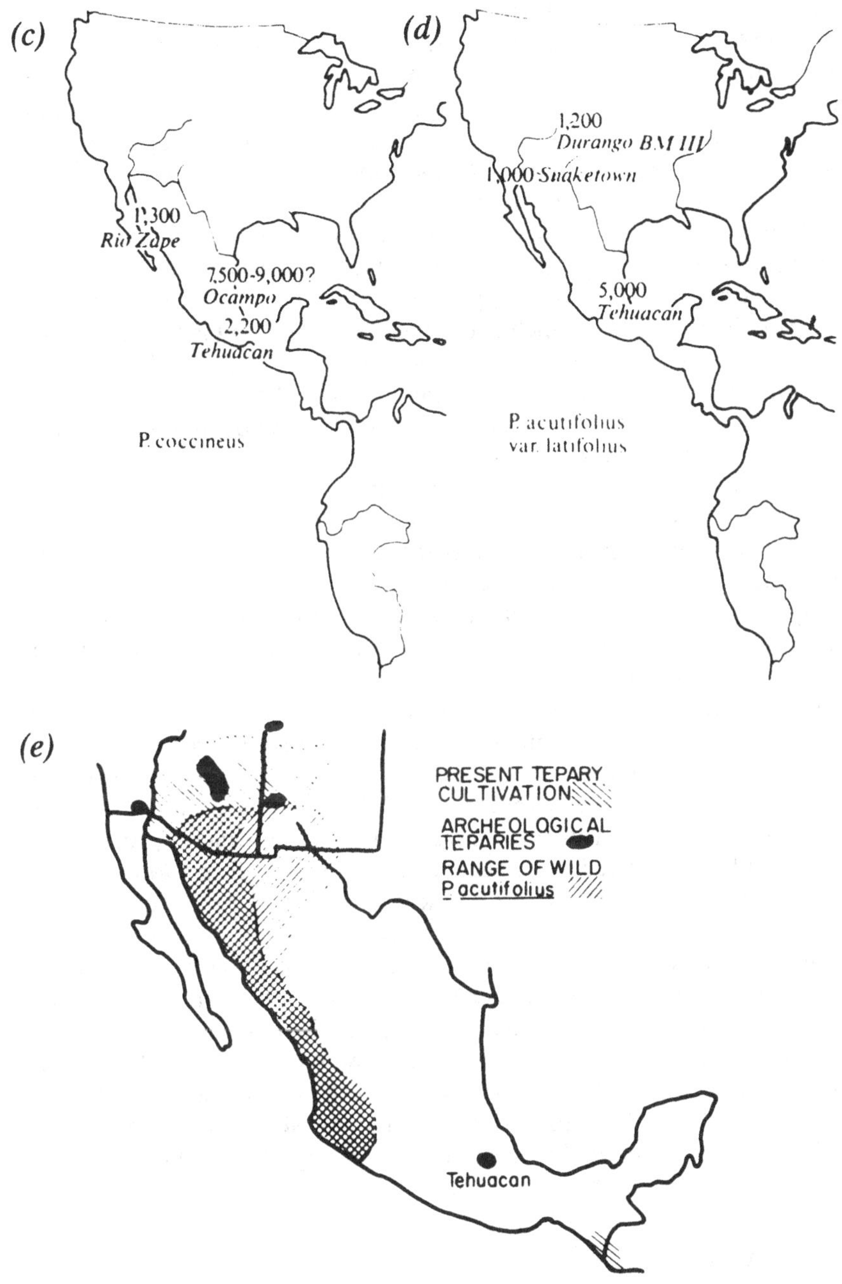
(c)
1,300
Rio Zape
7,500-9,000?
Ocampo
2,200
Tehuacan
P. coccineus
(d)
1,200
Durango BM III
1,000 Snaketown
5,000
Tehuacan
P. acutifolius
var. latifolius
(e)
PRESENT TEPARY CULTIVATION
ARCHEOLOGICAL TEPARIES
RANGE OF WILD P. acutifolius
Tehuacan

Recorded datings go back to almost 2000 years ago from the Ocampo caves in Mexico and to over 7000 years ago at Chilca in Peru. The Mexican material is of the 'Sieva' type while that from Peru is seed of the 'Large Lima' form. Records of *P. acutifolius* go back 5000 years at Tehuacan and of *P. coccineus* over 2000 years at this site for domesticates; some possibly wild *P. coccineus* material from the Ocampo caves has been dated at over 7000 years old (7500–9000) (Kaplan, 1965).

Kaplan (1981) observes that very ancient material of beans from Mexican sites is not particularly abundant and believes that the discovery of additional material would be particularly informative. This is certainly the case. The questions which particularly exercise Kaplan are the abrupt appearance of essentially modern forms of bean in the archaeological record without intermediates, and the origin of the genetic factors producing larger seed size in *P. vulgaris*. Too much should not be expected of the archaeological record, particularly that of earlier periods. It is of interest to note that even when contemporaneous written records are available it can be difficult to reconcile these with the nature of the materials recovered from archaeological sites (J. Renfrew, personal communication). The problem basically is one of sampling; this may not be truly representative of the materials of the time, and surviving written records may equally lead to biased interpretation because they may not be comprehensive in their coverage. The conclusions which can be drawn safely are that beans have been domesticated for 5000–7000 years in the Americas and that this enables us to make an extremely conservative estimate of the time they have been in cultivation. What we cannot estimate is the time scale for the development of the modern form of the beans observed at the sites. This could have been a fairly protracted process and may well have occurred some distance from sites of preservation.

The importance of the archaeological record of *Phaseolus* is considerable in that it is among the best for the grain legumes; only the Mediterranean pulses have a comparable record. Archaeological remains are very scarce (e.g. *Arachis hypogaea*) or non existent (cowpea, soyabean) for several grain legumes. What is clear from the grain legume archaeological record, such as it is, for both the New World and the Old, is that they are of comparable antiquity to that of the major staple crops. The evolution of cereal grains and grain legumes went hand in hand and helped to secure a better balance, in both farming systems and diets, than might otherwise have been the case.

4.3. Evolution

The nature of evolutionary changes in cultigens

Evolutionary studies of *Phaseolus* beans have centred very largely on questions of 'age and origin'; that is to say, the duration of the period under domestication and the geographical areas in which this occurred. Morphology has been considered only insofar as the characters have been expressed in preserved archaeological material; this is of course restricted to those of the seed and to a lesser extent the pod. Only rarely can the shoot system which bore them be studied (Kaplan, 1965). Kaplan is fully aware of this restriction of the value of archaeological materials in the study of *Phaseolus* evolution where extensive changes have been brought about in the shoot system. The discovery of Burkart and Brücher (1953) of wild *Phaseolus* beans in north-west Argentina which were obviously close relatives of cultivated *P. vulgaris* (Bergland-Brücher and Brücher, 1976) and the discovery of comparable wild material in Mexico by Miranda Colín (1968) and Gentry (1969) have enabled preliminary comparative studies of the morphology, physiology and biochemistry of wild and cultivated forms to be carried out (Smartt, 1969, 1976*a*, *b*; Kloz and Klozová, 1968; Kloz, 1971). These studies have been somewhat limited in scope but nevertheless useful generalisations can be made and conclusions drawn. It must be borne in mind that wild populations can be polymorphic as well as those of the cultigen, and the possibility of complex patterns of domestication must also be noted.

The special characteristics which broadly differentiate domesticated plants from their closest (and often conspecific) wild relatives have been considered by Schwanitz (1966) and Hawkes (1983) for the whole range of domesticated plants. Relatively few studies have been carried out on *Phaseolus* species but some work has been initiated (Smartt, 1969, 1976*a*). The character changes which have followed domestication can be shown to have produced the following effects, namely:

1 gigantism;
2 suppression of seed dispersal mechanisms;
3 changed growth form;
4 changed life form (*sensu* Raunkiaer, 1934);
5 loss of seed dormancy;
6 other physiological changes;
7 biochemical changes.

In the genus *Phaseolus*, as now defined (Maréchal *et al.*, 1978), there are no natural polyploids which have been domesticated. Experimentally produced allopolyploids are known (Dhaliwal *et al.*, 1962; Smartt and Haq, 1972; Prendota *et al.*, 1982; Thomas and Waines, 1984; de Tau *et al.*, 1986).

Domesticated *Phaseolus* species

There are five cultigens in the genus *Phaseolus*, namely *P. vulgaris* L., *P. coccineus* L., *P. lunatus* L., *P. acutifolius* A. Gray and *P. polyanthus* Greenm. The status of the last has been, as noted, somewhat problematical. Difficulties have arisen because it apparently has a wide geographic range from Mexico to Colombia and has been collected, described and named several times independently. Names which have been given, such as *P. dumosus* Macf. and *P. flavescens* Piper, are therefore synonymous with *P. polyanthus* Greenm. Mexican material of the same taxon has also been described as a subspecies of *P. coccineus* by Hernandez *et al.* (1959). The three species *P. vulgaris*, *P. coccineus* and *P. polyanthus* are indeed closely related and isolating mechanisms between them are not completely effective in preventing gene flow. The situation is complicated by the evolution of rather different isolating mechanisms in reciprocal crosses (e.g. between *P. vulgaris* and *P. coccineus*). In crosses on *P. vulgaris* the effective obstacle in most parental genotype combinations is hybrid sterility; however, there are combinations in which F1 hybrid inviability and loss of vitality are expressed at the seedling and early post-seedling stages. In *P. coccineus* as seed parent there is an apparently earlier development of post-zygotic barriers and most inter-specific hybrid embryos abort. The very small minority that survive are of a higher level of fertility (*ca.* 70%) at maturity than in the reciprocal cross (20–30%) (Smartt, 1970). This evolution of isolating mechanisms has apparently intensified under domestication; wild *P. coccineus* is much more readily cross-compatible with *P. vulgaris* than are modern cultivars (Miranda Colín, 1974). It is of interest to note that landraces of *P. coccineus* from Mexico are more cross-compatible with *P. vulgiaris* than are those cultivars developed in western Europe. This can be ascribed to two causes: both species are commonly grown together in north-west Europe, and *P. coccineus* has a higher incidence of out-crossing than *P. vulgaris*. This could be expected to intensify selection for the more efficient isolating mechanisms in *P. coccineus*. The spontaneous occurrence of inter-specific hybrids by *P. coccineus* on *P. vulgaris* is usually very low (but not unknown) owing to the predominance of self-pollination; the selection pressures for the establishment of early-acting isolating mechanisms are therefore less in *P. vulgaris*.

Evolutionary changes

We can now review the extent of our knowledge and understanding of the evolution of cultigens within the genus *Phaseolus*, identify the more serious areas of deficiency and, where appropriate, suggest strategies for remedying any such deficiencies. This can best be done by considering *Phaseolus* cultigen evolution in the light of the relevant criteria of Schwanitz (1966) and Hawkes (1983) listed earlier.

Gigantism

As human food *Phaseolus* species are exploited first and foremost for the seed, which may be stored dry for later consumption, or collected and used fresh as mature or nearly mature seed. Depending on the nature of the pod wall, the pod itself may be consumed. In some parts of the world, most notably in Africa, the leaves are used as spinach. As is general with domesticated crop plants it is almost an invariable characteristic that the parts exploited by man show gigantism. Depending on the nature of the plant itself there may or may not be a correlated response in other parts. For example, in the domesticated apple or cherry a similar shoot system architecture serves equally well for the small-fruited wild forms and the larger-fruited domesticates. Although this is commonplace in domesticated tree species, it is in the herbaceous crops that the most profound changes of vegetative shoot morphology and architecture are aparent. This is well shown in *Phaseolus* species and the common pattern of correlated responses has been discussed by Smartt (1976*b*).

The consequences of gigantism can all be related to the increase in seed size which can readily be seen in comparing domesticated *P. coccineus* and the wild forms. A similar contrast is clearly apparent between cultivated genotypes of *P. lunatus* and their wild counterparts. Clear discontinuities in seed size between wild and cultivated populations are apparent in both these species. In terms of seed weight, the largest domesticated beans commonly encountered (*P. coccineus* and *P. lunatus*) are approximately 1 g in weight, whereas the largest wild seeds rarely, if ever, exceed 100 mg and are commonly much smaller. Seed weight has therefore increased by a factor of at least ten and probably much more. A similar situation occurs in *P. polyanthus* but the position in *P. vulgaris* is different and that in *P. acutifolius* different again. This can be related to the rather different patterns of domestication which have occurred in the five species. *P. coccineus* and *P. acutifolius* both have a Middle American domestication centred on Mexico (Kaplan, 1965) whereas *P. vulgaris*, *P. lunatus* and possibly also *P. polyanthus* have been domesticated both in Middle and South America. According to Kaplan (and his view is generally accepted) *P. lunatus* has been domesticated independently in Peru and Mexico. There is a size differential between Mexican and Peruvian domesticates; the latter are frequently double the size of the former and have been designated *microspermus* (Mexican) and *macrospermus* (Peruvian) respectively. This geographically based size differential is being eroded by the action of plant breeders who are crossing the two forms. It is interesting to note, as Kaplan does, that the characteristic differences between Mexican and Peruvian cultigens are paralleled in the local wild races. For example, both wild and domesticated Peruvian *P. lunatus* are more pubescent than are their Mexican counterparts, as well as having larger seeds.

The pattern of domestication in *P. vulgaris* has probably been more complex. If one compares seed sizes on a geographical basis those of both wild forms and cultigens tend to be smaller in the north and larger in the south. A situation similar to that of *P. lunatus* is found in *P. vulgaris* with the important distinction that although there is (or has been) a clear discontinuity in seed size between Mexican and Peruvian lima beans (*P. lunatus*) no such discontinuity is apparent in *P. vulgaris* domesticates but possibly a N–S cline of increasing seed size (Evans, 1980). The smaller-seeded domesticates are of comparable size to the largest-seeded southern wild forms. However, the work of Gepts (Gepts *et al.*, 1986) and Debouck (1986) suggests the very strong possibility that separate domestication in at least three major centres is the basis for the present variation pattern.

Although *P. coccineus* and *P. acutifolius* were domesticated in the Mexican region, their respective increases in seed size are of different orders of magnitude, that of *P. coccineus* being among the greatest and *P. acutifolius* the least. This difference may well have been an important factor in determining the differences between the vegetative morphology of their respective cultigens.

Genetic control of seed size is poly-factorial (Cheah, 1973; Miranda Colín, 1974) with additive effects. Small-seededness tends to show dominance.

Increased seed size has the logical consequence of giving rise to increases in pod size. Since there is no very strong or consistent trend in the differences between the number of seeds produced in the pods of wild representatives and those of their domesticated counterparts, the weight of individual pods increases by a similar factor to that of seeds. In addition the number of pods matured per inflorescence may also be very little different between wild and domesticated individuals. The overall loading factor on the individual inflorescence will again show an increase of a similar order. It is at this juncture that correlated effects become apparent on vegetative architecture. In order to support effectively a heavier load of maturing pods, the inflorescence axis itself must be stronger and more robust, and the general vegetative structure must show similar development in order that a properly balanced shoot architecture is maintained. There are two ways in which this could be achieved: either the vegetative frame could maintain the same general form but be generally enlarged, or it could be re-structured. It is the latter which has apparently occurred. The overall vegetative structure produced by domesticated *Phaseolus* is not consistently more extensive than that of the wild counterpart; in fact the reverse is frequently the case. If for the purpose of discussion we assume that the two types of canopy are equally extensive and the dry matter is roughly equivalent, the structuring of the canopy and the deployment of dry matter is distinctly different. The wild strategy of canopy development entails the production of numerous weak stems which can readily infiltrate shrubby vegetation and expand their numer-

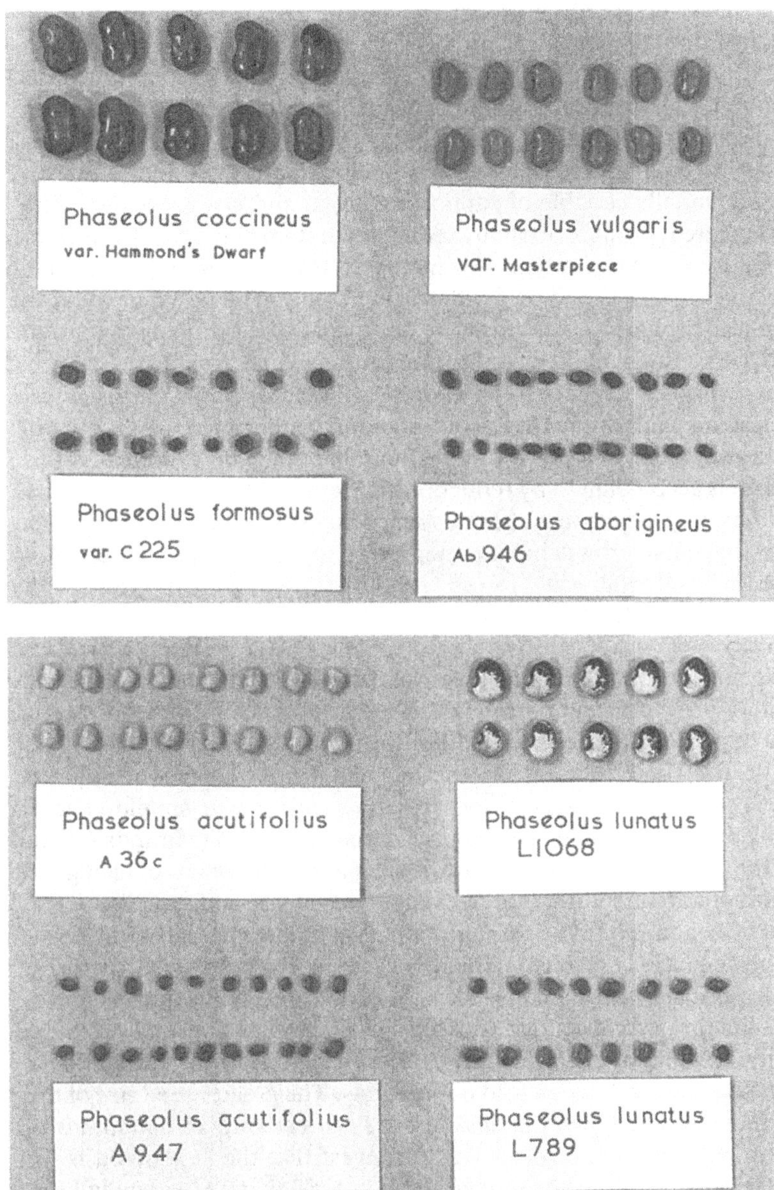

Fig. 4.12. Comparative seed sizes of wild and cultivated forms of *Phaseolus* species.

ous but relatively small leaves at the surface of the shrub canopy. In domestication support may be provided by an erect growing crop, such as maize, or by sticks, poles or trellises. The nature of the support is such as to encourage a less profuse system of branching, which therefore bears in total a reduced number of leaves. However, the more robust stems necessary to support the heavier pod loads established under domestication are equally capable of supporting larger and heavier leaves. The loss in leaf area, brought about by reduction in stem number, can be compensated for (at least partially) by increased leaf size. We thus can have fewer larger leaves borne on fewer thicker stems. The varying extent to which these evolutionary trends have occurred in the cultigen is apparent in the CIAT growth habit classification (Adams *et al.*, 1985).

Gigantism thus appears to be expressed in larger seeds, larger pods, thicker stems and larger leaves but no concomitant increase in overall size. Flower size also shows a comparable increase. Typically this gigantism is accompanied by reductions in stem and leaf numbers. These reductions may have come about by response to selection pressure for a more manageable growth habit better suited to management under cultivation. The well established mixed cropping system of beans grown with maize would be difficult to manage with the rampant growth habit of wild *P. vulgaris*.

The genetic bases for relatively few of these evolutionary changes have been studied in sufficient detail. Differences in branching pattern, determinate or indeterminate (i.e. whether the shoot axis is terminated by an inflorescence (determinate) or whether it remains vegetative (indeterminate)) have been analysed and are apparently under simple genetic control depending on allelic difference at one or two loci (Miranda Colín, 1969; J. Smartt, unpublished). The other major component of change in overall shoot architecture, that of changes in leaf size and stem diameter, has not been analysed. The presumption is that these characters are probably polygenically controlled. Changes in size of reproductive structures, flowers, pods and seeds are also likely to prove to be polygenic in nature. No full, integrated genetic analysis has been attempted here as yet. Such an analysis would in the long term entail some study of the variability of flower, pod and seed size in wild populations. The greater seed size of the wild South American populations of both *P. vulgaris* and *P. lunatus vis-à-vis* their Mexican counterparts would suggest that the genetic basis for this difference should be examined. It is possible that occasional hybridisation within both species plus selection could have resulted in the accumulation of polygenes with a positive effect on seed size in the early phases of domestication, resulting in a relatively rapid increase. It is also possible that polygenic mutations positively affecting seed size could have occurred and persisted in early stages of domestication. Since these processes are not necessarily mutually exclusive, they could also have occurred simultaneously.

Seed dispersal mechanism

Wild leguminous species have well-developed seed dispersal mechanisms, which typically consist of explosively dehiscent pods. Other dispersal mechanisms are exploited in genera such as *Desmodium*, in which the loments, into which the pod fragments, adhere to animal fur like burrs, and in *Arachis* and other geocarpic genera, in which soil movement by water appears to be the effective dispersal agent.

In the early days of domestication it was probably necessary to harvest crops when somewhat immature or in the morning when pods were damp with dew and before the dehiscence mechanism was activated by rapidly falling relative humidity and desiccation. This dehiscence is achieved as a consequence of the anatomical structure of the pod valves, especially the disposition of the fibrous tissues. These comprise (a) an inner parchment layer lining the pod cavity and containing fibres obliquely orientated; and (b) the longitudinal vascular bundles of the pod, especially those close to the sutures, which are the most strongly developed. In the primitive pod these tissues are highly lignified.

Loss of dehiscence potential can produce, from the dehiscent primitive or parchmented pod type, 'leathery' pods in which the valves of the pod can be separated readily at maturity but do not dehisce explosively. The ultimate suppression of the normal dehiscence mechanism is seen in the fleshy or 'stringless' pod type of some cultivars, in which lignification is virtually totally suppressed and the pod valves cannot readily be separated. The pod fragments into short segments enclosing single seeds rather than separating along the sutures to expose the seed (Fahn and Zohary, 1955; Roth, 1977).

The extreme development of producing a virtually indehiscent pod (fleshy or stringless) is seen in *Phaseolus vulgaris* where almost the entire crop grown for freezing, canning and freeze drying is of this type. Suppression of lignification in pod wall tissues extends the period during which the green pods are acceptable for eating. Although this situation is advantageous for exploitation as a vegetable, it poses some problems for use as a pulse. For the latter the leathery pod is more suitable. Leathery pods are also suitable in some species for use as green beans (Froussios, 1970).

In the commonly cultivated species edible green pods are produced by *P. vulgaris*, *P. coccineus*, and *P. polyanthus*. Those of *P. acutifolius* and *P. lunatus* are too tough and fibrous to have ever been used in this way. In recent times both *P. vulgaris* and *P. coccineus* have been subjected to selection pressures for further reduction in the lignification of fibrous tissue, resulting in strongly reduced lignification of the vascular bundles of the sutures (the 'strings') and the fibres of the pod walls, culminating in virtual suppression of lignification in tissues of the entire pod. The genetic control of the stringless character is apparently mono-factorial (Yarnell, 1965; Atkin, 1972).

As in many other domesticated plants, seed dispersal mechanism development can be suppressed by single mutations. The genetic control of the difference between the strongly dehiscent parchmented pod and the leathery pod types is less clear; this difference might be under polygenic control. However, in both leathery and stringless pod types the natural seed dispersal mechanism is rendered ineffective. This character is of high selective value since without it efficient harvest technologies would be more difficult to develop.

Changed growth form

Changed growth form is one of a complex of changes which have produced the more highly evolved cultivated beans of the present time. These differ so strikingly from their non-domesticated counterparts that, on the strength of these differences, conspecific wild and cultivated populations have at times, and by some authorities, been considered to be different species. Certainly there is an immediately obvious difference between a dwarf determinate cultivar and a rampantly growing indeterminate wild population. As has already been mentioned some change in growth form is dictated by the positive response of domesticated populations to selection for increased seed and pod size. The nature of the overall change in growth form is a little more complex.

The starting point for selection is obviously that of the wild form existing in its natural habitat. This is basically similar in all four species studied, which differ only in minor details of growth and physiology, Typically, germination of the seed produces an ascending shoot with indeterminate branching on which numerous lateral branches are produced. The growth of these may be horizontal or diageotropic for as much as 0.5 m before becoming ascending. This combination of erect and horizontal growth is a very effective strategy for exploiting shrubby vegtation as support. It is totally unsuited to production in cultivation unless some kind of a hedge is used as support. The association of maize and beans in cultivation demanded a somewhat more restrained growth of lateral branches, and a more strictly ascending mode of growth. The reduction in number of lateral branches developed and their more restrained growth produced an effective pair of crops in which maize provided support and the beans twined around the maize stems. This combination appears to have been very successful, not only in exploiting the soil very efficiently and meeting nutritional needs quite effectively, but also in providing rather less favourable conditions for pest and disease outbreaks than would be the case in pure stands.

The three *Phaseolus* species *P. vulgaris*, *P. lunatus* and *P. coccineus* have produced very similar wild and cultivated indeterminate, climbing forms. The behaviour of wild *P. coccineus* may be somewhat different from that of *P. vulgaris* initially in that growth of lateral branches in the open may produce a ground cover before becoming ascending even if provided with support. Domesticated climbing beans, because of their indeterminate growth, are able to maintain flowering and fruiting more or less indefinitely. The maize stems could provide support after harvest until they rotted or were consumed by termites. Maize and beans are not the only crops which could be grown together, but maize is one of the few American staple crops which could have provided effective support for climbing beans in early agricultural times. Mixed cultivation of climbing beans and squashes (*Cucurbita* spp.), for example, would be less effective than maize with beans. The need to produce compatible crop combinations involving beans with other crops could well have set in train a different kind of selection involving a further reduction in luxuriance of growth. This reduction can come about in two ways, the first by reduction of internode length, without reduction in node number, producing an indeterminate bush form, and the second by reduction in internode length coupled with a reduced number of nodes, the determinate bush form. It is therefore primarily the reduction in internode length which produces the bush form. Empirically a distinction is commonly made between 'indeterminate' and 'determinate' growth in *Phaseolus* beans. Main axes and lateral branches capable of producing indefinite growth in length and not terminated by an inflorescence are characteristic of 'indeterminate' forms, whereas such a termination occurs after production of usually less than ten nodes in the 'determinate' form. In *P. coccineus* it has been observed (J. Smartt, unpublished) that the 'determinate' forms produce main axes with not more than five nodes, usually bearing cotyledons, primary leaves and two nodes (occasionally three) producing trifoliolate leaves; the 'indeterminate' varieties produce at least 14–15 nodes. In *P. vulgaris* 'determinate' forms occur which may produce 3–6, 7–10 and 11–15 internodes (Evans and Davis, 1978). It seems probable that the distinction between 'determinate' and 'indeterminate' is not strictly valid in *P. vulgaris*. What probably is determined genetically is not whether the axis is terminated by an inflorescence or not but the number of nodes which are produced on the main axis before this comes about. In *P. coccineus* there appears to be a real discontinuity between what might be termed 'oligonodal' as against 'polynodal' growth forms but this does not seem to be so in *P. vulgaris*. True indeterminacy could be established experimentally by observing the number of nodes

(a) *(b)*

(c)

Fig. 4.13. Cultivated and wild forms of *Phaseolus* species. (*a*) *P. lunatus*: l, dwarf determinate; r, wild form. (*b*) *P. coccineus*: l, wild form; r, indeterminate climber. (*c*) *P. acutifolius*: l, dwarf indeterminate; r, wild form.

produced on a main axis which was maintained indefinitely in vegetative growth by continually being cut and rooted before apical growth ceased.

Reduction of stature by shortening internode length, whether in combination with production of fewer internodes or not, effectively suppresses stem twining and gives rise to an erect and bushy growth form. This is suitable for growing in pure stands or in mixtures with other low-growing crops. It is, however, the growth habit which *par excellence* is amenable to modern once-over or destructive harvesting techniques. It is easier with 'determinate' growth to produce a crop with near-simultaneous crop maturation than with 'indeterminate' growth. Cultivars which are developed for highly intensive, mechanised production systems are of the 'determinate' type. This notwithstanding, indeterminate growth has advantages in some circumstances, particularly under wet conditions when green pods or green mature seed may be produced and harvested over a relatively long period. This can obviate the necessity of storage, which can be difficult under humid conditions.

The full range of growth forms which have been observed can be found in the IBPGR descriptor lists, which are now available for all *Phaseolus* species (IBPGR, 1982–85).

(a) *(b)*

Fig. 4.14. Growth forms within cultigens of *Phaseolus*. (*a*) *P. vulgaris*: l, dwarf determinate; r, indeterminate climber. (*b*) *P. coccineus*: l, dwarf determinate; r, indeterminate climber.

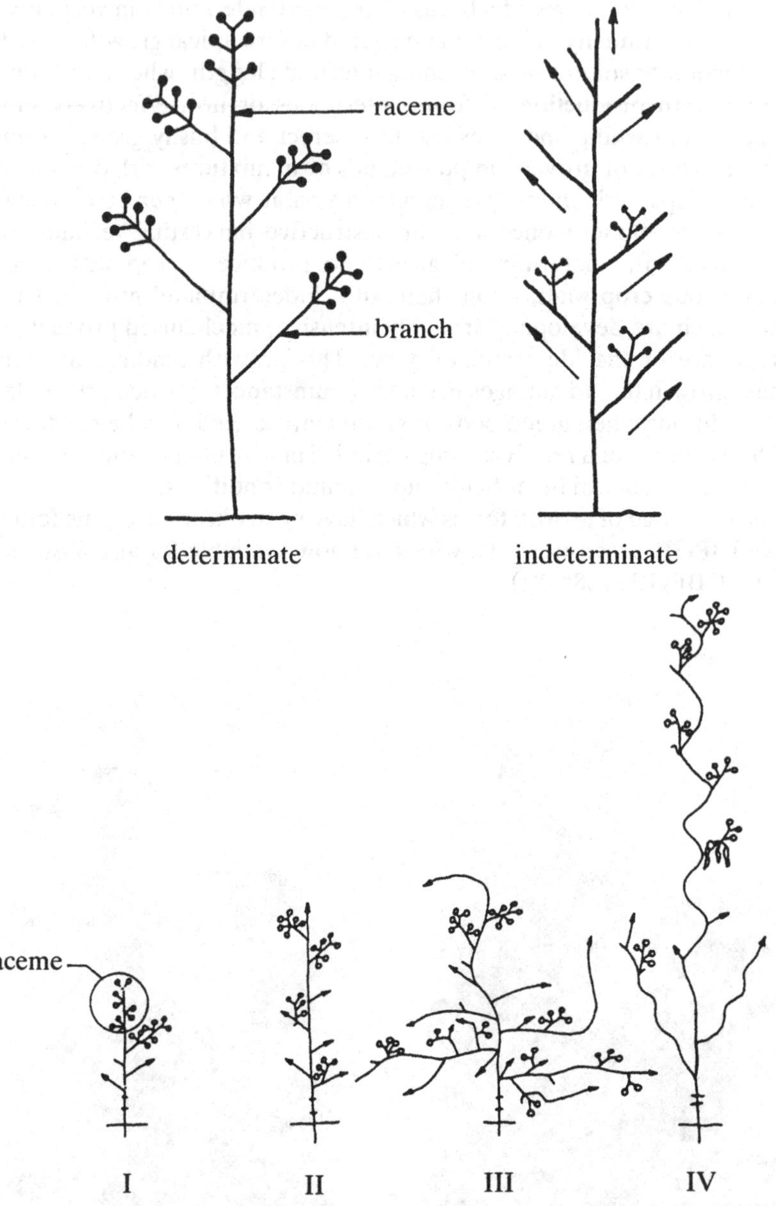

Fig. 4.15. Growth forms in *Phaseolus* recognised by CIAT. (*a*) Determinate vs indeterminate; (b) (I) determinate bush (dwarf); (II) indeterminate bush (dwarf); (III) indeterminate, straggling; (IV) indeterminate, climbing. (From Debouck and Hidalgo, 1984.)

Changed life form

Life form (*sensu* Raunkiaer, 1934) is defined in terms of overwintering strategy. In *Phaseolus* the perennial life form overwinters on reserves in a root tuber; annuals overwinter as seeds. These are both found in cultivated *Phaseolus* species and their wild progenitor types. Probably the primitive life form in the genus as a whole is the perennial. The annual life form may well have arisen in response to ecological selection pressures as in *P. acutifolius* which occurs naturally in semi-desert areas of north Mexico and the south-western United States. In *P. vulgaris* the annual life form also predominates. Although very occasionally wild perennial individuals are encountered, domesticated forms are typically annuals, as are most wild individuals. Primitive *P. coccineus*, *P. lunatus* and probably also *P. polyanthus* are perennials. Domesticated perennials and annuals are well known in *P. coccineus* and *P. lunatus*; *P. polyanthus* appears to be entirely perennial in domestication.

The change in life form from perennial to annual has arisen in response to rather different selection pressures in the wild and under cultivation. The annual life form probably arose in *P. acutifolius* in response to the ecological selection pressures of its desert environment. The majority of plant species in such a habitat are opportunistic annuals which take advantage of whatever rainfall occurs. The essentially mesophytic vegetative morphology of *P. acutifolius* would be inappropriate to a desert existence as a perennial. However, *P. acutifolius* does have some physiological adaptation to semi-desert conditions in its drought tolerance, which is greater than that of *P. vulgaris*. The annual life form has been established in perennial species under domestication for somewhat different reasons. The common feature is that the perennial life form has, perhaps, as little advantage in cultivated beans as it does for desert-inhabiting wild beans. The perennial *Phaseolus* species are relatively short-lived and their productivity falls off very rapidly after their first fruiting season. The result is that even though cultivated forms may potentially be perennial they are treated customarily as annuals. The establishment of the annual life form has also been favoured by the change from the climbing to the bush growth habit. This has encouraged more concentrated flowering and fruiting on a reduced shoot system which has apparently brought about exhaustion of assimilates, leaving no reserves for perennation. It is of interest to note that indeterminate cultivars of *P. coccineus* often develop a substantial root tuber; such root tubers are rarely if ever produced by determinate cultivars such as Hammond's Dwarf. The lima bean *P. lunatus* behaves similarly. Where a potential perennial was grown as an annual and selection for yield was practised, those genotypes which stored least in perennating organs would be better able to produce high seed yields. Such a selection régime could also result in a reduced ability to perennate.

Loss of seed dormancy

Seed dormancy is commonplace in *Phaseolus* species in the wild; equally, it is virtually absent in the cultigens. The loss of seed dormancy is a basic requirement for development of a successful crop plant which is seed propagated and harvested annually. Hard-seededness has obviously been successfully selected against in all important grain legume crops. Dormancy can be imposed by the development of an impervious coating on the testa and the sealing of hilum, raphe and micropyle by water-repelling waxes and the like (Kyle, 1959). In addition dormancy can be imposed by the presence of germination-inhibiting compounds in the seed. These may take a considerable time to disappear and permit germination. The presence of germination inhibitors is important in normal seed development. If immature embryos are excised from developing seeds (after differentiation of the cotyledons) and cultured *in vitro*, what occurs is effectively premature germination with development of a miniature shoot and root system. In normal development the embryo enlarges without any such extensive development of root and shoot systems. The point at which germination inhibition ceases can vary. In most bean cultivars germination does not occur until the seed is fully mature and the pod has dried. However, germination of the seed of cv. Kentucky Wonder can occur readily in the pod under high relative humidity conditions before the pod has dried out at maturity. Hard-seededness is under simple genetic control in *P. vulgaris*; the recessive allele at a single locus confers dormancy (Kyle and Randall, 1963).

Hard-seededness, in addition to imposing seed dormancy, has another consequence. Reduced penetration by water makes hard seed difficult to cook and prolongs cooking times as well as requiring more fuel. 'Cookability' is also inversely related to seed size and directly to protein content (Rutger, 1970). The character of 'cookability' is very important, particularly in Third World communities where fuel supplies are usually limited.

Other physiological changes

Apart from loss of seed dormancy the physiological change of greatest importance in the dissemination of *Phaseolus* beans has been the loss of photoperiodic sensitivity. The genus *Phaseolus* originated and evolved apparently in tropical and sub-tropical latitudes. The role of photoperiod sensitivity in low latitudes is interesting and appears to be important in synchronising flowering with growing season. In the equatorial zone day length varies very little during the year and the distribution of rainfall is such that it may fall at any time of the year. With movement away from the equator departures from a 12 hour day length occur, and distinctions can be drawn between wet and dry seasons. There may be two wet and two dry seasons in low latitudes, but closer to the tropics of Cancer and Capricorn a single wet and dry season alternate in the annual weather cycle. A longer day occurs during the rains and a shorter day in the dry season. Flowering and fruiting can be regulated in harmony with the annual cycle of day lengths if the critical photoperiod for flowering is appreciably shorter than the maximum day length. This ensures that vegetative growth can occur during the first half or two thirds of the growing season before the onset of flowering and fruiting. This will occur when rainfall is declining and conditions are favourable for fruit set and maturation. An appropriate photoperiod can harmonise fruiting with optimal seasonal conditions. In the tropics rainfall amounts and distribution are variable. If the rains are late the photoperiod will ensure that flowering and fruiting will still occur at the optimal time. Experiments in Zambia (Smartt, 1960) have shown that soyabeans sown successively over a two-month period all matured within the space of a fortnight. Photoperiodicity can therefore be a very effective means of regulating flowering and fruiting in the higher latitude tropics. It is not therefore surprising to find it established in many legume (and other) species. In strictly equatorial areas such photoperiodic control is superfluous or even disadvantageous. It is not surprising to find in the tropics as a whole both photoperiod-sensitive and day-neutral genotypes. It is these day-neutral genotypes which have in *Phaseolus* held the key to its wide dissemination into the temperate zones. The photoperiod requirements of adapted high latitude tropical and sub-tropical genotypes render them unsuitable for cultivation in temperate zones. The relatively short-day photoperiod requirement delays flowering and renders crops liable to frost damage. Thus the polymorphism for photoperiod sensitivity which can evolve in the tropics can materially assist dispersal of the crop beyond the tropics. This is of little significance in lowland tropical crops but of considerably greater importance in the montane tropical crops which effectively are growing under temperate conditions.

Biochemical changes

Seeds of the Leguminosae are known to develop a wide range of toxic materials which probably play a role in protecting them from predation. Some of these materials such as the alkaloids (e.g. in *Laburnum* spp.) are very effective. The most notorious example of a toxic material produced by a *Phaseolus* species is the hydrogen cyanide which can be generated in the seed of *P. lunatus*. The amount of HCN which can be produced is variable between genotypes. Those which produce the greatest amounts have been responsible for deaths, the earliest records for which came from Mauritius in 1884 but others have been recorded from Burma and Puerto Rico. The HCN is produced by hydrolysis of a cyanogenic glycoside (phaseolunatin) (Montgomery, 1964). Viehover (1940) found that the amount of HCN liberated on hydrolysis ranged from 10 to 300 mg per 100 g of beans. Contents in the range of 10–20 mg per 100 g of seed are regarded as safe and legal imits of cyanide content of lima bean seed have been established in the USA and elsewhere. It has been at times suggested (Smartt, 1976*a*) that white-seeded genotypes tend to produce lower HCN contents than those with highly pigmented testas. Since legal permissible limits have been set on HCN content there has apparently been little subsequent difficulty with clinical problems arising from lima bean consumption. However, if it ever proves to be necessary to make wide crosses within the species to wild or landrace material, cyanide content of selected segregants will require careful monitoring. Vanderborght (1979) has in fact demonstrated that cyanogenic glycoside contents are higher in wild than in cultivated material, sometimes exceeding 400 mg per 100g of seed. It appears that what might be termed a quantitative polymorphism has evolved in *P. lunatus* for cyanogenic glycoside content and this has given scope for selection of relatively innocuous genotypes which can be exploited in safety. Another important area of biochemical change is in pigmentation of the plant: vegetative parts, flowers, fruits and seeds. These pigments are flavonoids and include the anthocyanidins which produce red, violet and blue pigments and flavones which include the yellow pigments. Genetical and biochemical aspects have been reviewed by Feenstra (1960) and show that a complex system of colour variation has evolved especially as regards seed coat colours and patterning. Pigmentation patterns of stem, flower and fruit are relatively simple. Stem colours may be green, indicating absence or low levels of pigmentation; reddish or purplish colours indicate presence of red and/or blue anthocyanins. Pod pigmentation may be similar to that of the stem; it may occur as blotches, flecks or streaks of anthocyanin or may be uniformly distributed. Another different type of pod colour variant is produced in wax-podded cultivars, where chlorophyll initially present in the pod is lost and the pod then becomes yellowish. Flower pigmentation is typically lilac, but paler and colourless (white) flowers are found in all species. *P. coccineus* is unique among cultivated *Phaseolus* species in producing red flower pigments in many genotypes.

The production of pigmented stems, flowers and fruits is characteristic of the wild forms of all cultivated species. Reduced pigmentation and its loss is very much more common in domesticated forms. Pigmentation of stems, flowers and fruits is broadly correlated. There is also a general correlation with testa pigmentation. Genotypes with intensely pigmented seed coats will also tend to produce pigmented stems, flowers and pods.

However, the range of pigments developed and colours expressed in seed coats is considerably greater than that in other parts. Wild forms are in fact polymorphic for pigmentation of stems, flowers, fruit and seed but there is a great preponderance of lilac-flowered forms with flecked pods, pigmented stems and speckled seed. This character combination is found not only in the wild populations of cultigen species but equally in species which are not domesticated.

The significance of the common speckled pattern of seed coat coloration is probably related to seed predation. Self-coloured seeds, particularly of light and bright colours, are conspicuous against the majority of soil backgrounds whereas speckled seed are inconspicuous. Selection pressures in the wild would probably act very strongly in favour of speckled seed coats. Some self-colours might have superior selective values in specific environments. Black of buff seed could be inconspicuous against many backgrounds whereas bright yellow, red or white seed would almost invariably be conspicuous. With the harvesting of bean seed under cultivation the adaptive value of speckled testas was lost and a greater range of colours became established. Speckled patterns persist, albeit in a modified form, as in the pinto bean. The aesthetic appeal of colour and general appearance of the seed has become important under domestication.

In all cultivated bean species genotypes are established in which little or no pigmentation of the testa occurs. White-seeded cultivars are commonplace in *P. vulgaris*, *P. lunatus* and *P. coccineus*, for example. Loss of pigmentation produces both advantageous and disadvantageous effects. Flavour is probably improved by loss of testa pigmentation, but the seed loses the protection given by the flavonoid compounds against fungal attack. It is a matter of common observation that wet harvest conditions have a severer effect on the quality of white haricot or navy beans than of those with strongly pigmented testas such as cv. Red Kidney. Navy bean production is therefore dependent on reliably dry harvest conditions. This requirement precludes successful large-scale production in the United Kingdom and other oceanic climates where wet conditions at harvest time are probable.

Unrealised evolutionary potential

If the extent of apparent evolutionary advance under domestication is compared in the five species then it is clearly apparent that this is considerably greater in *P. vulgaris* than in the other *Phaseolus* species. If the range of known genetic resources is also considered, a similar broader range of genetic resources is apparent; these two observations are obviously correlated. The implication of these is that with a greater pool of genetic variability already established in *P. vulgaris* then the evolutionary potential for the future is also greater in this species.

It is worthy of note that although the other *Phaseolus* species show less evolutionary advance and a narrower range of genetic resources, the patterns of variability generated show marked similarities. These have been discussed by Smartt (1976*b*) and his summary conclusions are given in Table 4.1. The question can be raised as to why *P. vulgaris* should have outstripped other *Phaseolus* species. The reason is probably that the ecological preferences of mankind and this species largely coincide and that where the environment could support large human populations *P. vulgaris* could in many cases be grown successfully. Such large bean populations could generate and store considerably more genetic variability than, say, the comparatively small populations of *P. ιacutifolius* existing in a specialised and not very extensive semi-desert habitat.

It is worth noting that parallel and presumably homologous genetic variants can be generated in *Phaseolus* species which are of considerable complexity. The range of testa colour variants in *P. vulgaris* and *P. lunatus* are very extensive indeed, yet very similar. It is difficult to find a testa colour variant in one species without its parallel in the other. These parallel variants have totally independent origins from their respective wild forms; the gene pools of the two species are completely isolated from each other, thus ruling out any possibility of introgression.

For the purpose of the present discussion only *P. vulgaris* will be considered in detail. Comparable, if lesser, evolutionary potential obviously exists in other species; breeding success in improving *P. vulgaris* may be repeated in these other species.

Improvement in plant form

Plant breeders at the present time no longer tend to make crosses in the hope that something better will turn up by selecting the good qualities of only two parents, but rather to draw up detailed specifications of the ideal types towards which they are striving. The ideotype (ideal type) concept has gained ground enormously in the last two decades with the successes of the Green Revolution breeding strategies. Ideotypes are unlikely to be achieved other than by very complex crossing programmes and extensive search and evaluation of genetic resource material. In constructing a bean ideotype for shoot and canopy architecture it is necessary to consider the variability available in the components of the shoot system. The range in leaf size; stem thickness; node numbers of main axis and lateral branches; dwarf, climbing or intermediate growth habit (i.e. internode length); inflorescence length and carriage all need to be considered and the desired level of expression selected. Ideotypes would have to be constructed also bearing in mind ecological and agronomic considerations.

Having decided on the plant type or the desired vegetative structure, then other decisions will be required on pod type (leathery or fleshy), pod and seed size, seed shape and testa colour. In addition to these standards it may be necessary to specify protein content and quality, tolerance levels of toxic materials and anti-metabolites (i.e. lectins, protease inhibitors, flatus factors, etc.), cookability or tenderness of seed and digestibility of the protein. It would be possible to improve greatly the agronomic and consumer qualities of beans by using the ideotype strategy since the necessary genetic variability is probably available. Plant breeding skill of a high order would no doubt be required as well as considerable time in order to achieve the ideotype objective. Adams (1982) has achieved practical success in applying the ideotype concept and produced genotypes with desirable growth habit and improved yielding capacity.

Improvement in seed quality

At one time a great deal was written about the potential for improvement of yield and quality of protein produced by grain legumes (PAG, 1973). Nowadays much less emphasis is placed on such breeding objectives (Payne, 1978). The current view is to 'take care of the calories and leave the proteins to take care of themselves'. There is some justification for this view in circumstances where people have access to a reasonably mixed diet. It is open to serious question when staple diets providing the calorific intake are predominantly or exclusively low-protein foods such as cassava or plantains. The spectre of a world protein famine has receded somewhat but not vanished. The cost of animal proteins has risen considerably in real terms since World War II and is set to rise further. The substitution of plant for animal protein can also be justified by the fact that the production of animal protein involves consumption of many times that amount of plant protein. If direct consumption of plant protein could replace part, if not all, of the animal protein intake by the human population, a higher standard of nutrition could be maintained for more people. If one accepts the argument that less protein is required in the human diet than formerly believed, there is no doubt that a protein-rich diet (even if this comes largely from pulses) is more attractive and palatable than a low-protein, high-carbohydrate diet. If eating is to afford any pleasure at all, then a reasonable protein content is essential.

Substitution of plant for animal protein in human diets is not straightforward. Nutritionally animal protein is of higher quality than plant protein. The amino acid profiles of the former are commonly better balanced than those of the latter. Even when profiles are similar, as in raw lean beef on the one hand and soyabean protein on the other, digestibility of the animal protein is higher. This is due to the presence in the seed of the soyabean and a number of other pulses of toxic materials and anti-metabolites. The occurrence of cyanogenic glycosides in *P. lunatus* has already been noted.

There are present in the seeds of the commonly used *Phaseolus* species materials which can be regarded as anti-metabolites if not outright toxins. The seed of *P. vulgaris* has been most studied in this regard of all cultivated *Phaseolus* species. Two types of anti-metabolite or toxin have been found, namely protease inhibitors and phytohaemagglutinins or lectins. Although both are present in *P. vulgaris* seeds it is the lectins which create the greater problems in utilisation. These, possibly on account of their haemagglutinating activity, produce disturbance of the digestive tract. Both protease inhibitors and lectins are inactivated by heating and beans are rendered innocuous by adequate cooking; however, the uncooked beans are lethal when fed to rats as a sole food source. It would be useful none the less to produce beans free of anti-metabolites and toxic materials. Although protease-inhibitor presence or absence polymorphisms are known in the soyabean (Orf and Hymowitz, 1979) they have not been studied extensively in *Phaseolus*. However, lectin presence or absence polymorphisms have been observed (Jaffé *et al.*, 1972). These have been studied serologically by Klozová and Turková (1978) who have shown that the cultivar 'Krupnaya sakharnaya' is lacking the lectin proteins I and II. In place of these proteins, others of similar electrophoretic mobility but different serological specificity are produced. Clearly, from the point of view of the human consumer, reduction or elimination of anti-metabolites and toxic materials would be desirable. This might reduce the cooking time necessary to make beans wholesome; it could also enable raw beans to be used safely as a livestock food. It is possible that the elimination of lectins and protease inhibitors might entail a yield penalty. However, the work of Osborn and Bliss (1985) has given reassurance on this point; lectin-free genotypes without any apparent impairment of yielding capacity have been produced.

Other possible evolutionary developments in *Phaseolus*

There are two potential areas of evolutionary development which have received relatively little attention in *Phaseolus*; the first is the exploitation of polyploidy and the second is the exploitation of hybrid vigour or heterosis.

Some work has been carried out on the induction of allopolyploidy in inter-specific F1 hybrids, initiated by Lorz (1952) on progeny of the cross *P. lunatus* × *P. polystachyus* (L). B.S.P. The objective in producing this inter-specific hybrid and others (*P. vulgaris* × *P. coccineus* and *P. vulgaris* × *P. acutifolius*) has been to attempt the inter-specific transfer of desirable traits. Little or no work has been published of studies on the agricultural potential of the allopolyploids themselves. The results of some preliminary studies of the agricultural potential of the amphidiploid *P. vulgaris* × *P. coccineus* have shown it to be later-maturing than both parents, which is disadvantageous in the United Kingdom but might be of interest elsewhere. The fertility of some *Phaseolus* amphidiploids is not high but could improve to adequate levels under selection.

The exploitation of F1 hybrid vigour and uniformity has expanded enormously in both agriculture and horticulture since the end of World War II. It is appropriate therefore to consider the potential for exploitation of the phenomenon in the cultigens of this genus. In order to do this it is necessary to consider breeding systems in the genus. Two distinct breeding systems are found in *Phaseolus* species: the predominantly self-

Fig. 4.16. Wild *Phaseolus vulgaris* growing on a shrubby *Solanum* species, Michoacan, Mexico (from Gentry, 1969).

ing system in which self-pollination and self-fertilisation occurs naturally, and a more outbreeding system in which effective, spontaneous self-pollination does not occur. The first system is found in *P. vulgaris*, *P. lunatus* and *P. acutifolius* and the second in *P. coccineus* and *P. polyanthus*. In spite of the high selective advantage that the self-pollination system has under domestication, there is no evidence that breeding systems of cultivated *Phaseolus* species have changed in cultivation. Self-pollination can be maintained indefinitely in the three habitually self-pollinating species. If such a régime is imposed on *P. coccineus* a typical inbreeding depression syndrome develops after the S2 generation. Frequently, flower development becomes very irregular and pollen pro-

Fig. 4.17. Wild *Phaseolus vulgaris* from Durango, Mexico. Peduncle length is unusually long in this form (H. S. Gentry).

duction becomes poor. Many inbred lines cannot be maintained. However, as in other crop plants with similar breeding systems, e.g. maize and cotton, viable inbred lines have been produced, as they have in cv. Hammond's Dwarf Scarlet (J. Smartt, unpublished). It should be possible to produce synthetic varieties of scarlet runner bean by this method if it was thought worth while. F1 hybrids might also be developed if the male sterility which is occasionally found in *P. coccineus* proves amenable to management.

In both *P. coccineus* and *P. polyanthus* self-pollination is normally prevented by the inability of pollen to germinate on an intact stigmatic surface. When this surface is ruptured or abraded, germination of pollen

Fig. 4.18. Typical habitat of wild *Phaseolus vulgaris* in Michoacan, Mexico (from Gentry, 1969).

(that of the same or a different plant) can occur. No self-incompatibility mechanism as such appears to operate. A change in breeding system in either of these species would require a change in the nature of the stigmatic surface. If a change was to occur then there would be no obstacle to self-pollination as in other *Phaseolus* species. However, the population would experience a period of inbreeding depression during which the frequencies of deleterious recessive alleles could be expected to decline rapidly as they were exposed to selection as homozygotes and eliminated. This could have occurred in the evolution of *P. vulgaris* from its common ancestor with *P. coccineus* and *P. polyanthus*.

4.4. Future research objectives

In the course of any evolutionary advance it is inevitable that certain traits and characteristics can be lost. This frequently happens in the absence of positive selection for a particular character. In the absence of a specific pathogen, for example, there is no positive selection value for

Fig. 4.19. Wild *Phaseolus coccineus* at Nacaltepec, 20 miles north of Telixtlahuaca, Oaxaca, Mexico (from Gentry, 1969).

resistance to it. When it does appear the situation is of course totally reversed. This explains why reserves of genetic variability are continually being searched for genetic resistance to pests and pathogens. Extensive germplasm resources enable characters, which have been lost passively or even actively selected against, to be re-instated. This is a commonplace of crop plant evolution. However, a character of considerable current importance which appears to be diminishing in *Phaseolus* species, and especially *P. vulgaris*, is the ability to establish and maintain effective nitrogen fixation by *Rhizobium*.

It is a matter of common observation that plants of *Phaseolus vulgaris* at maturity are frequently poorly if at all nodulated. There are two possible major causes for this. There could perhaps have been unconscious selection against efficient symbiotic nitrogen fixation by constant pro-

Fig. 4.20. Bean cultivation in Middle America. An Indian farm in the Peten highlands of Guatemala. Bush beans are being cultivated beyond the fenced garden (from Gentry, 1969).

duction and selection on relatively nitrogen-rich soils. The selection of early maturity (and incidentally the dwarf growth habit) could substantially reduce the proportion of the lifespan during which effective nodulation is possible. These two effects have probably been operating in conjunction. As Sprent (1979) has noted, the climbing beans (which are closer in growth form to the ancestral type) are more effective in nitrogen fixation than are the dwarf beans. It is of course the latter which have been most extensively used for bulk production. It is perhaps opportune to consider whether some research effort might be devoted to the improvement of nitrogen fixation by dwarf beans. Selection for early establishment of effective nodules is obviously necessary.

There is undoubted scope for further research in the genetic control of those characters which have responded most to selection under domestication. While progress has undoubtedly been made in the genetic analysis of the differences between wild and domesticated populations of *Phaseolus* species, the surface of this very substantial task has only been scratched. Understandably greater progress has been achieved in the analysis of qualitative than in quantitative differences. Even here the

Fig. 4.21. Threshing of a common bean crop in the highland grassland zone of Durango, Mexico (from Gentry, 1969).

analysis has been piecemeal and unsystematic. The analysis of quantitative genetic differences, for example of pod and seed size, has only reached a preliminary stage (Cheah, 1973; Miranda Colín, 1974). The possible interactions between oligogenic and polygenic systems have scarcely been considered, if at all. No analysis has been attempted of correlated responses to selection for increases in size of seeds, pods, flowers and leaves together with the increased diameter of stems.

Our future ability to direct the further evolution of *Phaseolus* beans would undoubtedly be enhanced by progress in investigations such as those mentioned above. We cannot hope to influence future development of our crops without adequate knowledge of their present state and their evolutionary history.

Fig. 4.22. A market stall in Uruapan, Mexico, with dry beans for sale (from Gentry, 1969).

Table 4.2. *Comparative evolutionary development in* Phaseolus *species under domestication*

Characters	*P. vulgaris*	*P. lunatus*	*P. coccineus*	*P. acutifolius*
Increased size of pods and seeds	+	+	+	+
Reduction of hard-seededness	+	+	+	+
Reduction of parchment tissue	+	+	+	+
Production of 'stringless' pod types	+	–	+	–
Reduction of branching in polynodal forms	+	+	+	+
Occurrence of oligonodality	+	+	+	–
Production of polynodal dwarf forms	+	–	–	+
Life forms of cultigens				
(a) annual	+	+	+	+
(b) perennial	–	+	+	–
Life forms of wild forms				
(a) annual	+	–	–	+
(b) perennial	+	+	+	–
Flower colours in				
(a) wild forms				
lilac	+	+	+	+
white	+	–	–	–
red	–	–	+	–
(b) cultigens				
lilac	+	+	–	+
white	+	+	+	+
red	–	–	+	–
Testa colours in				
(a) wild forms				
wild speckled	+	+	+	+
self colours	+	+	+	–
patterns	–	–	–	–
(b) cultigens				
wild speckled	–	–	–	–
self	+	+	+	+
patterns	+	+	+	+
Wild growth habit scrub infiltrator	+	+	+	+
Significantly increased leaf size in cultigens	+	+	+	–

4.5. The present state of genetic resources

The changes which have been brought about as a result of domestication can be inferred from appropriate comparisons between cultigens and conspecific wild relatives. The assumption is made that the rate of apparent evolutionary change has been greater in the cultivated than in wild populations since the time of initial domestication. This can be supported by comparisons between wild populations of species with cultigens and those with none. Morphology of wild species and wild populations of cultivated species is very similar; changes from this wild common pattern in cultigens are inferred to be as a consequence of domestication. Changes affecting vegetative morphology and shoot architecture tend to follow a common pattern of more restrained vegetative growth and a compact form coupled with gigantism of leaves and more robust stems. Gigantism of seeds and pods also occurs and the latter are effectively supported by the thicker and stronger stems of the domesticates. Loss of seed dormancy and reduced effectiveness of dehiscence mechanisms have been advantageous in cultivation. The occurrence of day-neutral genotypes has also enabled *Phaseolus* to penetrate the temperate climatic zones very effectively. Biochemical changes under domestication are best exemplified by reduction in the cyanogenic glycoside content of cultivated *P. lunatus vis-à-vis* the wild form. Parallel evolution under domestication of cultigens in all domesticated species is clearly apparent with *P. vulgaris* having achieved the greatest advance and *P. acutifolius* the least. The genetic resources of *P. vulgaris* would also appear to be the most extensive of any of the cultivated forms. It has a large and extensive GP1A and probably an extremely large GP1B. There is an exploitable secondary gene pool and the tertiary gene pool is also extensive. Further work on collection of wild *Phaseolus* species and experimental hybridisation with all the cultigens is required before the full accessibility of wild genetic resources can be ascertained. The genetic resources available to the species of the common bean syngameon (*P. vulgaris*, *P. coccineus* and *P. polyanthus*) for use in crop improvement is virtually co-extensive with the syngameon itself. Those of the species *P. lunatus* and *P. acutifolius* are much less extensive. The most readily exploitable resources are in the primary gene pool (GP1A + GP1B); no secondary gene pool is known but presumably the tertiary gene pool comprises the bulk if not all of the genus. Future developments in genetic engineering and biotechnology may permit, eventually, the exploitation of potentially vast genetic resources.

5 The Old World pulses: *Vigna* species

The taxonomic revision of Maréchal *et al.* (1978) has brought within the confines of a single genus crops which had previously been distributed over three genera: *Vigna* itself, the genus *Voandzeia* (now merged with it) and *Phaseolus* (re-defined). The genus, in the revised sense, is distributed in the warmer parts of both the Old World and the New; all the cultigens are of Old World origin although the cowpea and the mungbean are now more widely distributed.

The members of the genus commonly cultivated are the cowpea itself (*Vigna unguiculata*), the groundbean or Bambara groundnut (*Vigna subterranea*) and the group of Asiatic forms (commonly known as grams) *Vigna radiata*, *V. mungo*, *V. angularis*, *V. umbellata*, *V. aconitifolia*, *V. glabrescens* and *V. trilobata*. The last two are of peripheral interest; *V. glabrescens* is a polyploid and *V. trilobata* can perhaps best be regarded as a semi-domesticate.

The extent to which these different species are exploited has been determined largely by economic factors. The cowpea is the subsistence legume *par excellence*; it is grown extensively in Africa and India in the Old World, and in Brazil in the New (Rachie, 1985). It is only in the USA that the cowpea is primarily a commercial crop, where it is grown to meet the demand for 'black-eyed peas' (dry mature seed). Green mature seed is also used both canned and frozen. With the increased popular use of pulses in high-fibre diets, the demand for cowpeas, along with that of other pulses, can be expected to increase in the near future.

The groundbean is a subsistence crop pure and simple. Commercial exploitation is negligible at the present time, although canning for local use has been attempted in some parts of Africa. Significant cultivation is largely confined to Africa south of the Sahara, in which it is widespread. The green mature seed is very palatable when eaten boiled but it is notorious for producing flatulence. The dry mature seed is excessively hard and difficult to use as a pulse. There have been periodic attempts to make use of the crop outside Africa but these have not been successful.

Of the Asiatic grams the green gram or mungbean (*V. radiata*) has achieved the widest distribution. This owes a great deal to the popularity of Asian cuisines in the Western world. The need to produce seed for the production of bean sprouts has brought about the cultivation of green gram in the USA to meet the relatively small but steady demand for the crop. The black gram (*V. mungo*), a very close relative of the green gram, has not achieved such a wide distribution and is still mainly cultivated in India. The adzuki bean (*V. angularis*) and the rice bean (*V. umbellata*), particularly the former, are becoming better known to western consumers. The moth bean (*V. aconitifolia*) also has potential in the increasingly pulse-conscious western world. While pulses of all descriptions are increasingly available to consumers in the developed world, there is a question regarding sources of supply. The demand at the present time is not particularly large but the potential for a steadily increasing demand exists. It is clearly apparent that, in an agricultural sector where cereal grain surpluses are a growing embarrassment and Third World famines are becoming increasingly frequent and severe, agricultural production strategies have to be re-considered. There is obviously extensive scope for diversification in cropping. Concentration on the improvement of carbohydrate staple crops and the almost complete neglect of other crops has made these almost totally uncompetitive with cereals. In Green Revolution economies one can find gluts of staple crops combined with shortages of legumes and other subordinate crops all too frequently. This balance urgently needs to be redressed.

The economic difficulties besetting improved production of legumes arise from a number of causes which result in generally low productivity. This can arise from an inherently low biological productivity of the species as a whole, or of particular genotypes, an unsatisfactory soil nutrient status and/or unsatisfactory conditions for symbiotic nitrogen fixation together with losses brought about by uncontrolled pest and disease incidence. The utilisation of some legumes is beset with nutritional problems arising from toxic or undesirable seed constituents; this fortunately (apart from the instance already mentioned) is not a serious problem with *Vigna* species (Bressani, 1985).

The problems of low production of *Vigna* crops are complex and need detailed analysis. Inherently low biological productivity does seem to be a major factor, especially in green gram. This might be overcome by development of alternative production strategies, such as catch-cropping with selected appropriate genotypes. If it does not prove possible to breed a satisfactory long-term variety, good total dry matter production might well be achieved with two consecutive crops of a short-term cultivar.

The economics of increased grain legume production require careful study. At the moment some increase in production could probably be absorbed in the areas of production. Relatively small increases might, in some areas at least, transform marginal deficiencies beyond the level of self-sufficiency to surpluses and even gluts. Increased efficiency of production need not necessarily greatly increase the total volume produced if this would produce surpluses, but could enable alternative commercial crops to be produced without jeopardising subsistence.

5.1. Classification and biosystematics of *Vigna*

Taxonomy of *Vigna*

The current view of *Vigna* taxonomy has arisen from the extensive studies of several legume taxonomists, most notably Verdcourt (1970*b*), and this has culminated in the revision of Maréchal *et al.* (1978), which has brought order out of chaos. This work has resulted in a very considerable amplification of the genus, a synopsis of which is presented in Table 5.1. Prior to the revision there was only one cultigen in the genus, namely the cowpea (*V. unguiculata*). The genus *Voandzeia* has now been merged with *Vigna* and the former Asian *Phaseolus* species have been transferred *en masse* to *Vigna*. This has ended a state of very considerable nomenclatural confusion because some but not all members of the present subgenus *Ceratotropis* had been transferred piecemeal to *Vigna*. This led certain authors (e.g. Smartt, 1976*a*) to retain the old nomenclature until the taxonomic revision of the genera *Phaseolus* and *Vigna* had been completed. However, some confusion does remain, as can be seen in Tindall (1983) where the green and black grams are correctly assigned to *Vigna* but the rice bean is still included in *Phaseolus* and not re-assigned to *Vigna*.

The genus *Voandzeia* has now formally been included within *Vigna* by Verdcourt (1981), implementing the suggestion of Maréchal *et al.* (1978) that this be done. There was a difficulty here in that the name *Voandzeia* had priority over *Vigna* but this problem has been satisfactorily solved by Verdcourt (1978, 1981). It was believed, in view of the strong resemblances in floral morphology to members of the genus *Vigna*, that the geocarpy of *Voandzeia* did not of itself constitute sufficient grounds for retaining its generic status. The present situation, therefore, is that we have three distinct groups of cultivated *Vigna* species, two of which are of African origin and the third Asiatic. The African groups each contain a single species: the cowpea, *Vigna unguiculata* (L.) Walp, is assigned to section *Catiang* of sub-genus *Vigna*, and the groundbean *Vigna subterranea* (L.) Verdc. is assigned to section *Vigna* of the same sub-genus. The Asiatic grams are assigned to sub-genus *Ceratotropis*; this contains eleven species, of which seven are domesticated or semi-domesticated forms (Table 5.1).

The revised taxonomy, although soundly conceived, has created a large and perhaps unwieldy genus (Smartt, 1981*b*) which Maréchal (1982) justifies on the grounds of strong palynological (and other) resemblances between sub-genera and sections. However, distinctive palynological features are less strongly developed in the Leguminosae than in some other families and it could be argued that in this instance perhaps too much weight has been attached to them. What is at stake here is the ranking of sub-generic taxa in the revision, whether in fact these would be better recognised as genera in their own right or not. There is a regrettable possibility in the near future that the genus as constituted will be dismembered and present sub-genera raised to generic rank. In some important ways the revised and enlarged genus *Vigna* does not seem to constitute an entirely satisfactory grouping. The difficulties probably arise because it appears to contain both ancient and more recently evolved phylogenetic lineages. The cowpea on this view would appear to belong to an ancient lineage in which virtually complete genetic isolation has evolved between it and its closest relatives in the genus. The Asiatic grams (sub-genus *Ceratotropis*), between which limited inter-specific hybridisation is possible, could well be of more recent origin. The question of inter-specific hybridisation will be discussed in more detail in relation to the question of genetic resources of individual species within the genus; these will be considered first separately and then comparatively.

5.2. The cowpea, *Vigna unguiculata* (L.) Walpers

It is now generally agreed that the cowpea is of African origin. Conspecific wild forms are found in Africa but are absent from Asia. Maréchal *et al.* (1978) state that the distribution of wild forms covers much of tropical Africa and that the greater part of the variability within the wild segment of the species is located in southern Africa. Verdcourt (1970*b*) considered this species to comprise five sub-species: two wild, *dekindtiana* (Harms) Verdc. and *mensensis* (Schweinf.) Verdc. and three cultivated, *unguiculata*, *cylindrica* (L.) Verdc. and *sesquipedalis* (L.) Verdc. The latter two cultivated forms have evolved in Asia; subspecies *unguiculata* is found in most areas where cowpeas are grown. The post-domestication evolution of cultivated *Vigna unguiculata* has two sequential components, an African followed by an Asian. The African dimension embraces primary domestication and the evolution of the typical *unguiculata* form, and the Asiatic the subsequent evolution of the sub-species *cylindrica* and *sesquipedalis*. It can be noted in passing that Verdcourt's cultivated sub-species were originally recognised by Linnaeus as distinct species within the genus *Dolichos*.

Table 5.1. *Synopsos of* Vigna *Savi*

Genus *Vigna* Savi
Sub-genus *Vigna*
Section 1 *Vigna*
1. *Vigna luteola* (Jacq.) Bentham
2. *Vigna marina* (Burman) Merrill
3. *Vigna fischeri* Harms
4. *Vigna bequaertii* Wilczek
5. *Vigna oblongifolia* Richard
var. *oblongifolia*
var. *parviflora* (Baker) Verdcourt
6. *Vigna lanceolata* Bentham
var. *lanceolata*
var. *filiformis* Bentham
7. *Vigna filicaulis* Hepper
var. *filicaulis*
var. *pseudovenulosa* Maréchal, Mascherpa & Stainier
8. *Vigna multinervis* Hutch. & Dalziel
9. *Vigna laurentii* De Wild.
10. *Vigna ambacensis* Baker
var. *ambacensis*
var. *pubigera* (Baker) Maréchal, Mascherpa & Stainier
11. *Vigna heterophylla* Richard
12. *Vigna hosei* (Craib) Baker
var. *hosei*
var. *pubescens* Maréchal, Mascherpa & Stainier
13. *Vigna parkeri* Baker
ssp. *parkeri*
ssp. *maraguensis* (Taubert) Verdcourt
14. *Vigna gracilia* (Guill. & Perr.) Hooker fil.
var. *gracilis*
var. *multiflora* (Hooker) Maréchal, Mascherpa & Stainier
15. *Vigna racemosa* (G. Don) Hutch. & Dalziel
16. *Vigna desmodioides* Wilczek
17. *Vigna subterranea* (L.) Verdc.
var. *subterranea*
var. *spontanea* (Harms) Hepper
18. *Vigna angivensis* Baker
19. *Vigna stenophylla* Harms
20. *Vigna gazensis* Baker
Section 2 *Comosae* Maréchal, Mascherpa & Stainier
21. *Vigna comosa* Baker
ssp. *comosa* var. *comosa*; var. *lebrunii* (Baker) Verdcourt
ssp. *abercornensis* Verdcourt
22. *Vigna haumaniana* Wilczek
var. *haumaniana*
var. *pedunculata* Wilczek

Table 5.1. (*cont.*)

Section 3 *Macrodontae* Harms
23. *Vigna membranacea* Richard
  ssp. *membranacea*
  ssp. *macrodon* (Robyns & Boutique) Verdcourt
  ssp. *caesia* (Chiov.) Verdcourt
  ssp. *hapalantha* (Harms) Verdcourt
24. *Vigna friesiorum* Harms
  var. *friesiorum*
  var. *angustifolia* Verdcourt
  var. *ulugurensis* (Harms) Verdcourt
Section 4 *Reticulatae* Verdcourt
25. *Vigna reticulata* Hooker fil.
26. *Vigna wittei* Baker
27. *Vigna radicans* Baker
28. *Vigna dolomitica* Wilczek
29. *Vigna kassneri* Wilczek
30. *Vigna tisserantiana* Pelleg.
31. *Vigna pygmaea* Fries
32. *Vigna phoenix* Brummitt
33. *Vigna platyloba* Hiern
Section 5 *Liebrechtsia* (De Wild.) Baker
34. *Vigna frutescens* Richard
  ssp. *frutescens* var. *frutescens*; var. *buchneri* (Harms) Verdcourt
  ssp. *incana* (Taubert) Verdcourt
  ssp. *kotschyi* (Schweinf.) Verdcourt
Section 6 *Catiang* (DC.) Verdcourt
35. *Vigna unguiculata* (L.) Walpers
  ssp. *unguiculata* cv. gr. unguiculata cv. gr. biflora cv. gr. sesquipedalis cv. gr. textilis
  ssp. *dekindtiana* (Harms) Verdcourt var. *dekindtiana* (Harms) Verdcourt
    var. *mensensis* (Schweinf.) Maréchal, Mascherpa & Stainier
    var. *pubescens* (Wilczek) Maréchal, Mascherpa & Stainier
    var. *protracta* (E. Mey.) Verdcourt
  ssp. *tenuis* (E. Mey.) Maréchal, Mascherpa & Stainier
  ssp. *stenophylla* (Harvey) Maréchal, Mascherpa & Stainier
36. *Vigna nervosa* Markotter
Sub-genus *Haydonia* (Wilczek) Verdcourt
Section 1 *Haydonia*
37. *Vigna monophylla* Taubert
38. *Vigna triphylla* (Wilczek) Verdcourt
39. *Vigna juncea* Milne-Redhead
  var. *juncea*
  var. *major* Milne-Redhead
Section 2 *Microspermae* Maréchal, Mascherpa & Stainier
40. *Vigna microsperma* Viguier

Table 5.1. (*cont.*)

41. *Vigna richardsiae* Verdcourt
42. *Vigna schimperi* Baker
Section 3 *Glossostylus* Verdcourt
43. *Vigna nigritia* Hooker
44. *Vigna venulosa* Baker
Sub-genus *Plectotropis* (Schumach.) Baker
Section 1 *Plectotropis*
45. *Vigna vexillata* (L.) Richard
var. *vexillata*
var. *macrosperma* Maréchal, Mascherpa & Stainier
var. *angustifolia* (K. Schum. & Thonn.) Baker
var. *dolichonema* (Harms) Verdcourt
var. *yunnanensis* Franchet
var. *pluriflora* Franchet
46. *Vigna davyi* Bolus
47. *Vigna hundtii* Rossberg
48. *Vigna kirkii* (Baker) Gillett
Section 2 *Pseudoliebrechtsia* Verdcourt
49. *Vigna nuda* N.E. Br.
50. *Vigna lobatifolia* Baker
51. *Vigna longissima* Hutch.
Sub-genus *Ceratotropis* (Piper) Verdcourt
52. *Vigna mungo* (L.) Hepper
var. *mungo*
var. *silvestris* Lukoki, Maréchal & Otoul
53. *Vigna radiata* (L.) Wilczek
var. *radiata*
var. *sublobata* (Roxb.) Verdcourt
var. *setulosa* (Dalzell) Ohwi & Ohashi
54. *Vigna glabrescens* Maréchal, Mascherpa & Stainier
55. *Vigna trilobata* (L.) Verdcourt
56. *Vigna khandalensis* (Santapan) Raghavan & Wadhwa
57. *Vigna aconitifolia* (Jacq.) Maréchal
58. *Vigna angularis* (Willd.) Ohwi & Ohashi
var. *angularis*
var. *nipponensis* (Ohwi) Ohwi & Ohashi
59. *Vigna umbellata* (Thunb.) Ohwi & Ohashi
var. *umbellata*
var. *gracilis* (Prain) Maréchal, Mascherpa & Stainier
60. *Vigna minima* (Roxb.) Ohwi & Ohashi
61. *Vigna dalzelliana* (O. Kuntze) Verdcourt
62. *Vigna bourneae* Gamble
Sub-genus *Lasiospron* (Bentham emend. Piper) Maréchal, Mascherpa & Stainier
63. *Vigna lasiocarpa* (Bentham) Verdcourt
64. *Vigna longifolia* (Bentham) Verdcourt
65. *Vigna juruana* (Harms) Verdcourt

Table 5.1. (*cont.*)

Sub-genus *Sigmoidotropis* (Piper) Verdcourt
  Section 1 *Sigmoidotropis*
  66. *Vigna speciosa* (H.B.K.) Verdcourt
  67. *Vigna candida* (Vellozo) Maréchal, Mascherpa & Stainier
  68. *Vigna halophila* (Piper) Maréchal, Mascherpa & Stainier
  69. *Vigna elegans* (Piper) Maréchal, Mascherpa & Stainier
  70. *Vigna antillana* (Urban) Fawcett and Rendle
  Section 2 *Pedunculares* Maréchal, Mascherpa & Stainier
  71. *Vigna peduncularis* (H.B.K.) Fawcett & Rendle
        var. *peduncularis*
        var. *clitorioides* (Bentham) Maréchal, Mascherpa & Stainier
        var. *pusilla* (Hassler) Maréchal, Mascherpa & Stainier
  72. *Vigna firmula* (Bentham) Maréchal, Mascherpa & Stainier
  Section 3 *Caracallae* (DC.) Maréchal, Mascherpa & Stainier
  73. *Vigna caracalla* (L.) Verdcourt
  74. *Vigna hookeri* Verdcourt
  75. *Vigna linearis* (H.B.K.) Maréchal, Mascherpa & Stainier
        var. *linearis*
        var. *latiofolia* (Bentham) Maréchal, Mascherpa & Stainier
  76. *Vigna vignoides* (Rusby) Maréchal, Mascherpa & Stainier
  Section 4 *Condylostylis* (Piper) Maréchal, Mascherpa & Stainier
  77. *Vigna venusta* (Piper) Maréchal, Mascherpa & Stainier
  Section 5 *Leptospron* (Bentham) Maréchal, Mascherpa & Stainier
  78. *Vigna adenantha* (G. F. Meyer) Maréchal, Mascherpa & Stainier
Sub-genus *Macrorhyncha* Verdcourt
  79. *Vigna macrorhyncha* (Harms) Milne-Redhead
  80. *Vigna grahamiana* (Wight & Arn.) Verdcourt
  81. *Vigna praecox* Verdcourt

After Maréchal, Mascherpa and Stainier (1978).

(a)

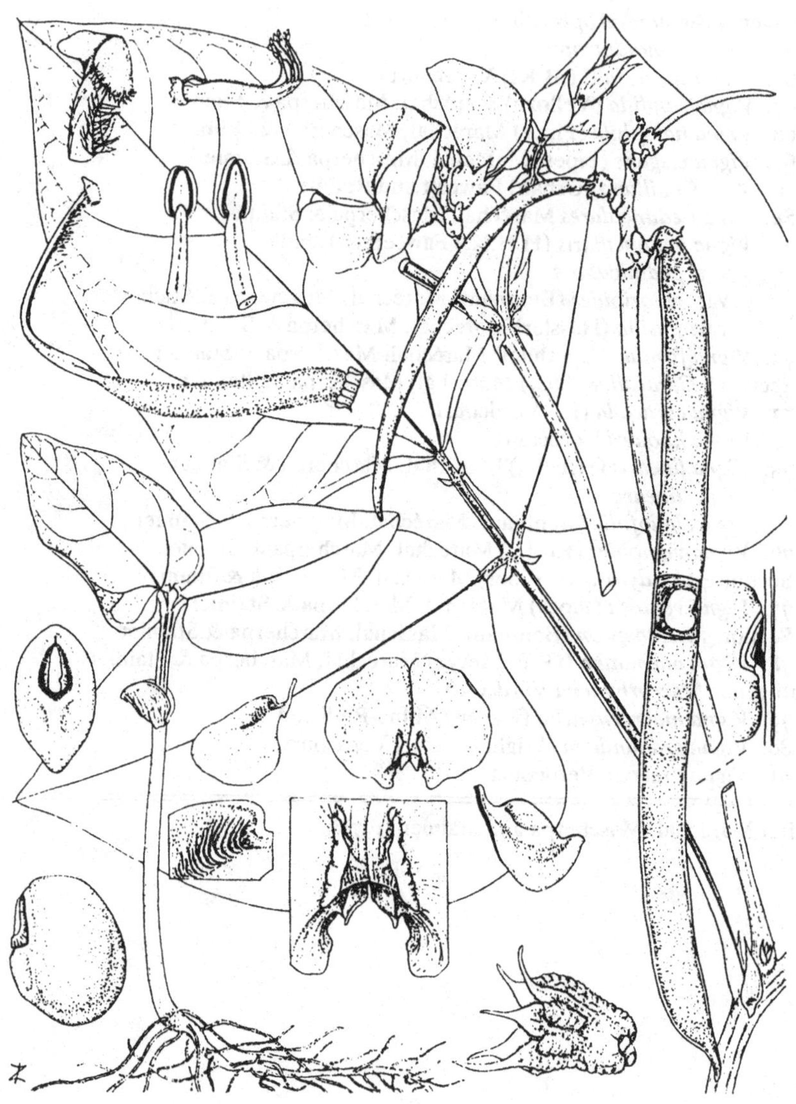

Fig. 5.1. Forms of *Vigna unguiculata*, the cowpea: (*a*) *unguiculata*; (*b*) *cylindrica*; (*c*) *sesquipedalis*. ((*a*) from Westphal, 1974, (*b*, *c*) from Purseglove, 1974.)

(*b*)

(c)

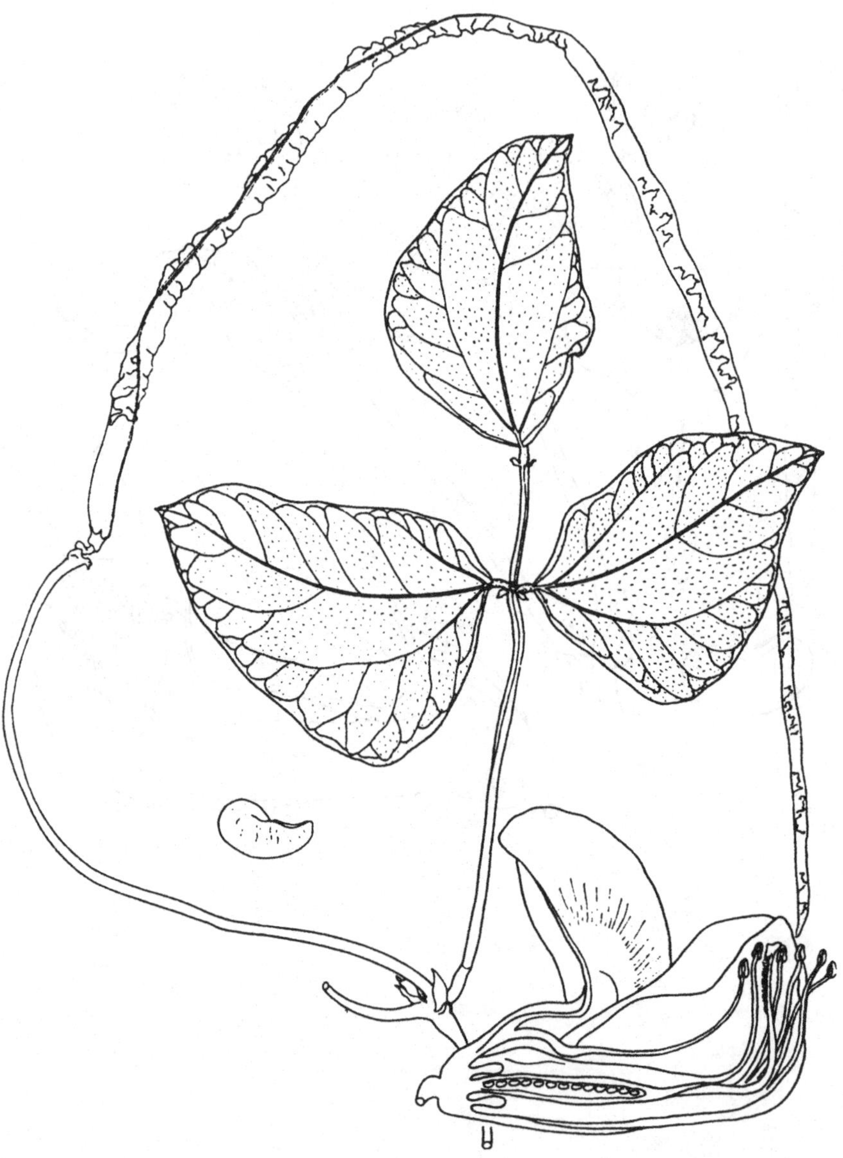

Fig. 5.1. (*cont.*)

Maréchal *et al.* (1978) view the species somewhat differently and assign all cultivated forms to sub-species *unguiculata* but follow Westphal (1974) in recognising four distinct cultivar groups within it, namely unguiculata, biflora (= cylindrica), sesquipedalis and textilis. In their scheme sub-species *dekindtiana* is enlarged to include the earlier subspecies *dekindtiana* and *mensensis* together with the former *Scytalis protracta* E. Mey. and *Vigna pubescens* Wilczek, all of which are assigned varietal rank only. Two new subspecies, *tenuis* (E. Mey.) Maréchal, Mascherpa & Stainier (from Natal) and *stenophylla* (Harvey) Maréchal, Mascherpa & Stainier (from the Transvaal), are also recognised. The specimens cited for sub-species *dekindtiana* come from a considerably wider area than that indicated by Steele (1976) and Steele and Mehra (1980), var. *dekindtiana* ranging as far south as Zimbabwe and Mozambique. The total range of wild forms of *V. unguiculata* extends as noted into the South African provinces of Natal and the Transvaal (subspp. *tenuis* and *stenophylla* respectively); this total range covers much of sub-Saharan Africa, therefore.

Origin and domestication

Faris (1965) concluded that the cowpea arose from the domestication of wild *Vigna unguiculata dekindtiana* forms in West Africa. Steele (1976) noted that variability in sub-species *dekindtiana* is greater in Ethiopia than in West Africa, and suggested that domestication could actually have occurred in Ethiopia and dissemination was westwards across the landmass of Africa, and eastwards across the Indian Ocean (cf. guar (Hymowitz, 1972)). Steele also noted the possibility of a diffuse centre of origin, consistent with the distribution of wild *dekindtiana* cowpeas. He suggested that the cowpea could have had a diffuse origin like that of the common bean in the New World. The origin of the cowpea can with confidence be located in a broad sub-Saharan belt. Agriculture in Africa appears to have originated here and was carried southwards by the Bantu migrations. Lush and Evans (1981) consider that var. *dekindtiana* is a more probable progenitor type than var. *mensensis* (recognised as sub-species by these authors) because it is closer morphologically to the cultivated cowpea. However, this similarity is also explicable if var. *dekindtiana* arose by introgression of cultigen germplasm into var. *mensensis*. It is of some interest to note that these two botanical varieties have different geographical ranges. That of var. *mensensis* is relatively restricted, whereas that of *dekindtiana* covers much of Africa south of the Sahara (Maréchal *et al.*, 1978). The weedy var. *dekindtiana* type could have originated from hybrids of var. *mensensis* with sub-species *unguiculata* and perhaps spread with the crop since var. *mensensis* is an outbreeder (Ng and Maréchal, 1985); this is not only possible, but probable. This hypotheses would quite clearly support the case for a West African origin, advanced by Faris (1965). Because var. *mensensis* is morphologically farther from the cultigen than var. *dekindtiana*, which is somewhat intermediate, it would seem preferable to regard var. *mensensis* as the most probable ultimate progenitor type. It must be acknowledged that this phytogeographic and morphological evidence is somewhat equivocal; however, the establishment of strong 'weedy' tendencies in a population of a species in which other populations are 'non-weedy' does suggest contact with cultivation and that the 'weedy' tendency is a consequence of this. Maréchal *et al.* (1978) also noted that the range of variation within the species as a whole is greatest in southern Africa. If one were to equate high levels of variability with centre of origin or domestication this might suggest a southern African domestication (cf. Steele, 1976). This, however, is highly improbable, as the direction of migration of agricultural peoples in pre-colonial and early imperial days was from north to south. The practice of agriculture came south with the Bantu migrations, which had not reached the Cape of Good Hope at the time of European settlement there.

It is unfortunate that there is so little archaeological record of crops of sub-Saharan African origin, both in Africa and elsewhere, e.g. in Asia. Steele (1976), however, related the spread of the cowpea to the development and spread of a cereal culture system based on sorghum and pearl millet. The spread of the cowpea to Asia probably occurred in a similar fashion and possibly at about the same time as sorghum. There are some implements from archaeological sites in West Africa (the upper Niger and Ghana) which suggest that a cereal-based agriculture was practised at about 3000 BC (Steele and Mehra, 1980) and it is entirely possible that the cowpea could have been an element in this agricultural system. We have alternative hypotheses, based on the somewhat meagre evidence available, of either a West African origin of the cowpea or of a more diffuse origin in northern sub-Saharan Africa. At the present time the evidence marginally favours the former view.

The movement of the cowpea from Africa to Asia, principally the Indian sub-continent, appears to have occurred by about 2000 years ago. The much greater diversity of agricultural systems established in India subjected the crop to much more disruptive selection pressures than in Africa. The African use as a pulse continued, but use as a forage apparently resulted in the establishment of the biflora type by selection, while utilisation of green pods as a vegetable resulted in the establishment of the long-podded sesquipedalis type in the Far East. There can be little doubt that the intensive and widespread cultivation of the cowpea in India generated a large gene pool; this has responded to strongly disruptive selection pressures by producing distinct evolutionary lines, once regarded as different species. The richer diversity generated in Asia persuaded some authorities (e.g. Darlington, 1963) that the cowpea had an Indian origin.

Subsequent dispersion of the crop has been documented by Ng and Maréchal (1985). They believe that the crop was established in southwestern Asia by about 2300 BC and in southern Europe by 300 BC. Dissemination of the crop to the Americas occurred in the sixteenth and seventeenth centuries from southern Europe and more particularly from West Africa with the slave trade.

Archaeology at present sheds little light on the origin and domestication of the cowpea. The historical documentary evidence is more informative (Steele and Mehra, 1980) on dissemination in Asia and southern Europe (Burkill, 1935; Purseglove, 1974). More information from archaeological and historical sources would be most valuable and hopefully may eventually be forthcoming.

Evolution of form

There now appears to be general agreement that the typical cowpeas, cv. gr. unguiculata, evolved in Africa. These have a trailing growth habit but are generally grown without support, as the inflorescences grow well above the leaf canopy and the pods mature satisfactorily since they are held well out of contact with the soil. In the African context the cowpea's role is predominantly that of a pulse, although it may be exploited to a minor degree as a leaf vegetable or spinach, as are *Phaseolus* beans. It is apparent that the crop in Asia (India in the first instance) was subjected to quite a different range of selection pressures. As Steele and Mehra (1980) indicate, there is no shortage of good indigenous pulse crops in India and Asia generally, such as green gram, pigeonpea, chickpea etc., but few produce entirely satisfactory pods for consumption as a green vegetable. It is probable that selection was practised among introduced unguiculata lines for succulent, fleshy pod types and this eventually culminated in the sesquipedalis cultivars and landraces which have reached their greatest development in the Far East. The most satisfactory growth habit for this form is that of a twining climber rather than trailing or erect. With a pod length of 50–100 cm satisfactory production demands that developing pods be kept out of direct contact with the soil and the hazard of soil-inhabiting pathogens.

A very different line of selection was also practised in India where the cowpea came to be used as an animal fodder. It seems entirely probable that this line of selection led to the establishment of the erect growth habit, with reduced node number and internode length. It is manifestly easier to harvest an erect growing plant as fodder than either a trailing or climbing form. There would probably be absolutely no advantage to a dwarf erect growth habit for production of seed crops until the advent of large implements for cultivation. Mechanised cultivation, particularly in the southern USA, has led to the development of numerous erect cowpea cultivars from crosses between biflora lines and those with good seed quality.

Changes in pod morphology have had repercussions on that of the seed. It is commonplace for the part of a crop plant actually exploited by

Fig. 5.2. Growth habits in *Vigna unguiculata*: (*a*) prostrate; (*b*) climbing; (*c*) erect ('60 day') (R. J. Summerfield).

man (especially for food) to show gigantism. This is clearly shown in the yard-long bean. Gigantism is also apparent in seed size. Cultivars have seed weights up to 34 times those of wild forms, with mean seed weights ranging from 10 mg to 340 mg. There appears also to be a correlation between seed size and pod carriage. In ssp. *dekindtiana* and cv. gr. biflora types this is erect, but in cv. gr. sesquipedalis and cv. gr. unguiculata (with longer and/or heavier pods) it is pendent. Pods are usually borne only in twos and threes and so the flowering stems are not usually over-burdened with crop, maintaining pods reasonably free of soil contact in both erect and trailing types.

The difference in the nature of the selection pressures in Asia and Africa has produced some interesting effects on seed size and shape. If a crop is to be grown for forage, then large seed size is of no advantage, probably quite the reverse. It is not surprising, then, to find that biflora cultivars have on average smaller seed than those of unguiculata. The seed of both these cultivar groups (particularly biflora) may be crowded in the pod, producing oblong or cylindrical seeds. The long pods of some unguiculata and all sesquipedalis lines do not necessarily contain more seeds than those of shorter-podded forms, but they are more widely separated, particularly in sesquipedalis. It is in the latter that the most elongated and even kidney-shaped seeds are found.

As is commonly found in domesticated legumes, pods tend to become less dehiscent at maturity, particularly where the green pod is used as a vegetable. The cowpea has produced a range of pod types comparable with that of *Phaseolus vulgaris*: strongly dehiscent in the wild forms with progressive reduction of the parchment tissue in domestication, culminating in indehiscent pods in the most advanced cultivars (Lush and Evans, 1981). In common with most other pulses, wild forms show seed dormancy imposed by impermeable seed coats (Lush and Evans, 1980). In domesticates the testa may be permeable all over its surface or only in specific areas such as the hilum. Hard-seededness has been selected against very effectively in the course of cowpea domestication. Lush and Evans found no physiologically determined dormancy in the samples of wild cowpea they examined; seed impermeability would thus seem to be the prime agent in imposing dormancy.

Some mention should be made of the rather curious group of cowpeas in cv. gr. textilis. This group (Westphal, 1974) is known from northern Nigeria where a strong fibre is obtained from the erect peduncles, which may be up to 60 cm in length. No exploitable fibre is apparently produced by any other part of the plant and the seed of this type is apparently not eaten, according to Dalziel (1937).

The morphological features of different cultivar groups and the wild forms are well summarised in Table 5.2, from Ng and Maréchal (1985).

Table 5.2. *Characteristics of various cv. gr. of sub-species* unguiculata *and varieties of sub-species* dekindtiana

Character	cv. gr. unguiculata	cv. gr. biflora	cv. gr. sesquipedalis	cv. gr. textilis	var. *mensensis*	var. *dekinditiana*	var. *pubescens*
Flower colour	white, purple	white, purple	white, purple	white, purple	purple	purple	purple
Standard petal							
width (mm)	24–30	24–28	28–35	24–27	33–42	23–42	28–32
length (mm)	18–23	18–24	21–24	19–21	—	17–34	19–21
Pod							
length (cm)	6.5–25	7–13	15–90	7–14	—	6.0–11.6	—
width (mm)	3–12	4–6	5–11	—	—	2.8–7	—
orientation	mostly pendent, vertical	mostly vertical	all pendent	vertical	vertical	vertical	vertical
Texture	fibrous, hard, firm; not inflated when young	fibrous, hard firm; not inflated when young	succulent; inflated towards maturity; shrinking after maturity	fibrous, hard, firm; not inflated when young	fibrous, hard firm; not inflated when young	fibrous, hard, firm; not inflated when young	fibrous, hard, firm; not inflated when young
Dehiscence	nil	nil to moderate	nil	nil to moderate	shatters	shatters	shatters
Locules/pod	7–23	12–16	15.8–23	—	16–19*	14–17	—
Calyx lobe (mm)	<5	<5	<5	<5	>5	<5	<5
Seed							
length (mm)	6–11	5–7	7–11	5.1–7.6	—	3–6	—
width (mm)	4–9	3–5	5–8	4–5.6	—	2–4	—
orientation in pod	crowded	crowded	far apart	crowded	crowded	crowded	crowded
Breeding system	inbreeder	inbreeder	inbreeder	inbreeder	outbreeder*	inbreeder	inbreeder
Growth habit	erect, prostrate, climbing	prostrate, climbing	semi-erect, climbing	prostrate	prostrate, climbing	semi-erect, prostrate, climbing	semi-erect, prostrate
Shoots	glabrous	glabrous	glabrous	glabrous	glabrous	glabrous	pubescent
Inverted V-shaped pigmentation on leaves	nil	nil	nil	nil	some	some	nil

*Information quoted from Lush and Evans (1981).

Cytogenetics and hybridisation

The chromosome complement of the cowpea, $2n = 2x = 22$, is typical of the tribe *Phaseoleae* (Figure 5.3). This possibly is why it has been comparatively little studied (Faris, 1964; Frahm-Leliveld, 1960; Yarnell, 1965). Meiotic chromosomes at pachytene have been studied by Mukherjee (1968), who classified the complement as comprising 1 short, 7 medium and 3 long chromosomes. He noted that the distribution of chromomeres along the chromosome arms was not uniform. Somatic C-metaphase chromosomes are relatively small and the longest do not usually greatly exceed 3 μm in length. Induced tetraploidy has been studied by Sen and Hari (1956) and Sen and Bhowal (1960). As is common with induced autopolyploids, meiosis was disturbed and fertility impaired. It seems highly improbable that any practically useful autotetraploid could be produced.

Crosses have been made (Rawal, 1975; Lush, 1979) between var. *mensensis*, var. *dekindtiana* and the domesticates which suggest that some genetical crossing barriers may have developed between these different forms. These could have a cytological basis. However, Steele (1976) reported that all subspecies of *V. unguiculata* are interfertile. The com-

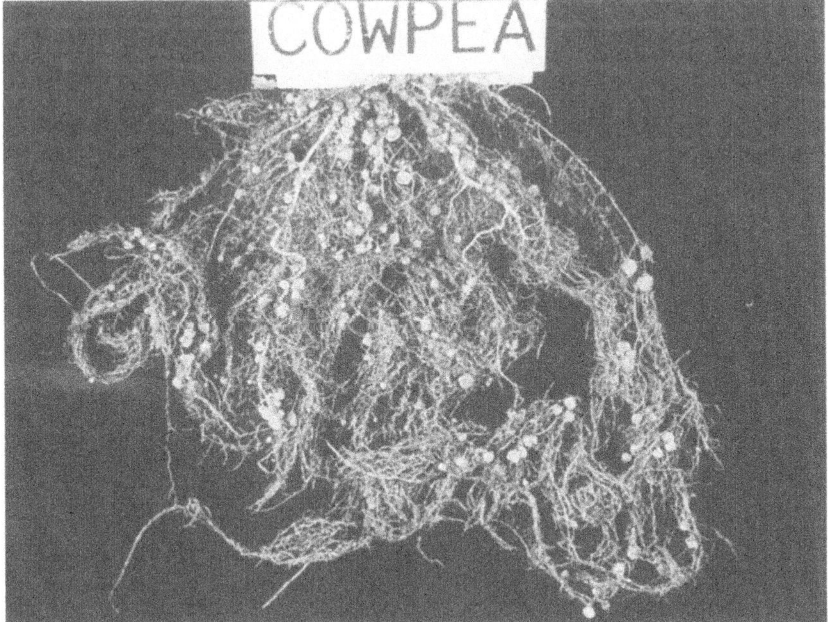

Fig. 5.3. Nodulated root system of cowpea (R. J. Summerfield).

mon explanation for such contradictory reports is the occurrence of chromosome structural polymorphism. There are alternative possibilities, such as complementary lethal or deleterious gene action, which could be tested. No viable inter-specific hybrids have been reported with other species of *Vigna* (Singh *et al.*, 1964; Evans, 1976) such as *V. umbellata*, *V. vexillata*, *V. mungo*, *V. radiata*, *V. aconitifolia* and *V. angularis*. The inter-generic cross *V. unguiculata* × *Phaseolus coccineus* (Ballon and York, 1959) also failed.

Photoperiod sensitivity

Short-day and day-neutral genotypes of the cowpea are known. This question has not been studied in depth by geneticists but Sene (1967) con-

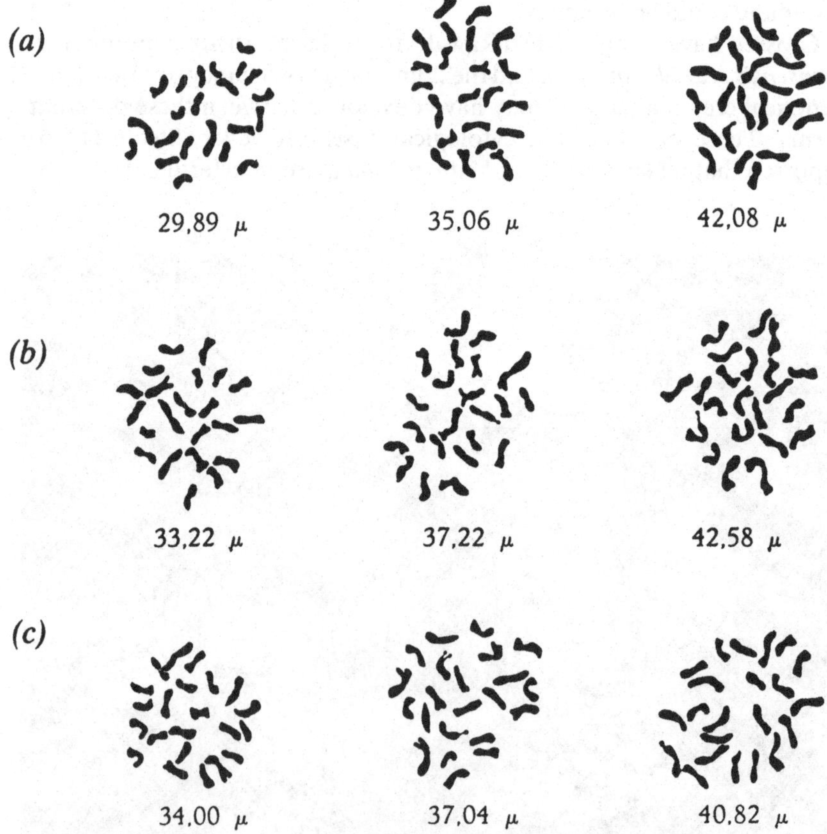

Fig. 5.4. Chromosome complements of *Vigna unguiculata* forms (× 2500): (*a*) *unguiculata*; (*b*) *cylindrica*; (*c*) *sesquipedalis* (from Maréchal, 1969).

sidered that a pair of major genes was involved, with short-day photoperiod being dominant to day-neutral.

Biochemistry

The use of cowpea seed does not apparently present any serious nutritional problem, although protease inhibitors have been found in the seed (Liener, 1982). Lectins have not as yet been reported in significant quantity. Biochemical goals do not figure largely in breeding objectives for the crop (Rachie, 1985). The low content, or apparent absence, of many common legume anti-metabolites may partly explain the extreme susceptibility of cowpeas to attack by storage pests. Protein content and amino acid profile are broadly similar to those of the common bean, *Phaseolus vulgaris*, except that the sulphur amino acid content is higher (Mossé and Pernollet, 1982). Selection for improved amino acid profile (i.e. for increased sulphur amino acid and tryptophan content) is an objective common to many pulse crops. It is possible that selection for increased resistance to seed storage pests, if successful, could increase contents of anti-metabolites and toxic materials which are present at low levels, a consequence which should be borne in mind. Neutralisation of these could increase cooking times, which again is undesirable; rapid cooking (cookability) is desirable (less than 30 min) because of limiting fuel supplies in many areas of consumption.

Genetic resources and future developments

The cowpea has a large primary gene pool (GP1) (Harlan and De Wet, 1971) with a good balance between the domesticated and wild components (GP1A and GP1B). The crop is extensively cultivated in Africa, Asia, Brazil and some southern states of the USA. There is thus extensive development of landraces and cultivars which together constitute a substantial genetic resource which has not yet been seriously undermined by genetic erosion through introduction of 'improved varieties'. The wild component of the gene pool is very extensively distributed in sub-Saharan Africa and constitutes a genetic resource of great potential value. This resource is obviously worthy of further collection and investigation. It could well be a source of important germplasm in the future evolution of the crop. There is a great necessity, as Rachie (1985) stated, for improvement in resistance to a wide range of pests and pathogens which at the present time assail the crop.

There does not appear, at the present time, to be a secondary gene pool (GP2) since no viable and fertile inter-specific hybrids have been reported and it is quite probable that none can be produced by conventional means. The tertiary gene pool (GP3), such as it is, would appear to be genetically remote from the cowpea. More investigation of the nature of the failures recorded in attempts at inter-specific hybridisation is necessary to determine the stage at which the failure occurs and whether there is any possibility of rescue for the inter-specific hybrid embryos.

The cowpea has proven very responsive to human selection in producing a range of plant and seed types. There is no good reason to presume that its potential for further evolutionary change has been exhausted. Its present use is closely similar to that of *Phaseolus vulgaris*; however, it is much better adapted to climatic conditions in the semi-arid, sub-humid and humid tropics. It therefore performs a similar function under a rather different range of environments and is therefore best regarded as a complement to, and not a competitor of, the common bean. The general similarity of the seed in the two species and the fact that they can be prepared in similar ways has in a sense pre-adapted the cowpea for culture in the New World and the common bean in the Old.

A potentially useful line of investigation which could well be taken further, not only in the cowpea but in grain legumes generally, is the study of the genetics of nitrogen fixation in the host (Miller *et al.*, 1986). Experience tends to show that host genotype is more easily manipulated than that of *Rhizobium* populations and should receive more detailed study.

5·3· The groundbean (Bambara groundnut) *Vigna subterranea* (L.) Verdc.

From time to time the groundbean generates interest as an under-exploited crop, but inevitably the difficulties in its economic exploitation cause interest to subside. One of the crop's attractive features is that, probably owing to its wide geographic dispersion in Africa, there appears to be a wide range of genetic variability which could give scope for selection. This is most obviously manifested as variation in seed size and the colour and patterning of the testa. Morphology does not show much apparent variability, but there may be some in response to attacks of pests and pathogens; this aspect has not been investigated to any great extent.

Biosystematics

In their revision of the genera *Phaseolus* and *Vigna*, Maréchal and co-workers (1978) have clearly demonstrated that the groundbean belongs in the section *Vigna* of the genus *Vigna*. Since *Voandzeia* had priority over *Vigna*, the transfer of the groundbean to *Vigna* raised a difficulty of nomenclature, resolved by conserving the name *Vigna* (Verdcourt, 1981). Section *Vigna* contains no other cultigens. This name change is, perhaps, something of a surprise, if not an irritation, to interested agricultural scientists. However, the new name gives a very much clearer indication of the biosystematic affinities of the groundbean; the value of this is probably not unappreciated.

The species was known to science initially in its cultivated form; however, Hepper (1963) collected the wild form in West Africa where domestication is presumed to have occurred. This is currently recognised as var. *spontanea* (Harms) Hepper and differs principally from the cultivated form by its more spreading growth habit and smaller pods and seeds.

The groundbean is geocarpic. The inflorescences are aerial but after pollination and fertilisation of ovules tropic growth buries the developing pods in the surface layer of soil. Inflorescences are borne on a number of rather short horizontal branches arising from a taproot; the branches bear trifoliolate leaves on relatively long petioles. As the internodes are quite short, the plant at maturity has a definite tufted appearance. In the wild groundbean internode length is longer and the petiole length is shorter so that the characteristic tufted appearance of the cultigen is not in evidence.

Fig. 5.5. *Vigna subterranea*: the plant (from Purseglove, 1974).

Phytogeography and ecology

The groundbean is of widespread African distribution. The finding of the wild form in West Africa (Hepper, 1963) pinpoints this part of the continent as the probable area of origin. It has been transported in post-Columbian times to Brazil, the Far Eastern tropics and the Philippines (Smith, 1976), where it survives as a subsistence crop. Its ecological preferences have not been studied in detail, but it seems to be very well

(*a*)

Fig. 5.6. The Bambara groundnut (*Vigna subterranea*). (*a*) The wild form; (*b*) the cultivated form; (*c*) pods and seeds of wild and cultivated forms. (From Hepper, 1963. British Crown Copyright. Reproduced with permission of the Controller, Her (Britannic) Majesty's Stationery Office and the Trustees, Royal Botanic Gardens, Kew, © 1963.)

and broadly adapted to the African sub-Saharan zone. Although there is no good ecological reason why it should not be extensively cultivated in other semi-arid tropical areas, the fact is that its ecological preferences are broadly similar to those of the groundnut. Cultivation of the latter is tending to inhibit and even supplant that of the groundbean in Africa, more especially when there is a market for surplus production. It is possible that cultivation of the groundbean paved the way for the success of the groundnut after its introduction to Africa.

(b)

(c)

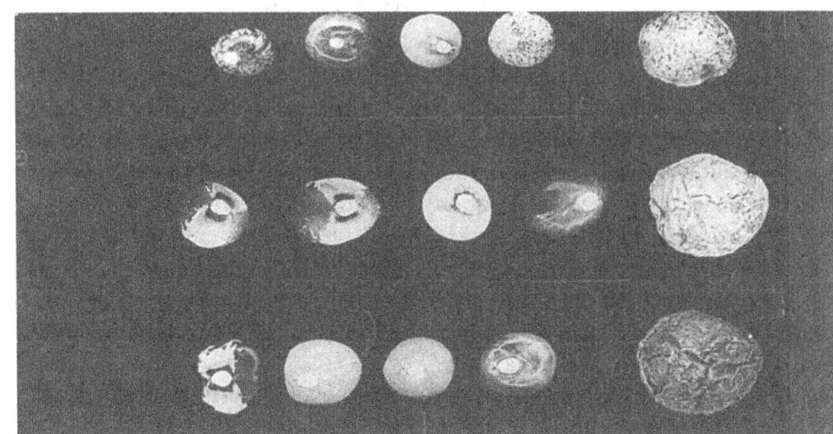

Hybridisation and cytogenetics

Very few if any taxonomic studies involving hybridisation have been carried out on the groundbean. Since it belongs to a different sub-genus to the Asiatic grams and a different section of sub-genus *Vigna* to the cowpea, it is highly unlikely that any viable inter-specific hybrids would result from experimental crosses. The groundbean has some very distinctive features and could well, like the cowpea, have developed very effective mechanisms to isolate it genetically from even those species in the same section as itself. The chromosome number, $2n = 2x = 22$, is, however, quite typical of the *Phaseoleae*.

Biochemistry

No significant chemotaxonomic studies have been carried out on this species but some studies of the biochemistry of its seed have been undertaken. The nature of carbohydrate seed storage materials has been investigated (Bailey, 1971), the absence of lectins demonstrated (Toms and Western, 1971), the protein content and amino acid profiles determined (Mossé and Pernollet, 1982) and the presence of a trypsin inhibitor has also been reported. The protein content is relatively low for a legume, 16–21%, and typically the sulphur amino acid content is limiting, relative to reference protein.

Archaeology

In common with all other pulses of sub-Saharan origin there is no known archaeological record.

Domestication, evolution and future prospects

There can be no doubt that the groundbean was domesticated in Africa and almost certainly in West Africa. The chief effects of selection under domestication, which have already been noted, are an increase in pod and seed size and a general shortening of internode length to produce a tufted form, similar to the change which has occurred in the evolution of the bush marrows (*Cucurbita pepo*). In West Africa there appears to be continuing inter-crossing between wild and domesticated populations of the groundbean. This tends to erode the effects of selection and establish a near continuum between the wild and cultivated segments of the species. In areas away from the centre of origin, in Central Africa for example, the effects of selection on seed size are clearly apparent. There is a very definite discontinuity in size between local Central African landraces and the West African wild forms.

Persistence of the groundbean in cultivation is perhaps under threat. There is no doubt that it is not a strong commercial competitor to the groundnut. However, it may well persist in cultivation in areas where specific nutrient deficiencies, such as those of calcium and boron, for example, limit groundnut production. It is highly unlikely that any significant expansion of its cultivation will occur under prevailing economic and agricultural conditions.

5.4. The Asiatic grams

This group of pulses can be expected to excite increasing interest if the present enthusiasm for health foods and diets with reduced meat content continues. The green gram, *Vigna radiata*, already is a popular food in the form of bean sprouts; present levels of consumption can be expected to increase rather than diminish. The black gram, *V. mungo*, although very similar to green gram, is not as palatable and is less likely to generate sufficient demand to stimulate production significantly outside its traditional areas. The adzuki bean, *V. angularis*, has generated interest as a pulse outside traditional areas of production and consumption and demand for it could increase in the near future. Perhaps the most interesting future exists for the rice bean, *V. umbellata*, which reputedly has a high food value (Herklots, 1972). It is cultivated in the general area of Indo-China and Thailand. It possibly has the highest yielding capacity of any of the Asiatic grams and if a sizeable consumer demand were built up it could become a useful crop. Two other species can be mentioned briefly, *V. aconitifolia* and *V. trilobata*, since they are frequently confused although they are quite distinct species most easily distinguished by the short inflorescence rachis of *V. aconitifolia* and the long peduncle of *V. trilobata*. The moth bean (*V. aconitifolia*) produces a seed comparable in size with green gram which can be used as a pulse, while that of *V. trilobata* is considerably smaller. The latter is probably most useful as a forage crop in semi-arid conditions.

Biosystematics

The five species of Asiatic pulses belonging to the genus *Vigna* are closely related and are characteristically small-seeded. They are all assigned to the sub-genus *Ceratotropis* (Maréchal *et al.*, 1978). Previously, four of them had been assigned to a new genus, *Azukia*, by Ohwi (1965).

Taxonomically, cultigens and conspecific wild forms are recognised in all species except *V. aconitifolia* (Table 5.3), although the position in this respect has only been clarified recently (Maréchal *et al.*, 1978; Lukoki *et al.*, 1980).

The four more important pulse species have a number of common characteristics; typically they have flowers of similar size and relatively much larger than those of *V. aconitifolia* (and *V. trilobata* (L.) Verdc.).

(a)

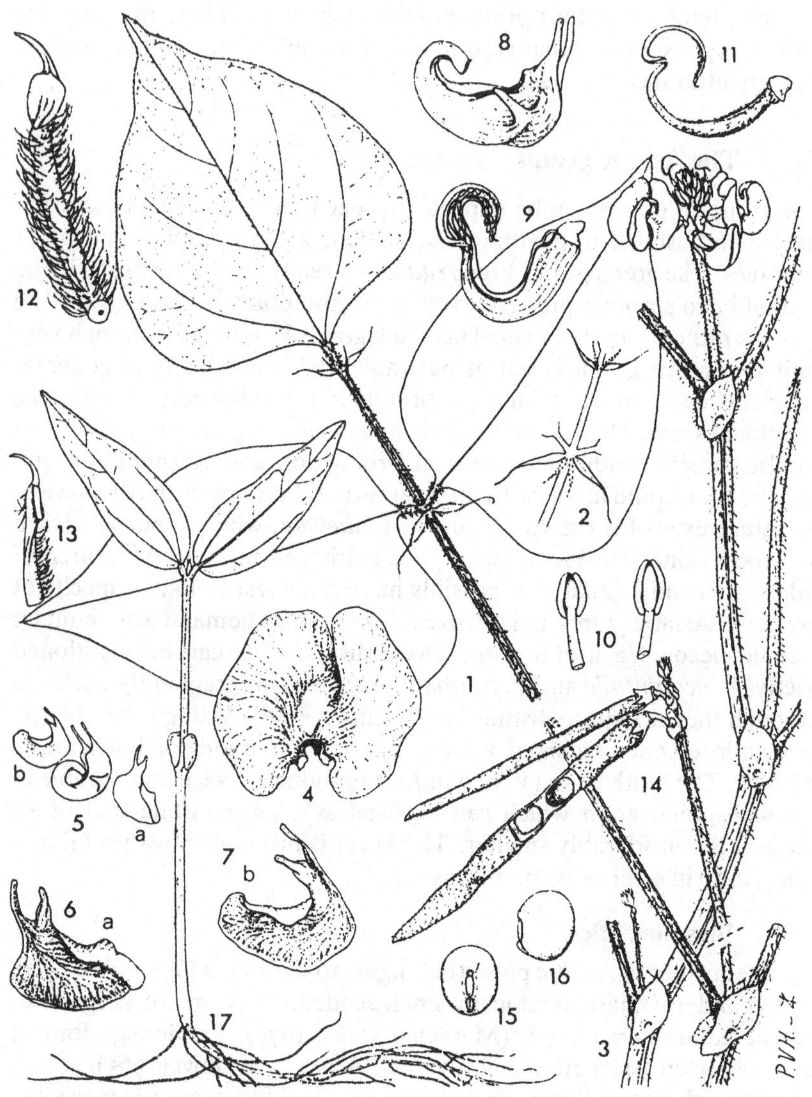

Fig. 5.7. Morphology of Asiatic *Vigna* species. (*a*) *V. radiata*, green gram or mungbean; (*b*) *V. mungo*, black gram or urd; (*c*) *V. angularis*, adzuki bean; (*d*) *V. umbellata*, rice bean. ((*a*) from Westphal, 1974; (*b–d*) from Duke, 1981.)

(b)

(c)

(d)

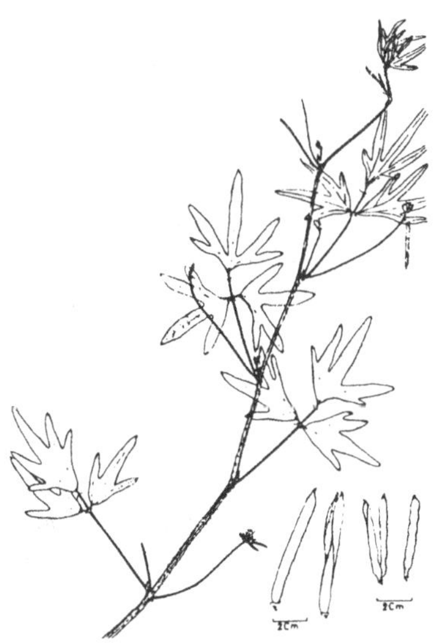

Table 5.3. *Nomenclature of taxa in domesticated Asiatic* Vigna *species*

Species	Cultigen	Wild form
Vigna angularis (Willd.) Ohwi & Ohashi	var. *angularis*	var. *nipponensis* (Ohwi) Ohwi & Ohashi
Vigna mungo (L.) Hepper	var. *mungo*	var. *silvestris* Lukoki, Maréchal & Otoul
Vigna radiata (L.) R. Wilczek	var. *radiata*	var. *sublobata* (Roxb.) Verdcourt
Vigna umbellata (Thunb.) Ohwi & Ohashi	var. *umbellata*	var. *gracilis* (Prain) Maréchal, Mascherpa & Stainier
Vigna aconitifolia (Jacq.) Maréchal		no distinction recognised

They show a very similar evolution of form from the trailing to the erect. It is not customary to provide support for the trailing forms; growth is decumbent rather than climbing. In any event, even in cultivated forms, the burden of seed and pods does not tax the strength of the peduncles unduly and they are able to maintain the crop out of soil contact quite satisfactorily.

Pods in cultivated forms are less dehiscent at maturity than those of the conspecific wild relatives. Green pods are usually tender enough for satisfactory use as a green vegetable. They are relatively small in size and perhaps less used on this account than they might otherwise be when cowpeas are also grown. The seed size relative to that of cowpeas and the common bean is small in all the grams. However, relative to the size of the wild forms (e.g. *V. r. sublobata*), the seed does exhibit gigantism and a fivefold size increase has probably been achieved. In *V. trilobata* (L.) Verdc. (a semi-domesticated species often confused with *V. aconitifolia*) no increase in seed size is apparent, a good distinction from the five fully domesticated species.

Hybridisation and cytogenetics

The taxonomic status of the six species considered has been studied experimentally. They behave as good, distinct species. Viable hybrids can be obtained between several species; *V. radiata* is probably the most satisfactory seed parent. It crosses with the wild *V. r. sublobata* reciprocally, and with *V. angularis*, *V. umbellata*, *V. mungo* and *V. trilobata* as seed parent only. *V. trilobata* crosses as pollen parent successfully with *V. mungo*, *V. radiata* and *V. aconitifolia*; reciprocal crosses fail. *V. umbellata* is cross-compatible with *V. angularis* as seed parent only, and with both *V. radiata* and *V. mungo* as pollen parent. *V. angularis* is cross-compatible as pollen parent with *V. umbellata* and *V. radiata*, while *V. mungo* is cross-compatible as seed parent only with *V. trilobata* and *V. umbellata* and as pollen parent only with *V. radiata*. Cross-compatibility of *V. aconitifolia* has not been widely investigated but it has been crossed as seed parent with *V. trilobata*. These results are summarised in Table 5.4.

Chromosome complements in *Ceratotropis* are typical of the tribe *Phaseolae* with $2n = 2x = 22$, with the exception of *V. glabrescens* Maréchal, Mascherpa and Stainier. This is unusual in that it is polyploid; $2n = 4x = 44$. It is probably an amphidiploid combining the genomes of *V. radiata* and *V. umbellata* (Dana, 1964). It is apparently the only natural amphidiploid in the sub-tribe *Phaseolinae* (Maréchal *et al.*, 1978). Artificially produced amphiploids include one derived from the hybrid (*V. radiata* × *V. mungo*) (Singh *et al.*, 1986).

A considerable amount of cytogenetic study has been carried out on Asiatic *Vigna* species and inter-specific hybrids. The first species to be investigated was *V. radiata* (Bose, 1939; Singh and Mehta, 1953; Sen and Ghosh, 1961) with the most detailed studied being carried out by Krishnan and De (1965). These authors have subsequently investigated meiotic behaviour in other inter-specific hybrids and attempted to determine chromosome homologies (De and Krishnan, 1966; Krishnan and De, 1968*a*, *b*). The conclusion which can be drawn from the work of these and other authors (Dana, 1966*a–d*; Biswas and Dana, 1975, 1976*b*; Machado *et al.*, 1982; Satyan *et al.*, 1982) is that chromosome structural rearrangements are a significant part of the genomic differentiation of these species. Even the two closest relatives, *V. radiata* and *V. mungo*, have some structural differentiation of their genomes.

These two species are morphologically very similar, so much so that Verdcourt (1970*b*) has expressed the view that they might better be regarded as a single species. On morphological grounds this is very true; however, breeding tests show that the two grams have almost completely separate gene pools. Gene flow can be induced between them, but only with some difficulty, and on balance they can satisfactorily be viewed as separate species. It is interesting to note that subsequent generations of the hybrid *V. radiata* × *V. mungo* tend to revert to the morphology of *V. radiata* (cf. hybrids *Phaseolus vulgaris* × *P. coccineus*) (Shanmugam *et al.*, 1985). At one time the wild form *Vigna sublobata* was thought to be the ancestral type of both cultigens. However, it has been shown subsequently (Arora *et al.*, 1973) that what was regarded as *V. sublobata* included two distinct wild forms which were not freely cross-compatible with each other but which could be crossed easily with cultigens, one with *V. radiata*, the other with *V. mungo*. This situation has been reviewed

Table 5.4. *Results of inter-specific hybridisation between Asian* Vigna *cultigens*

Seed parents:	*V. aconitifolia*	*V. angularis*	*V. mungo*	*V. radiata*	*V. trilobata*	*V. umbellata*
Pollen parents						
V. aconitifolia		—	—	—	no viable embryos[2]	—
V. angularis	—		inviable seedlings (e.c.)[1]	viable seedlings (e.c.)[1]	—	viable seedlings (e.c.)[1]
V. mungo	—	inviable seedlings (e.c.)[1]		viable seedlings[1]	no viable embryos[4]	no viable embryos[1]
V. radiata	—	inviable seedlings (e.c.[1]	no viable embryos[1]		no viable embryos[3]	no viable embryos[1]
V. trilobata	viable seedlings[2]	—	viable seedlings[4]	viable[3] seedlings[3]		—
V. umbellata	—	no viable embryos[1]	inviable and viable seedlings (e.c.)[1]	viable seedlings[1]	—	

(e.c.): by embryo culture
[1]Ahn and Hartmann (1978)
[2]Biswas and Dana (1976*a*)
[3]Dana (1966*a*)
[4]Dana (1966*b*)

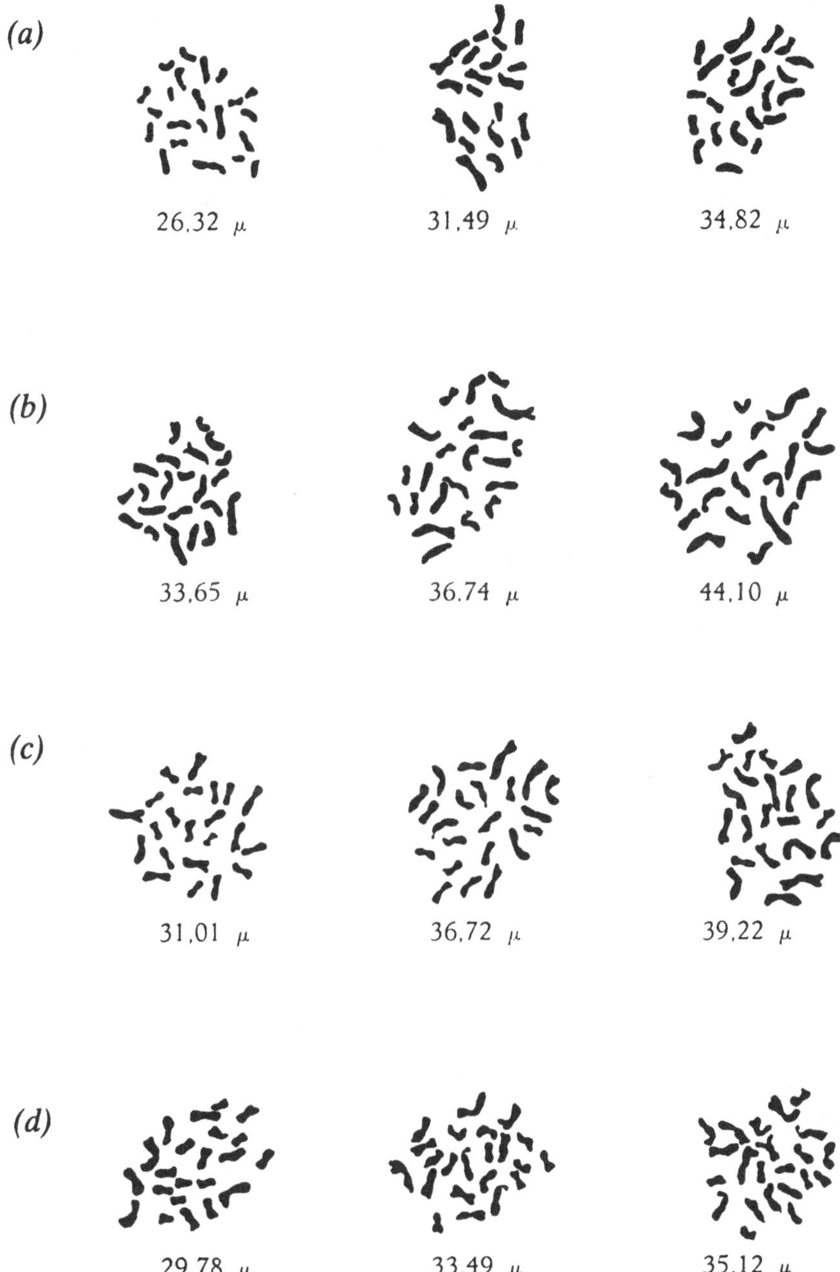

Fig. 5.8. Chromosome complements of Asiatic *Vigna* species (× 2500). (*a*) *V. radiata*; (*b*) *V. umbellata*; (*c*) *V. aconitifolia* (from Maréchal, 1969).

definitively by Lukoki *et al.* (1980) and suitable nomenclature for the wild and cultivated forms of the two species produced. The wild form of *V. radiata* is included within this species as var. *sublobata* while that of *V. mungo* is designated var. *silvestris*. Perhaps these might be more appropriately given sub-specific rank following the suggestions of Smartt and Hymowitz (1985). Both species are of undoubted Indian origin as is borne out by their occurrence in archaeological sites in the sub-continent. The domestication of *V. aconitifolia* is apparently Indian, whereas that of *V. angularis* and *V. umbellata* is Far Eastern. Wild *V. angularis* is distributed in Japan and also parts of China, Manchuria and Korea; wild *V. umbellata* has its origin in Indo-China and South-East Asia (Maréchal *et al.*, 1978). The situation in all species but *V. aconitifolia* is comparable: distinctly different wild and domesticated populations have become established which are still apparently conspecific where this has been tested (Lukoki *et al.*, 1980). In *V. aconitifolia* the domesticated and wild populations are not recognised as distinct taxonomic entities.

Chemotaxonomy and biochemistry

Like the cowpea, the gram species considered appear to contain very little by way of anti-metabolic factors, with the possible exception of a flatus factor in *V. mungo* (Gupta, 1982). An interesting study of seed proteins in *V. radiata*, *V. angularis*, *V. umbellata* and *V. trilobata* was carried out by Klozová (1965) (and also later by Turková and Klozová (1985)). Using immunochemical methods, she was able to demonstrate serological affinities which very closely agree with those established on morphological grounds by Maréchal *et al.* (1978). It is perhaps unfortunate that *V. mungo* was not included in this study since this would have provided additional data on the closeness (or otherwise!) of the *V. radiata* – *V. mungo* relationship.

Archaeology

The Indian grams *V. radiata* and *V. mungo* are alone in the genus in having an archaeological record. Jain and Mehra (1980) note records of finds aged 3500–3000 BP from archaeological sites at Navdatoli in Central India. The cultivated and wild forms *V. radiata* and *V. mungo* are also mentioned in ancient Hindu medical texts (2800 BP) in addition to *V. aconitifolia*. No comparable remains or records are reported for *V. angularis* or *V. umbellata*.

Domestication and evolution

The Asiatic grams have been domesticated in Asia from the Indian subcontinent to the Far East. They have responded very similarly to selection pressures under domestication to produce growth forms similar to those of the cowpea and *Phaseolus* species. From the primitive trailing and twining forms, erect, free-standing types have been established. Pods have reduced dehiscence and seeds are non-dormant. Both day-neutral and short-day genotypes can be found in all species; the limits of successful cultivation are determined by prevailing temperature régimes. They do not succeed in cooler temperate regions as do the common and scarlet runner beans (*Phaseolus* spp.).

Further evolutionary potential

Probably the greatest drawback to the utilisation of the *Vigna* pulses to a greater extent, in tropical and warm temperate agriculture, is the rather low yielding capacity. In spite of this green gram (*V. radiata*) is widely cultivated in the Old World and the New for the production of bean sprouts. Black gram (*V. mungo*), although widely cultivated in India, is less popular elsewhere. Adzuki bean (*V. angularis*) is a popular crop in Japan and Korea. The rice bean (*V. umbellata*) is cultivated from India to Japan, most extensively in the Far East (Indo-China and Thailand). Of all the gram species introduced to Central Africa this appeared to show the greatest promise and appears to have very useful yield potential.

In common with many pulses of the *Phaseoleae* the grams may be photoperiod-sensitive, in which case they require short days for flowering. However, day-neutral forms are commonly found and enable field cultivation to be practised in high latitudes e.g. Oklahoma. This polymorphism is extremely important in extending the range of crops which originated in lower latitudes. Photoperiod sensitivity is a very effective device in locally adapted forms for synchronising flowering and fruiting with the optimum period of the growing season in semi-arid climates. This is less important in humid equatorial areas where crop growth is possible all the year. It can be a very considerable drawback when cultivation is attempted in the summer under warm temperate conditions when flowering may only be induced when frost damage is a real hazard.

Probably the greatest advance that could be achieved in the Asiatic grams is improvement of yielding capacity. Duke (1981) quotes yields of 1125 kg ha^{-1} of dried beans with *V. radiata* in the USA; selections made in Taiwan have yielded from 800 kg ha^{-1} up to 2100 kg ha^{-1}. Yield stability is an important factor here. Yield levels of the other grams are comparable. Unless there is a premium price (as there is for green gram), production is unlikely to be economically attractive. The situation might be improved by selection for a higher leaf area. This might be achieved by selection for increased leaf number or greater area of individual leaves. At the present time there is a very active mungbean improvement project under way in Taiwan (AVRDC Annual Reports) which it is hoped will improve the competitive position of this crop.

5.5. General prospects for cultivated *Vigna* species

The genus *Vigna* contains seven species of pulse of which, from the point of view of their utilisation, the chief weakness appears to be in their relatively low yields as previously noted. It cannot be over-emphasised that when comparisons of yields are made, with cereals for example, these are not made on the crude dry matter basis. Due allowance should be made for the fact that pulses often have double the cereal protein content. None the less, even bearing this in mind, the yielding capacity of most *Vigna* species is low on their present showing. Consideration should be given to the possibility of improving this by modifying shoot architecture, selection for higher harvest indexes and more effective nodulation, to name but three possibilities. Much effort has been devoted to the selection of *Rhizobium* strains with high nitrogen fixing capacity. The practical results of this work are often disappointing. It is perhaps significant that the best results have been achieved with introduced legume species and introduced compatible *Rhizobium* strains. These introduced *Rhizobium* strains do not have to compete with established strains in establishing the symbiosis with their hosts and there is no problem in achieving very effective nitrogen fixation in the crop. The solution to this difficulty, especially in *Vigna*, might be to select for host genotypes capable of establishing effective symbiosis with a wide range of *Rhizobium* genotypes. The cowpea group of *Rhizobium* has a very wide host range and therefore this might well be the most effective strategy.

The evolutionary pattern in *Vigna* species has been greatly clarified in the past twenty years and most especially in the past ten years. A great deal of clarification has emerged from the work of Verdcourt (1980) and Maréchal and co-workers in the period. The taxonomic revision of Maréchal *et al.* (1978) has given us a firm foundation upon which to rest evolutionary hypotheses. The taxonomists have recognised and named wild and domesticated taxa within all species but *V. aconitifolia*. Although this species has apparently responded to selection for larger seed size, there has been no apparent increase in the size of the vegetative parts. One might say that it has retained a wild type vegetative morphology. This is in contrast to the situation in *V. radiata* and the other species, where cultigens are much more robust in growth, with large vegetative parts and often of erect growth. It is of interest to note that in some species of the genus the primitive growth habit is often straggling rather than twining. This is clearly expressed in the cowpea where var. *unguiculata* is straggling but var. *sesquipedalis* is twining and the differ-

Table 5.5. *Secondary and tertiary gene pools in the cultivated* Ceratotropis *species of* Vigna

Species	Secondary pool	Tertiary pool
Vigna radiata (L.) R. Wilczek	*V. mungo*	*V. umbellata*, *V. angularis*, *V. glabrescens*, *V. trilobata*
V. mungo (L.) Hepper	*V. radiata*	*V. umbellata*, *V. angularis*, *V. glabrescens*, *V. trilobata*
V. umbellata (Thunb.) Ohwi & Ohashi	*V. angularis*	*V. radiata*, *V. mungo*
V. angularis (Willd.) Ohwi & Ohashi	*V. umbellata*	*V. radiata*, *V. mungo*
V. aconitifolia (Jacq.) Maréchal	*V. trilobata*	—

ence in growth is clearly related to the great pod length of var. *sesquipedalis* and the necessity of keeping the pod out of contact with the soil.

Responses to domestication in terms of changed morphology are remarkably similar in the five more important species (*V. unguiculata*, *V. mungo*, *V. radiata*, *V. angularis* and *V. umbellata*) and they provide yet another exceptionally clear illustration of Vavilov's principle of homologous variation (cf. *Phaseolus*).

In common with most grain legume crop species, the wild related species and the other cultigen species do not form a particularly extensive or accessible genetic resource (Table 5.5.). Wild conspecific forms obviously should be collected and conserved carefully in addition to the fullest possible range of landraces and cultivars. The genus contains species with valuable characters; the fullest possible use will need to be made of available genetic resources to make extended cultivation economically attractive.

6 Pulses of the classical world

The grain legumes which evolved in the Mediterranean basin have a particular claim to the attention of students of crop plant evolution. Not only have they played an important supporting role to that of the cereals in sustaining the development of the classical civilisations of the area, but it is arguable that the scientific study of crop evolution began here. The crops of the Mediterranean region were among the most familiar to Linnaeus (1753) and to de Candolle (1886), the father of the scientific study of crop origins. De Candolle appreciated that non-biological disciplines could contribute valuable information on the evolutionary history of crop plants. For example, records of contemporary crops in classical writings such as Virgil's Georgics are readily accessible and useful. Representations in art and artefacts are equally valuable. The evidence shedding light on crop histories varies widely from crop to crop in quantity and quality. This depends on the significance of the crop in economic, social and religious life, and also on what records or materials have survived and been discovered. The evidence, although from a wide variety of sources, can only be fragmentary but it can nevertheless be very informative. In any event we have, perforce, to do the best we can with it.

In the present treatment I propose to treat each crop individually in the first place and then conclude with a brief comparative consideration of this group of pulses.

6.1. The pea (*Pisum sativum* L.)

Introduction

Originally the pea was cultivated as a winter annual crop in the Mediterranean basin. Its adaptation to relatively cool conditions has enabled its cultivation to spread far beyond the area of initial domestication. It is a major pulse crop of the temperate zones of the world, especially in northern Europe and most particularly in the Soviet Union, where it is the major grain legume crop (Makasheva, 1983; Khvostova, 1983). There are two major types of pea in cultivation: the garden pea, produced primarily for human consumption, and the field pea, grown for feeding livestock. The culture of the field pea is probably declining in at least some parts of its range but that of the garden pea is going from strength to strength. Traditionally the garden pea could be used as a mature seed or pulse ('blue' peas) or green mature or immature as a vegetable. Certain

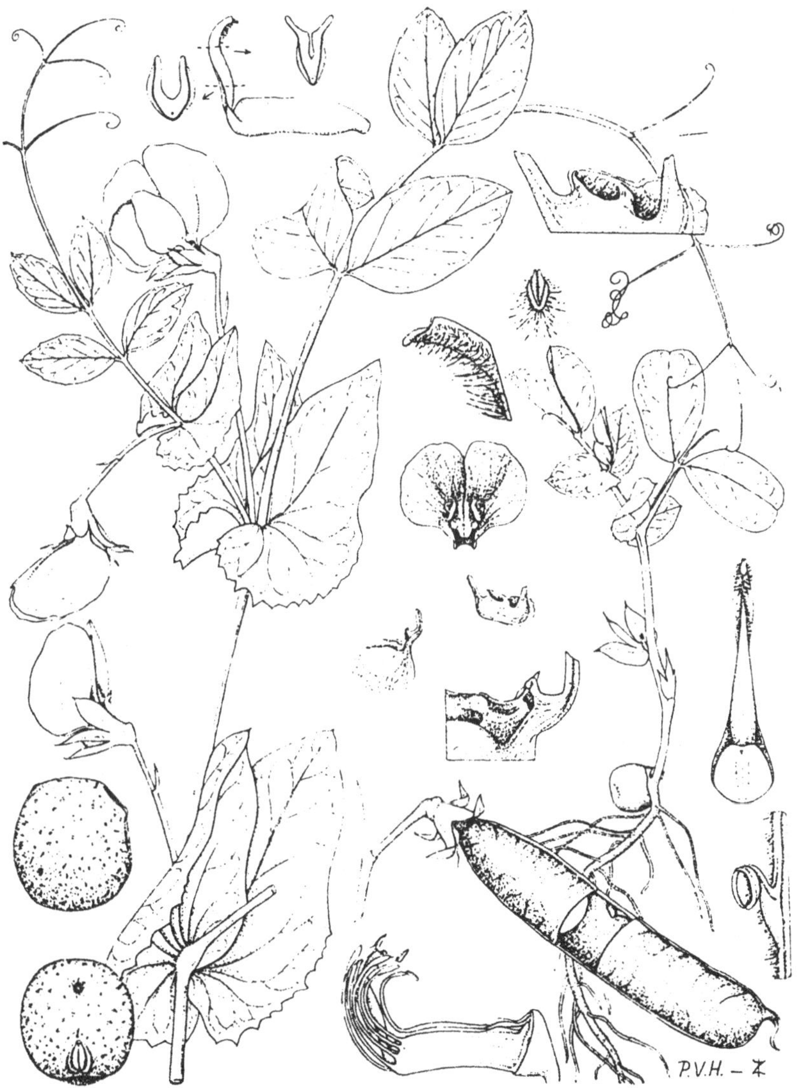

Fig. 6.1. The pea, *Pisum sativum*, cv. speckled Shoa (from Westphal, 1974).

cultivars had also been developed with pods which could also be consumed as a green vegetable: the 'mangetout' or sugar peas. Much of current pea production and consumption in Europe and North America is of cultivars specially developed for the frozen pea trade. The marketed product is of consistently high quality, due to meticulous attention to production, processing and merchandising. Peas for domestic shelling now have a very small market share but continue to be produced where freezer technology is less in evidence.

The crop is very palatable and nutritious. There are no problems attending its use in the way of significant contents of toxic materials or anti-metabolites. It ranks in the four major grain legume crops of the world.

Biosystematics

Pisum is a very small genus and at the present time comprises only two species, *P. sativum* itself and *P. fulvum* Sibth. & Sm. Other species formerly assigned to it have been transferred to other genera or reduced to synonymy, most frequently with *P. sativum*. The rationalisation of *Pisum* taxonomy has in a sense transferred the problem of classification from the infra-generic to the infra-specific level. There is a very real problem in recognising most appropriately the enormous range and wealth of variation which has developed within the cultivated forms. Perhaps the most reasonable approach is to view the question phylogenetically. The wild populations from which the domesticates probably arose were initially described as species in their own right, *P. elatius* Bieb. and *P. humile* Boiss. & Noe (syn. *P. syriacum* (Berger) Lehm.). As a consequence of Ben-Ze'ev and Zohary's work (1973), these species are now regarded as part of the biological species *P. sativum* in the broadest sense. The ranking of these two wild groups within the biological species creates a problem because they are morphologically and ecologically distinct and taxonomic ranking as sub-species could be justified. In the cultivated populations there is a somewhat similar problem. In its diffusion as a cultivated plant the pea has spread from the eastern Mediterranean into Europe, Asia and Africa. It has been subjected to different natural and artificial selection pressures in these areas. A distinctive Ethiopian form has been recognised as a species, *P. abyssinicum* A. Br., for example. Should this and other distinctive forms be recognised at the sub-species level? The European peas are clearly differentiated into the garden peas (*hortense*) and field peas (*arvense*); at what taxonomic level should this distinction be recognised? Some authorities, such as Gentry (1971) believe in recognising them all as sub-species of *P. sativum*. The alternative argument is that the prime distinction, at the sub-species level,

Table 6.1. *Taxonomy of* Pisum *(after Davis, 1970; Polhill and van der Maesen, 1985)*

P. sativum L.		
	ssp. *sativum*	var. *sativum*
		var. *arvense* (L.) Pair
	ssp. *elatius* (M. Bieb.)	Aschers. & Graebn.
		var. *elatius* (M. Bieb.) Alef
		var. *pumilio* Meikle (*P. humile* Boiss. & Noe)
		var. *brevipedunculatum* Davis & Meikle
P. fulvum Sibth. & Sm.		

between the members of the species is whether they are wild or domesticated. One could recognise a single wild sub-species containing the three distinct wild forms, and a single cultivated sub-species containing the range of the different domesticated forms. This is the scheme devised by Davis (1970) and reproduced in Polhill and van der Maesen (1985).

The only reservation one might have about this scheme is that arguably it does not give sufficient recognition to the geographic differentiation which has come about in the domesticated populations. This problem can be easily surmounted by description of such geographic variants as cultivar groups within var. *sativum*. In the wild subspecies *elatius*, morphological extremes are represented by var. *elatius* and var. *pumilio* with var. *brevipedunculatum* being intermediate.

In a widely cultivated species exposed to highly disruptive selection processes, extreme and even bizarre morphological variants can arise and be perpetuated. To the taxonomist more attuned to natural ranges of variability, such variants have on occasion seemed to merit taxonomic rank as species. A good example of this is the mangetout type of pea, which has been recognised as *P. macrocarpum* Ser. ex Schur. Genetic analysis shows that such differenccs arc undcr vcry simplc gcnctic control, and if these were to merit distinct binomials, then every recognisably distinct genotype could be similarly named.

What appears to be very much more sensible in cultivated species is to produce informal groupings of cultivars and landraces, perhaps on the basis of major common morphological features and/or geographic origin. Some formal schemes of cultivar classification have been devised (Parker,1978) and used in the pea by Makasheva (1983) but these can be extremely complex and confusing to use and do not find much favour. In cultivated species, the pattern of variation which has evolved often dictates the nature of workable and useful classification schemes.

Cytotaxonomy, hybridisation and genetic resources

An extremely valuable cytogenetic study was undertaken by Ben-Ze'ev and Zohary (1973) on the genus *Pisum*, which has a chromosome complement $2n = 2x = 14$. Their observations and conclusions have important implications for taxonomists and students of crop plant evolution. On the basis of morphological evidence Davis (1970) had already concluded that *P. sativum*, *P. humile* and *P. elatius*, previously recognised as species, together constituted a single biological species. With the reassignment of *P. formosanum* (Stev.) Alef. to a monotypic genus *Vavilovia* A. Fed., the genus then contained, as noted, only two species, *P. sativum* and *P. fulvum*. The studies of Ben-Ze'ev and Zohary (1973) of intra- and inter-specific hybrids within the genus not only supported Davis' conclusions but provided much interesting and additional information. By and large, crosses within the '*elatius*', '*pumilio*' and '*sativum*' botanical varieties and between different accessions of *P. fulvum* were made without undue difficulty and produced fertile progeny. There was, however, evidence of chromosome structural differentiation between northern and southern var. '*pumilio*' populations with reduced fertility in the translocation heterozygotes produced by crossing them. Crosses between groups showed that, cytologically, crosses var. *pumilio* × var. *elatius* had a normal meiosis when the *pumilio* parent was from southern Israel and was a translocation heterozygote when the *pumilio* parent was from northern Israel. Comparable crosses var. *sativum* × var. *elatius* showed similar meiotic figures to those given by northern *pumilio* × *elatius*, i.e. five bivalents and one translocation quadrivalent. This resulted, as in the previous instance, in a reduction of fertility. The crosses of var. *sativum* with var. *pumilio* gave results depending on the origin of the '*pumilio*' parent. Hybrids between '*sativum*' and the northern forms of '*pumilio*' produced F1 hybrids with normal meiosis (7 bivalents) and high pollen and seed fertility; those F1 progeny produced with southern '*pumilio*' parents showed a characteristic translocation quadrivalent and appreiciably reduced pollen and seed fertility. High production of normal pollen (over 90%) could be produced in *elatius* × *pumilio* and *sativum* × *pumilio* F1 hybrids when no chromosome structural heterozygosity was apparent. This clearly demonstrated that loss of pollen and seed fertility was a consequence in large part of this reciprocal translocation and not of fundamental divergence at the genic level. Detailed karyotype studies showed that the short arms of chromosomes IV and VI were involved in the interchange. Karyotype studies of *P. fulvum* showed an even greater divergence in chromosome structure. Hybridisation studies showed that crosses between *P. fulvum* and members of the three groups within *P. sativum sens. lat.* could be made without difficulty but hybrids were only viable when *P. fulvum* was pollen parent. This suggests that there is cytoplasmic differentiation as well as that of the karyotype.

On the basis of these studies Ben-Ze'ev and Zohary (1973) have also suggested that the northern Israeli '*pumilio*' populations represent the extant common pea prototype. The basic gene pool of the cultivated peas can sensibly be considered to have originated from the forbears of the northern Israeli '*pumilio*' population. It should be noted that the IV–VI translocation has been recorded in *P. sativum* cultivars and this could have introgressed from var. *elatius* and/or southern var. *pumilio*. Introgression is also possible from *P. fulvum* via hybrids with *P. sativum* and backcrosses since the F1 hybrids which survive are somewhat fertile. This obviously constitutes a secondary gene pool.

The bulk of genetic resources available for improvement of *Pisum sativum* reside within the biological species itself. The range of variation found within the cultivated sub-species is very considerable; the conspecific wild forms also constitute a notable genetic resource. Outside the biological species the only exploitable resource is *P. fulvum*, as has already been noted. It is possible that there is an extensive tertiary gene pool, comprising parts at least of the related genera, *Vicia*, *Lathyrus* and *Vavilovia*. The work of Gritton and Wierzbicka (1975) provides some support for this view.

Chemotaxonomy

As Waines (1975) pointed out, prior to this date there had been little thorough chemosystematic work carried out. The chief shortcoming of many such studies is that, by and large, small and probably unrepresentative samples had been studied and only tentative conclusions could be drawn safely from the results. This is frequently a drawback of such studies, when initiated by biochemists who have had no guidance in their selection of materials and frequently fail to secure reasonably representative collections. Some studies have been carried out on anthoxanthins and anthocyanidins (Harborne, 1971) suggesting that more complex compounds may be produced, in the case of the leaf anthoxanthins, in domesticated than in wild forms, and a greater diversity of these in the case of domesticates. However, the great bulk of chemosystematic work on peas has involved study of seed storage proteins. Kloz (1971) has reviewed his work (and that of his co-workers) on the serology of pea seed storage proteins. He has made an informed effort to secure a cross-sectional range of material: wild species, landraces etc. He found that with the exception of *P. fulvum* and '*P. abyssinicum*' all the other forms were indistinguishable immunoelectrophoretically. These two both had in common certain distinctive proteins of *P. fulvum*. This result is interesting if puzzling; more chemotaxonomic information on '*P. abyssinicum*' would be desirable. Apart from this curious feature these results are broadly consistent with the views of Davis (1970) and Ben-Ze'ev and Zohary (1973) in that *P. fulvum* is distinctive and that all peas except *P. fulvum* form a single group. Fox *et al.* (1964) have shown that the albumin fraction of the seed storage protein can be resolved by electrophoresis into as many as 23 components and that differences between genotypes of *P. sativum sens. lat.* can be demonstrated. Boulter and Derbyshire (1971) studied the globulin fraction of the seed protein and noted its broad similarity in properties to those of other members of the Vicieae.

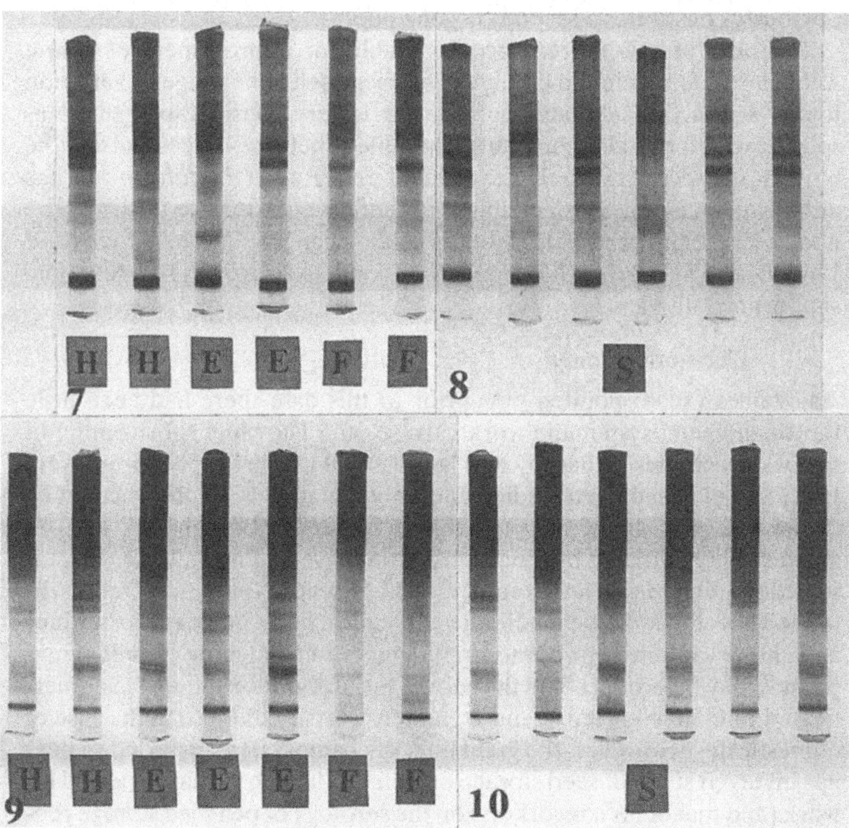

Fig. 6.2. Electrophoretic patterns (PAGE) of seed protein from *Pisum* spp.: 7 and 8, albumin proteins; 9 and 10, globulin proteins. Key: H, *Pisum pumilio* (*P. humile*); E, *P. elatius*; F, *P. fulvum*; S, *P. sativum*. (Note *P. pumilio* and *P. elatius* are currently thought to be conspecific with *P. sativum*). (From Waines, 1975.)

Waines (1975) carried out polyacrylamide gel electrophoretic (PAGE) analysis of the albumin and globulin fractions of a range of *Pisum* material. He found variation between accessions in both albumin and globulin fractions. He identified his material as wild or cultivated and attempted to find, without success, correlations between PAGE pattern changes and the morphological and other changes which have occurred in the course of domestication.

Domestication

The studies of Ben-Ze'ev and Zohary (1973) and Zohary and Hopf (1973) have provided the basis of our understanding of pea domestication. The wild populations with the closest affinity to the cultigen are clearly those resembling the northern Israeli populations. The distribution of var. *pumilio* is more restricted than that of var. *elatius* and confined to the eastern Mediterranean, Turkey and the Fertile Crescent. It seems likely that this is the area in which domestication occurred on phytogeographic grounds. This conclusion is strongly supported by the available archaeological evidence, the most ancient finds of peas in archaeological sites are in precisely this area, dating back to 7000–6000 BC. The remains are of carbonised seeds and have been obtained from Jarmo (north Iraq), Çayönü (south-east Turkey) and Jericho (Israel). More abundant remains of more recent date have been obtained from Çatal Hüyük (5850–5600 BC), Can Hasan and Hacilar (5400–5000 BC) in south-east Turkey. This is consistent with the view that domestication was initiated in the Fertile Crescent itself and westward diffusion to Turkey and Greece occurred (Nea Nikomedeia, ± 5500 BC) fairly soon afterward, beyond the present area of *P. humile* distribution (Zohary and Hopf, 1973). It is significant that some of the pea remains accompany finds of cultivated cereals in south-east Turkey. By 4400–4200 BC peas were common in the Rhine Valley, beyond the distribution range of wild

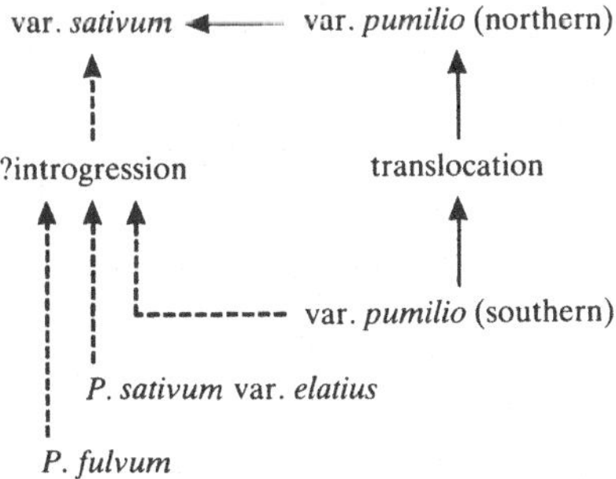

Fig. 6.3. Possible evolutionary pathway of the cultivated pea, *Pisum sativum* L.

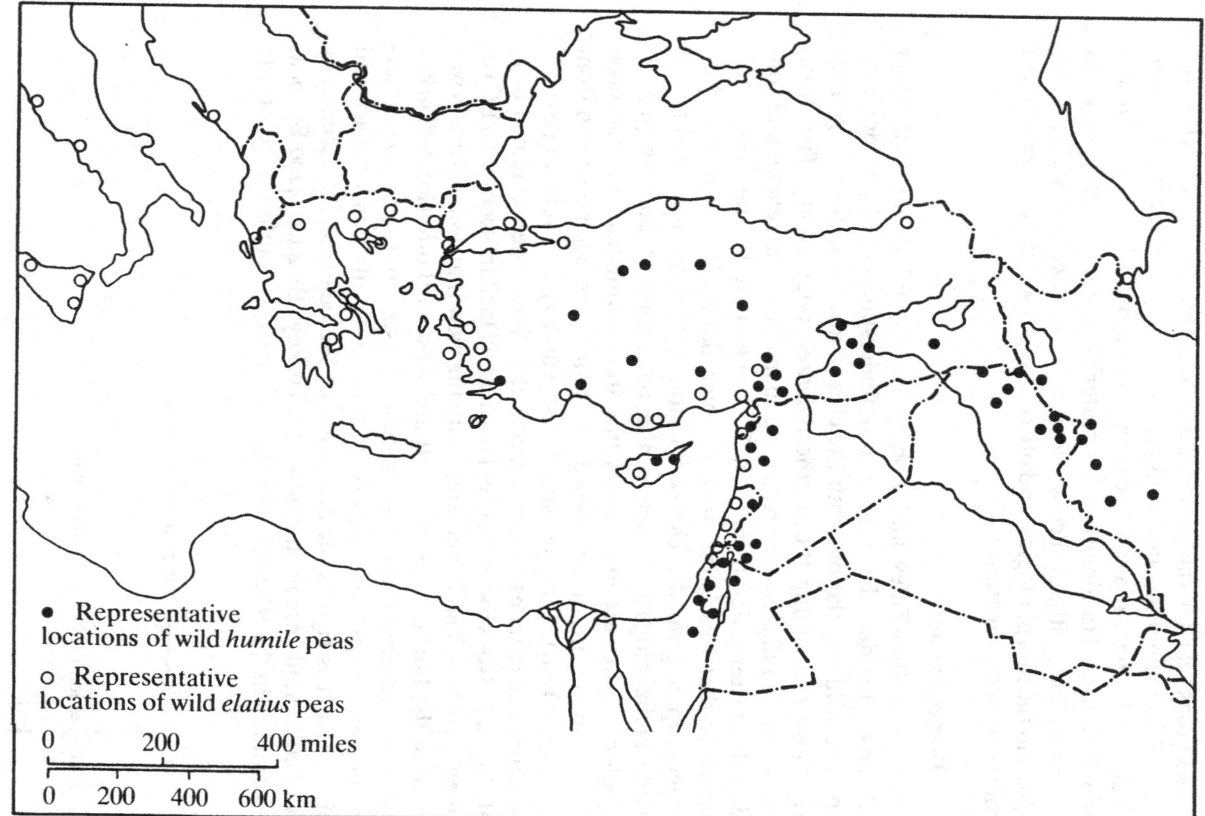

Fig. 6.4. Distribution of wild races of *Pisum sativum* (*P. elatius* and *P. pumilio* forms). (From Zohary and Hopf, 1973.)

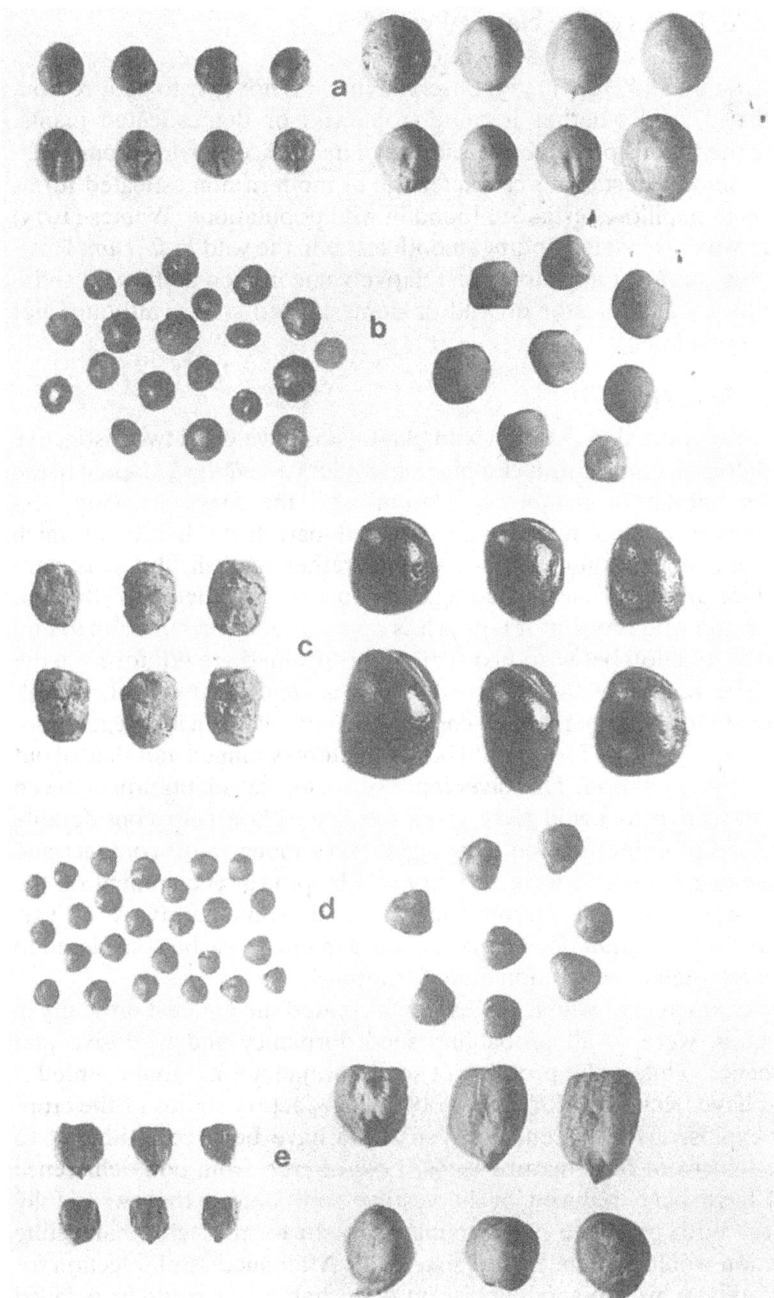

Fig. 6.5. Archaeological seed remains (left) and for comparison, seed from modern cultivars (right) of Mediterranean pulses. (a) Pea: carbonised seed from Early bronze Age, Arad, Israel. (b) Lentil: carbonised remains from Late Bronze Age, Manole, Bulgaria. (c) Faba bean: carbonised seed from Copper Age, Chibanes, Portugal. (d) Bitter vetch: remains from Late Bronze Age, Arad, Israel. (e) Chickpea: carbonised seed from Early Bronze Age, Arad, Israel (× 1.8) (from Zohary and Hopf, 1973).

peas. Zohary and Hopf (1973) observed that it is not easy to deduce from carbonised seed whether it came from wild or domesticated plants. Where the testa is preserved the nature of its surface provides some indication: smooth testas are characteristic of modern domesticated forms and rough papillose testas are found in wild populations. Waines (1975) has, however, reported finding smooth testas in the wild *P. fulvum*. However, this character appears to be relatively uncommon in the wild and is probably a fair indicator of wild or domesticated status, although not totally reliable.

Evolution

The pea is unusual in that as a wild plant it has developed two distinctive growth forms, the vigorous climbing scrambler var. *elatius*, adapted to the scrubby habitat of maquis vegetation, and the lower growing, less rampant var. *pumilio* found in the drier oak park forest habitat in which competing vegetation consists of grasses rather than shrubs. It is commonplace among domesticated legumes to find that their growth habit *vis-à-vis* the wild progenitor type is less rampant; it is less common to find a clear distinction between luxuriant and restrained growth forms in the wild. The reason in the case of the pea is clearly ecological. A tall-growing tendril-climber is at no competitive advantage in low vegetation, just as a low-growing form would be completely swamped and shaded out in shrubby vegetation. This divergence of ecological adaptation between the two wild peas could have given var. *pumilio* a very considerable advantage in domestication in being so very much more compact and manageable in growth form. Zohary and Hopf (1973) comment on the strong morphological resemblance between some cultivars of var. *sativum* and var. *pumilio*. The latter could in one sense be considered to be pre-adapted to cultivation quite fortuitously.

The characters of wild legumes which created the greatest difficulty in cultivation were in all probability seed dormancy and explosive pod dehiscence. Unless the problem of seed dormancy was circumvented it would have been very difficult to obtain satisfactory stands of the crop. With explosively dehiscent pods, it would have been very difficult to harvest crops of fully mature seeds. Loss of crop from pod dehiscence could have been reduced by harvesting pods before they were fully mature. With primitive indeterminate growth forms such a harvesting operation would have been very extended. After successful selection for less readily dehiscent and indehiscent pods, harvesting could be delayed until the majority or all pods were fully mature. The quality and storage properties of naturally matured seed would, other things being equal, be superior to that of seed harvested prematurely. Mature seed might have been harvested from plants with dehiscent pods in the morning before

Fig. 6.6. *Pisum elatius* (D. Zohary).

evaporation of dew and pod desiccation occurred; the establishment of indehiscence or delayed dehiscence would remove this restriction and give the cultivator much greater freedom of action.

It is apparent that, in common with other pulse crops, gigantism of pods and seeds has occurred in peas. The increase is perhaps less spectacular than in other pulses but real none the less, probably not in excess of a factor of ten in the most extreme cases. Gigantism of pods and seeds has evoked a correlated response of the vegetative parts in which the stems became more robust and overall leaf size larger.

The use of peas as food does not present any problems (cf. *Phaseolus* beans); they are perfectly wholesome in the uncooked state although both a trypsin inhibitor and a lectin have been identified from the seed (Liener, 1982). Three distinct pea types find use as human food: the starchy peas (the round peas of Mendel) which can be used green mature or dried; the sweet or sugary peas (Mendel's wrinkled peas) in which a higher concentration of sucrose and a lower content of starch appears, which are used immature, fresh or frozen; and the mangetout (or sugar) pea, which produces an edible parchment-free pod and is highly regarded as a gourmet vegetable although overall consumption and demand are not high.

The divergence between 'field' and 'garden' peas possibly came about in response to selection for improved palatability. Field peas can be used for human food and undoubtedly have been so used in the past and recently in times of scarcity. These have pigmented vegetative parts, flowers and seeds. Loss of anthocyanin pigments is associated with improvement in palatability and quality in many other pulses as it is in peas. Pigmentation of testas is virtually eliminated in 'garden' peas and the only apparent colouring matter is that of the photosynthetic pigments, in yellow- and green-seeded types; these may be absent in white-seeded cultivars.

Further evolutionary potential

A number of leaf-form variants are to be found in the cultigen, many more than are to be found both in wild vars. *elatius* and *pumilio*. The leaflets in the former are usually 6–8 in number, relatively small, rhomboid and with entire leaflets; the leaf terminates in three tendrils. In the latter the length of the leaf rachis is reduced and the four (usually) leaflets have a toothed margin. The leaf terminates in five tendrils. The total number of appendages on the leaf rachis in the cultigen can be as high as 11 or 13. It is possible to have leaves produced in which all appendages are leaflets and equally those in which all appendages are tendrils. The tendril-less form is termed 'acacia' leafed, while the forms with tendrils instead of leaflets are termed 'leafless' or 'semi-leafless', depending on whether they have stipules of reduced or normal size. The stems of peas are relatively weak and when carrying a substantial crop lodge easily. This tendency is less marked in leafless genotypes which can, in spite of the reduction in photosynthetic area, produce surprisingly good yields. There may well be a yield penalty entailed in making use of the leafless character but opinion on this point is not unanimous (Davies *et al.*, 1985).

Fig. 6.7. *Pisum pumilio (P. humile)* (D. Zohary).

Another leaf-form variant which might have some value in future breeding programmes is the multiple-imparipinnate type, which has a thrice-pinnate form with very small leaflets. There does thus appear to be considerable scope for the genetic manipulation of leaf area. It might well prove possible to maximise effective photosynthetic (leaf) area and combine this with a canopy less likely to produce lodging. The other variable which must be included in this consideration is the degree to which main-stem branches are produced. This could also be manipulated to produce an effective light-intercepting, lodging-resistant canopy.

6.2. The grasspea (*Lathyrus sativus* L.)

Introduction

The role of the grasspea in those areas in which it is grown as a subsistence crop is interesting. It is highly drought-resistant and produces a yield when other crops fail completely. The value of this capacity is offset by the occurrence of the pathological state of lathyrism when excessive quantities of this pulse are consumed (Rutter and Percy, 1984). The seed is nutritionally valuable and contains ±28% protein; it also contains the water-soluble, non-protein amino acid, β-*N*-oxalyl-L-α,β diaminopropionic acid (ODAP). This free amino acid, which is widely distributed throughout the plant, probably exercises its toxic effect as an antagonist in protein amino acid metabolism. Attempts at controlling the problem by prohibiting cultivation have not been successful since no satisfactory alternative crops are available. The most satisfactory approach is likely to be selection for reduced ODAP content or better still its elimination. Some progress has been achieved but a yield penalty is, however, entailed.

Classification and biosystematics

The grasspea is a member of the genus *Lathyrus*, which according to most authorities contains about 150 species. The breakdown of the genus according to Kupicha (1983) is given in Table 6.2. As a monographer of both this genus and *Vicia* her comments on the breakdown of the two genera are of interest. The genera are of about equal size but the structure of *Lathyrus* is more straightforward than that of *Vicia*. There appears to be a better overall correlation of vegetative characters in *Lathyrus*, which shows at the same time a comparable range of variation in floral structure to that found in *Vicia*. As a result a simple breakdown of the genus into 13 sections suffices.

The distribution of the genus is cosmopolitan. It is found naturally in Eurasia, North America, temperate South America and East Africa

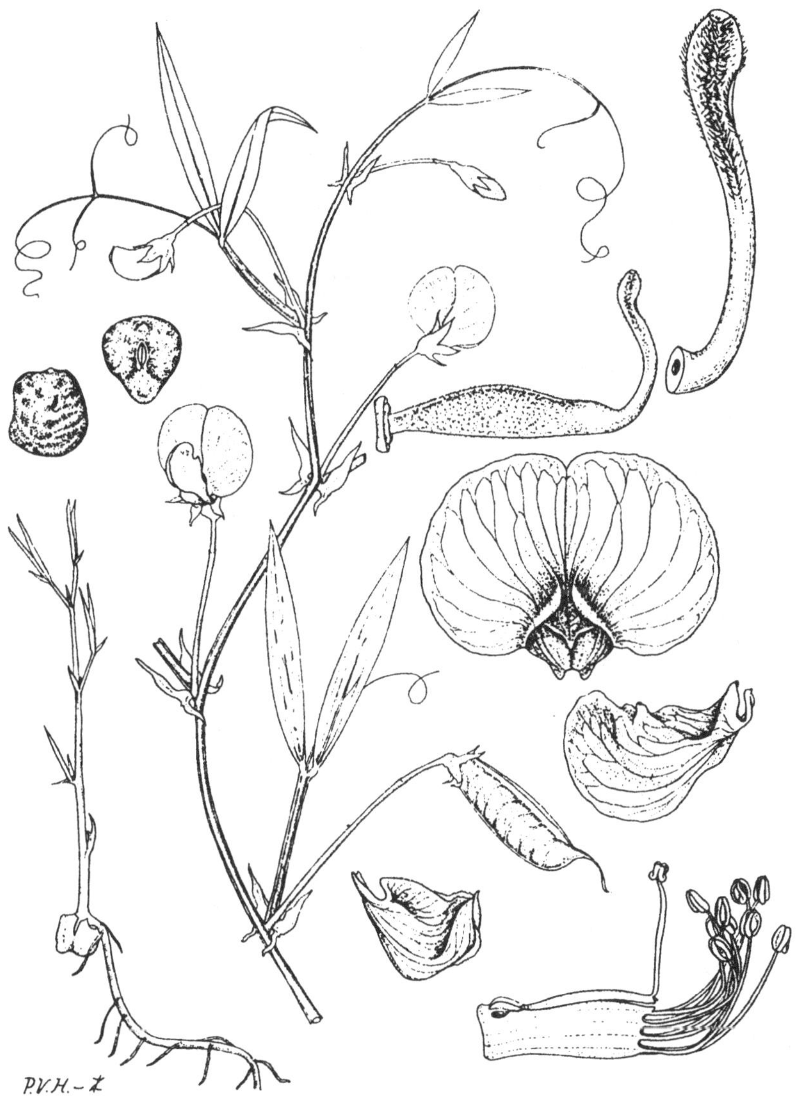

Fig. 6.8. The grasspea, *Lathyrus sativus*, cv. Gondar Marble (from Westphal, 1974).

Table 6.2. *Taxonomic conspectus of genus* Lathyrus *L.*

Sections	
1. *Orobus*	
Old World members	
L. davidii Hance	*L. komarovii* Ohwi
L. gmelinii Fritsch	*L. venetus* (Miller) Wohlf.
L. krylovii C. Serg.	*L. alpestris* (Waldst. & Kit.) Kit.
L. emodi Fritsch	*L. incurvus* (Roth) Willd.
L. vaniotii Léveillé	*L. niger* (L.) Bernh.
L. libani Fritsch	*L. japonicus* Willd.
L. aureus (Steven) Brandza	*L. pisiformis* L.
L. occidentalis (Fischer & Meyer) Fritsch	*L. palustris* L.
L. laevigatus (Waldst. & Kit.) Gren.	*L. dielsianus* Harms
L. transilvanicus (Sprengl) Reichb.f.	*L. winsonii* Craib
L. humilis (Ser.) Sprengel	*L. quinquenervius* (Miq.) Litv.
L. vernus (L.) Bernh.	*L. linifolius* (Reichard) Bassler
L. frolovii Rupr.	*L. dominianus* Litv.
New World members	
L. arizonicus Britton	*L. ochroleucus* Hook. f.
L. bijugatus T. White	*L. parvifolius* S. Watson
L. brachycalyx Rydb.	*L. pauciflorus* Fern.
L. delnorticus C. Hitchc.	*L. polymorphus* Nutt.
L. eucosmus Butters & St. John	*L. polyphyllus* Nutt.
L. graminifolius (S. Watson) T. White	*L. rigidus* T. White
L. hitchcockianus Barneby & Reveal	*L. splendens* Kellogg
L. holochlorus (Piper) C. Hitchc.	*L. sulphureus* Brewer
L. jepsonii E. Greene	*L. torreyi* A. Gray
L. laetiflorus E. Greene	*L. tracyi* Bradshaw
L. lanszwertii Kellogg	*L. venosus* Muhlenb.
L. leucanthus Rydb.	*L. vestitus* Nutt.
L. littoralis (Nutt.) Endl.	*L. whitei* Kupicha
L. nevadensis S. Watson	*L. zionis* C. Hitchc.
2. *Lathyrostylis*	
L. ledebouri Trautv.	*L. digitatus* (M. Bieb.) Fiori
L. pannonicus (Jacq.) Garcke	*L. armenus* (Boiss. & Huet.) Čelak.
L. pallescens (M. Bieb.) K. Koch	*L. nivalis* Hand.-Mazz.
L. pancicii (Jurišić) Adamović	*L. atropatanus* (Grossh.) Širj
L. brachypterus Čelak	*L. tukhtensis* Czeczott
L. bauhinii Gentry	*L. variabilis* (Boiss. & Kotschy) Čelak
L. filiformis (Lam.) Gay	*L. spathulatus* Čelak
L. satdaghensis P. H. Davis	*L. elongatus* (Bornm.) Širj.
L. karsianus P. H. Davis	*L. cilicicus* Hayek & Siehe
L. cyaneus (Steven) K. Koch	*L. boissieri* Sirj.
3. *Lathyrus*	
L. mulkak Lipsky	*L. chrysanthus* Boiss.
L. cirrhosus Ser.	*L. trachycarpus* (Boiss.) Boiss.
L. grandiflorus Sibth. & Smith	*L. lycicus* Boiss.
L. rotundifolius Willd.	*L. phaselitanus* Huber-Mor. & P. J. Davis
L. tuberosus L.	*L. sativus* L.
L. undulatus Boiss.	*L. amphicarpos* L.

Table 6.2. (*cont.*)

	L. heterophyllus L.	*L. cicera* L.
	L. latifolius L.	*L. stenophyllus* Boiss. & Heldr.
	L. sylvestris L.	*L. marmoratus* Boiss. & Blanche
	L. tingitanus L.	*L. blepharicarpus* Boiss.
	L. tremolsianus Pau	*L. ciliolatus* Rech. f.
	L. annus L.	*L. hirticarpus* Mattalaia & Heyn
	L. hierosolymitanus Boiss.	*L. basalticus* Rech. f.
	L. cassius Boiss.	*L. lentiformis* Plitm.
	L. odoratus L.	*L. gorgoni* Parl.
	L. hirsutus L.	*L. pseudo-cicera* Pampan
	L. chloranthus Boiss.	
4.	*Orobon*	
	L. roseus Steven	
5.	*Pratensis*	
	L. binatus Pančić	*L. laxiflorus* (Desf.) Kuntze
	L. czeczottianus Bassler	*L. layardii* Ball ex Boiss.
	L. hallersteinii Baumg.	*L. pratensis* L.
6.	*Aphaca*	
	L. aphaca L.	*L. stenolobus* Boiss.
7.	*Clymenum*	
	L. clymenum L.	*L. ochrus* (L.) DC.
	L. gloeospermus Warb. & Eig.	
8.	*Orobastrum*	
	L. setifolius L.	
9.	*Viciopsis*	
	L. saxatilis (Vent.) Vis.	
10.	*Linearicarpus*	
	L. angulatus L.	*L. tauricola* P. H. Davis
	L. hygrophilus Taubert	*L. vincalis* Boiss. & Noë
	L. inconspicuus L.	*L. woronowii* Bornm.
	L. sphaericus Retz.	
11.	*Nissolia*	
	L. nissolia L.	
12.	*Neurolobus*	
	L. neurolobus Boiss. & Heldr.	
13.	*Notolathyrus*	
	L. berterianus Colla.	*L. nervosus* Lam.
	L. cabrerianus Burkart	*L. nigrivalvis* Burkart
	L. campestris Philippi	*L. paraguayensis* Hassler
	L. hasslerianus Burkart	*L. paranensis* Burkart
	L. hookeri G. Don	*L. parodii* Burkart
	L. linearifolius Vogel	*L. pubescens* Hook. & Arn.
	L. lomanus I. M. Johnston	*L. pusillus* Elliott
	L. longipes Philippi	*L. subandinus* Philippi
	L. macropus Gillies	*L. subulatus* Lam.
	L. macrostachys Vogel	*L. tomentosus* Lam.
	L. magellanicus Lam.	*L. tropicalandinus* Burkart
	L. multiceps D. Clos	

From Kupicha (1983).

(Kupicha, 1981). The main centre of distribution is apparently the Mediterranean Basin and the Near East. In contrast to the common pea the nomenclature of the grasspea (also known as khesari dhal, chickling pea and blue vetchling) has been remarkably stable since the time of Linnaeus; the synonymy is not extensive and is given by Westphal (1974). Few infra-specific taxa have been described and it is interesting to observe that Townsend and Guest (1974) do not find it necessary to recognise more than were described by Boissier (1872) a century earlier. The two forms so described are typical *Lathyrus sativus* and *L. sativus* var. *stenophyllus*. Westphal (1974) comments that many infra-specific breakdowns proposed do not work well in practice. The patterning of the variability within the species is such that morphological discontinuities which exist are not distributed in such a way that distinct forms can be recognised on any reasonable geographical basis. Even more surprising is the lack of a consistent pattern of differences between wild populations and those that are domesticated (Townsend and Guest, 1974). It is possible that this lack of taxonomic divisions within the species may be a reflection of neglect by taxonomists on the one hand and its lack of economic significance as a human food outside some areas of the Third World on the other. Certainly descriptions of the plant from its whole geographic range are sparse and features such as presence or absence of pod dehiscence, for example, are rarely if ever recorded.

Geography and ecology

Townsend and Guest (1974) consider that the natural distribution of *L. sativus* has been completely obscured by cultivation even in south-west and central Asia, its presumed centre of origin. At the present time it is distributed widely in southern and south-central Europe, the Near East, Ethiopia and India; it has also been introduced to Australia. This wide dispersion is due to its extensive utilisation as a fodder crop. The seed is exploited as a pulse principally in India and sporadically elsewhere. The great value of the grasspea as a crop is its tolerance of adverse environmental conditions, tolerating not only drought but also waterlogging (Purseglove, 1974). In India it is one of the most reliable grain crops and may be the only food available in some areas when famines occur (Rutter and Percy, 1984). This can lead to excessive consumption and may provoke the neurological form of lathyrism previously noted.

Cytogenetics and genetic resources

The chromosome complement is diploid $2n = 2x = 14$ in common with most species of the genus. Polyploidy has been reported, most extensively in North America, where complements of $2n = 4x = 28$ are to be found (Hitchcock, 1952). Chromosome morphology has been studied by Senn

(1938*a*, *b*) and Davies (1958). The karyotype is on the whole symmetrical with chromosomes having median–sub-median chromosomes for the most part. The small section *Clymenum* is characterised by a markedly higher level of karyotype asymmetry.

Comparatively little has been carried out in the way of experimental inter-specific hybridisation in the genus. Senn (1938*b*) reported that hybridisation was difficult between species of the genus. However, Saw Lwin (1956) produced hybrids *L. sativus* × *L. cicera* and Davis (1958) crossed *L. odoratus* × *L. hirsutus*. It is apparent that only very closely related species will cross. Adequate knowledge of inter-specific hybridisation capability of *L. sativus* is an essential pre-requisite to effective evaluation of the broader range of genetic resources. Initially a comprehensive study of inter-specific hybridisation between *L. sativus* and other members of section *Lathyrus* should be undertaken. It might then be considered worth while to extend the study to other sections. Cytogenetic studies of any hybrids produced would be important, as would a study of the possibility of inducing gene flow between species capable of producing fertile hybrids. The production of sterile hybrids only would indicate that more complex genetic engineering techniques would be required to effect gene transfer. Present indications are that simple gene transfers between species of the same section might be possible but intersectional gene transfer may well not be so without genetic engineering techniques.

Biochemistry

Biochemical studies of *Lathyrus* species have tended to focus on the non-protein amino acids for two reasons. In the first place they may have profound effects on human amino acid metabolism if they are ingested (Aykroyd and Doughty, 1964) and secondly they have a value in taxonomy (Bell, 1964). The condition of neuro-lathyrism, an irreversible paralysis, has been linked to excessive consumption of grasspea seed. Outbreaks have occurred most frequently in India (Rutter and Percy, 1984) but have also been recorded from Spain (Aykroyd and Doughty, 1964). The incitant of this condition appears to be β-*N*-oxalyl-L-α,β diaminopropionic acid (ODAP) which probably acts as an antagonist in the metabolism of essential amino acids (Adiga *et al.*, 1962). This material has been identified in other *Lathyrus* species, namely *L. clymenum* and *L. latifolius*. It is of some interest to note that whereas *L. latifolius* is a member of the same section of the genus as *L. sativus*, *L. clymenum* belongs to a different taxon, the small section *Clymenum*. Some alleviation of neuro-lathyrism has been reported from administration of methionine supplements in the diet. Apparently detoxification is a relatively simple process (Liener, 1978) which, if more widely adopted, could make safer exploitation of this pulse possible. In the long term it would be desirable, if it is feasible, to select for absence or low seed concentration of the toxin. A further point of interest is that male susceptibility to the condition is higher than that in the female. The most susceptible females are pre-pubertal and post-menopausal; this observation has led to the suggestion (Rutter and Percy, 1984) that the female sex hormone somehow inhibits development of the disease.

In some ways the development of the condition osteo-lathyrism is rather better understood than that of neuro-lathyrism. The toxic material which incites this condition, β-(γ-L-glutamyl)aminoproprionitrile (BAPN), produces skeletal abnormalities rather than neurological damage. Osteo-lathyrogens apparently block cross-link formation in the collagen molecule and inhibit normal development of skeletal elastic fibres. The occurrence of compounds such as ODAP and BAPN in the seeds of *Lathyrus* species enjoins caution in the use of other *Lathyrus* species in the improvement of *L. sativus*.

Archaeology

The archaeological record of the grasspea and other *Lathyrus* species goes back nearly 10000 years. *Lathyrus* remains from Ali Kosh have been dated at 9500–7600 BP and from Tepe Sabz at 7500–5700 BP and were apparently commonplace foods. The oldest remains of *L. sativus* itself have been obtained from Jarmo (Iraq) with a dating of 8000 BP (Helbaek, 1965). Remains from Indian sites have been dated at 3800–3200 BP, Navdatoli (Allchin, 1969) and 4000–3500 BP, Atranjikha (Saraswat, 1980). Renfrew (1973) notes that other *Lathyrus* species have been identified from archaeological sites, namely *L. cicera*, *L. aphaca* and *L. nissolia*. These species belong to different sections of the genus: sect. *Lathyrus*, sect. *Aphaca* and sect. *Nissolia* respectively.

Elsewhere, seed of *L. sativus* has also been recovered from sites in Israel, Italy, Hungary and Switzerland (Renfrew, 1973). This is consistent with the view that this crop is a native of southern Europe and southwest Asia (Duke, 1981). The archaeological remains of the grasspea are of comparable age to those of the other Mediterranean pulses which have subsequently eclipsed it. The probable reasons for this are worthy of some consideration.

Domestication and evolution

At first sight it is puzzling that a crop which apparently has been domesticated for at least 8000 years should have made so little evolutionary progress as a grain crop in this time. It is possible that the association of neuro-lathyrism with excessive consumption of the seed in times of food shortage has been generally appreciated and has tempered enthusiasm for the crop. This may have proved to have been a disincentive to cultivation. Equally it could have provided a positive incentive to improve the crop, since it is a reliable yielder under adverse conditions. Two additional factors probably operated to inhibit the development of the grasspea as a crop comparable to the other Mediterranean pulses. The first is that the grasspea is not very highly regarded as a pulse. It is perhaps primarily its role as a crop in drought and famine-producing conditions that has enabled it to persist as a pulse. The second factor is that its more important use has been as a forage. The selection pressures imposed on forage crops are in many ways the opposite of those on grain crops. For instance large-seededness, an advantage in grain crops, is unnecessary and undesirable in a fodder plant, whereas luxuriant vegetative growth, desirable in a forage crop, is less so in a grain crop, especially if this is achieved at the expense of seed production. These considerations may have nullified the selection pressures that could have been expected to act and favour progressive evolution as a grain crop. The status quo may thus have been maintained over the millennia. This suggests that there may be unrealised potential for the development of the grasspea as a pulse. The establishment of a more compact (and erect) growth habit, combined with some increase in seed size and elimination of the neurotoxin, could transform this neglected pulse crop into one of great value in climatically suitable, semi-arid areas of the Third World.

In order to gain a better understanding of the evolution of this crop more information is needed in several areas. More extensive collections from areas of cultivation and natural occurrence are needed. The variability in collections needs to be studied and evaluated. Fortunately, work on this has already been initiated (Jackson and Yunus, 1984). The characters of major importance are: variation in neurotoxin content of the seed; useful morphological variation in growth habit, such as compact growth; delayed dehiscence or suppression of the pod dehiscence mechanism; variation in number of flowers per inflorescence with a view to selecting for an increased number. Improvement in our understanding of the reproductive biology of thc crop would be helpful in the management of selection programmes and maintenance of improved lines. There is apparently no obstacle to self-pollination in the greenhouse (M. T. Jackson, personal communication), but it is possible that significant cross-pollination could occur in the field if the appropriate pollinators are present.

There is little doubt that in agriculturally marginal areas of the Third World there is a niche for a crop with the useful characters of the grasspea. Certainly some effort directed towards its improvement would seem to be justified where it is an accepted crop. Whether it could be more widely promoted in other suitable areas would be determined by its acceptability. Consumer preferences can be notoriously difficult to influence.

The neurotoxin problem has been a peculiarly difficult stumbling block to surmount. The measures which have been taken, such as prohibiting cultivation and sale of the crop in some areas of India, have not been very helpful. Such legislation has apparently been totally ineffective (Rutter and Percy, 1984). Had it been so it would have condemned many more people to starvation than would have been affected by lathyrism. Lack of enforcement of this prohibition has effectively allowed to crop to maintain its position in agriculture.

6.3. The faba bean (*Vicia faba* L.)

Introduction

The economic importance of the faba bean, while less than that of the common pea world wide, is nevertheless considerable. According to FAO statistics (FAO, 1981) annual world production is in excess of 4 million tonnes (approximately half that of the pea). Production is very heavily concentrated in China, which produces 60% of the world's crop. It is a widely disseminated crop with significant levels of production in Africa, Europe and Latin America.

Faba beans are commonly used both for human and livestock food and have been a major protein source for animal feeding stuffs. The pre-eminence of the crop in this role has, in the present century, been strongly challenged in Europe by the imported oilseed cakes from expression of soyabeans, groundnuts and cottonseed. Nevertheless, faba beans are now attractive in the economic context of import saving. Yield potential is considerable; Duke (1981) records yields in the United States of over 6 tonnes ha^{-1}, although those in the UK average only half this level. In western Europe and elsewhere it is the instability of yield which inhibits wider cultivation. Poor seasonal growing conditions can probably bring about poor yields in at least two ways, in addition to slower growth and development of the crop itself. At low temperatures and in excessively wet soils it is probable that *Rhizobium* symbiosis is not operating effectively; while in cloudy and overcast conditions pod set may be adversely affected by low levels of bee pollinator activity.

Selection under domestication has produced three distinctive types of faba bean which differ principally in seed size, which are commonly classed as small, medium and large. While there is no reason why all should not equally be used for human and livestock food, where different types are grown the smaller tend to be used in livestock foods and the larger for human food. This is understandable in that the larger-seeded

Fig. 6.9. The faba bean, *Vicia faba*, cv. Abyssinica (from Westphal, 1974).

forms have been selected the most strongly for improved palatability and cooking quality. They tend to have, as a result, lower tannin contents in the testa, which could produce such an improvement.

In common with the other pulsesof Mediterranean origin there are few problems in the utilisation of faba beans as food, from the point of view of lectin and protease inhibitor content. These are low and do not require heat or microbiological inactivation as they do in *Phaseolus* and soya-beans. The major problem in their use is that they can incite the condition of favism, a haemolytic anaemia, in susceptible individuals, having a congenital deficiency of the enzyme glucose-6-phosphate dehydrogenase. The onset of the anaemia is thought to be incited by the glucosides vicine,

Fig. 6.10. Extreme range of seed size in *Vicia faba* (D. A. Bond).

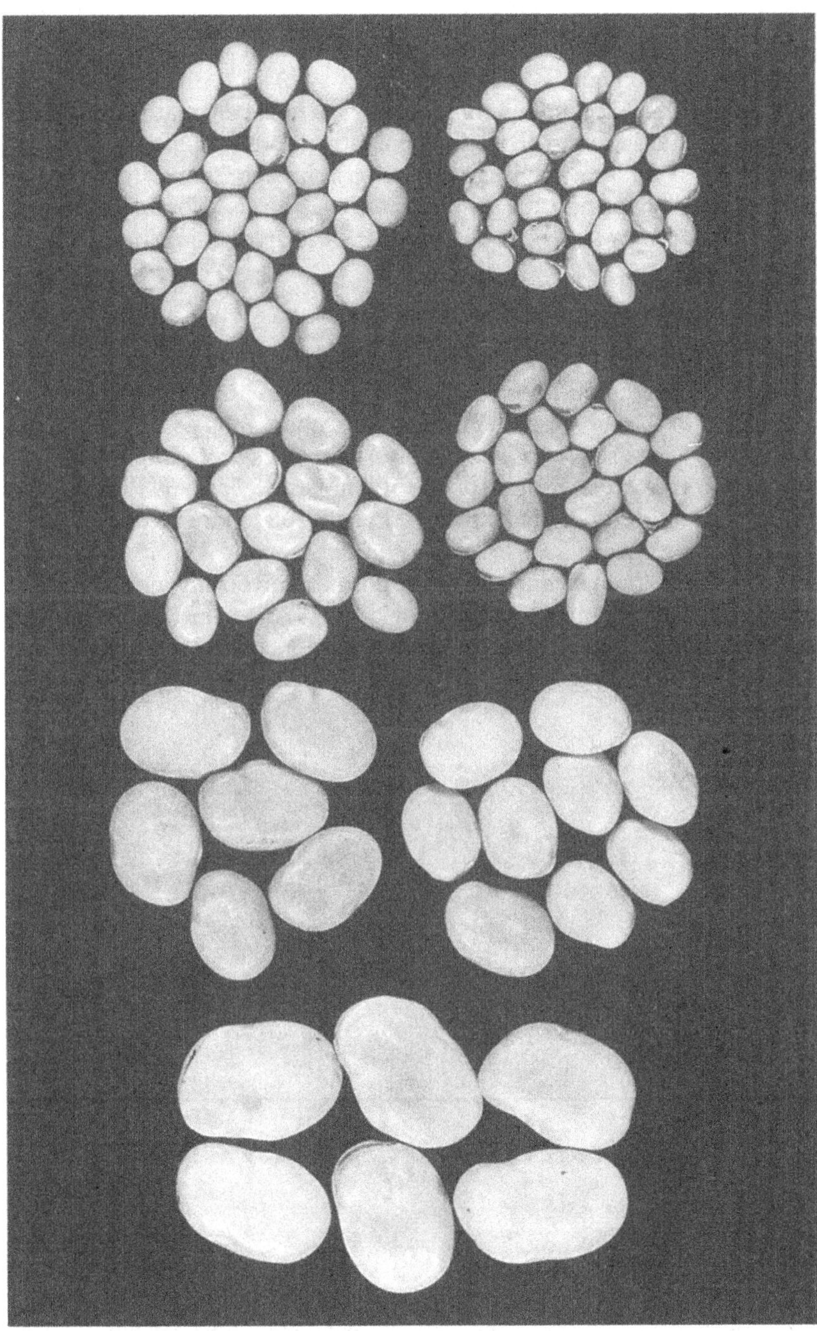

Fig. 6.11. Variation of seed size within *Vicia faba* (D. A. Bond).

convicine and DOPA glucoside (Liener, 1982). It is possible that judicious selection could reduce contents of these compounds and alleviate this problem.

Although predominant use of the crop is of its seeds (mature and green mature), limited use can be made of immature pods (cf. *Phaseolus* beans) and the tops of plants may also be used as a spinach. In livestock feeding and in green manuring the whole plant may be used before maturity. Crop residues may also be used as fodder or for ploughing in.

Classification and biosystematics

Taxonomy

Vicia faba is a member of a large genus with more than 130 members. It is assigned to the sub-genus *Vicia* and the section *Faba* of that sub-genus (Table 6.3.). The section *Faba* contains six species according to Kupicha (1976), but the status of these taxa is still not entirely clear. Pickersgill *et al.* (1983) consider there to be three groups of related forms conveniently recognised as species: *V. faba*, *V. narbonensis* and *V. bithynica*. The inclusion of *V. bithynica* in this section is not universally agreed but the section is clearly defined morphologically. *Vicia faba* (*sens. lat.*) is a highly variable species and there is no agreement on the relative status of variants within it, recognised as *paucijuga*, *major*, *equina* and *minor*.

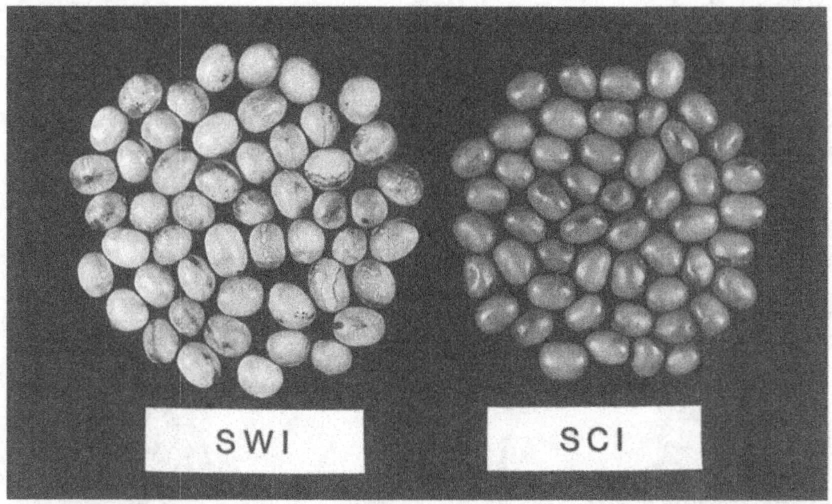

Fig. 6.12. Flower colour – testa colour correlation in isogenic lines SW1 and SC1 of *Vicia faba*. SW1, pale testa, white flowers; SC1, darker testa, coloured flowers (D. A. Bond).

These can be given equal status as botanical varieties or alternatively *paucijuga* can be recognised as a sub-species with the other three forms combined as sub-species *faba* and each assigned varietal status (Lawes *et al.*, 1983). The latter perhaps better reflects phylogenetic relationships within the species. There is much less of a consensus regarding *V. narbonensis* and its closest allies, *V. galilaea*, *V. hyaeniscyamus* and *V. johannis*. Plitmann (1970) has questioned the separate species status of *V. galilaea* and a form *V. serratifolia* not recognised by Kupicha (1976). Schäfer (1973) is not convinced that *V. hyaeniscyamus* and *V. galilaea* should be regarded as distinct species. *Vicia johannis*, however, is regarded as satisfactorily distinct from the other species in section *Faba*. There appear to be thus three species, whose status is not in doubt, and a complex which can be regarded either as a single variable species or three which cannot be sharply delimited.

In contrast to the great majority of other pulse species there is no clearly identifiable wild progenitor type or prototype of the faba bean. Comparative morphological studies suggest that of the various forms of *V. faba*, var. *minor* and ssp. *paucijuga* are probably closest to the ancestral form. This has not been found in spite of considerable exploration activity. However, since the critical area coincides largely with that of the present Middle East conflict, the obstacles to effective exploration are considerable. It may even be that the crucial material has already been collected and not recognised, regarded perhaps as a component of the *V. narbonensis* complex.

Cytotaxonomy

The chromosome complement of *V. faba* clearly differentiates it from the other members of the section. Chromosome number is $2n = 2x = 12$ in contrast to $2n = 14$ characteristic of other members of the section. They also differ in DNA content, size and symmetry (Chooi, 1971*a*, *b*). In addition the Giemsa banding patterns of *V. faba*, *V. narbonensis* and *V. bithynica* show no apparent homologies between chromosome arms in the three forms (Pickersgill *et al.*, 1983; Ramsay, 1984). The karyotype of *V. faba* is markedly different from that of *V. narbonensis*, *V. galilaea* and *V. hyaeniscyamus* while those of the three wild species are very closely similar (Ladizinsky, 1975*b*). The *V. faba* karyotype is extremely asymmetric with one large satellited metacentric pair; the remaining five pairs are acrocentric. The wild species have a small pair of sub-metacentric–metacentric satellited chromosomes and six larger pairs of sub-metacentric–sub-acrocentric chromosomes. These karyotypic differences suggest that differentiation between *V. faba* and the other species is of long standing and that common ancestry is relatively remote.

Table 6.3. *Taxonomic conspectus of genus* Vicia *L.*

I. Subgenus VICILLA

Sections

1. *Vicilla*
   - *V. unija* A. Braun
   - *V. crocea* (Desf.) B. Feditsch
   - *V. venosa* (Willd.) Maxim
   - *V. nipponica* Matsum.
   - *V. kulingiana* L. H. Bailey
   - *V. pseudo-orobus* Fischer & C. A. Meyer
   - *V. hirticalycina* Nakai
   - *V. venulosa* Boiss. & Hohen.
   - *V. dichroantha* Diels
   - *V. amoena* Fischer
   - *V. amurensis* Dettel
   - *V. japonica* A. Gray
   - *V. pisiformis* L.
   - *V. sylvatica* L.
   - *V. dumetorum* L.
2. *Cassubicae*
   - *V. abbreviata* Fischer ex Sprengel
   - *V. cassubica* L.
   - *V. dadianorum* Sommier & Levier
   - *V. montenegrina* Rohl.
   - *V. multicaulis* Ledeb.
   - *V. nigricans* Hook. & Arn.
   - *V. orobus* DC.
   - *V. semiglabra* Rupr. ex Boiss.
   - *V. sparsifolia* Ten.
3. *Perditae*
   - *V. dennesiana* H. C. Watson
4. *Cracca*

   Old World perennials
   - *V. cracca* L. agg.
   - *V. pinetorum* Boiss. & Spruner
   - *V. sibthorpii* Boiss.
   - *V. ochroleuca* Ten.
   - *V. atlantica* Pamel
   - *V. splendens* P. H. Davis
   - *V. kotschyana* Boiss.
   - *V. glareosa* P. H. Davis
   - *V. sicula* (Raf.) Guss.
   - *V. alpestris* Steven
   - *V. ciceroidea* Boiss.
   - *V. rafigae* Tamamschian
   - *V. multijuga* (Boiss.) Rech. f.
   - *V. glauca* C. Presl.

   Old World annuals
   - *V. villosa* Roth. agg.
   - *V. benghalensis* L.
   - *V. scandens* R. P. Murray
   - *V. cirrhosa* Webb & Berthel
   - *V. chaetocalyx* Webb & Berthel
   - *V. filicaulis* Webb & Berthel
   - *V. monantha* Retz
   - *V. leucantha* Biv.
   - *V. palaestina* Boiss.
   - *V. hulensis* Plitm.
   - *V. disperma* DC.
   - *V. durandii* Boiss.
   - *V. vicioides* (Desf.) Cont.
   - *V. hirsuta* (L.) Gray
   - *V. terronii* (Ten.) Lindb. f.

   New World members
   - *V. acutifolia* Elliott
   - *V. caroliniana* Walter
   - *V. exigua* Nutt.
   - *V. floridans* S. Watson
   - *V. hugeria* Small
   - *V. leavenworthii* Torrey & A. Gray
   - *V. ludoviciana* Nutt.
   - *V. mexicana* Hemsley
   - *V. minutiflora* Dietr.
   - *V. pulchella* Munth.
   - *V. reverchonii* S. Watson
5. *Variegatae*
   - *V. argenta* Lapeyr.
   - *V. vanescens* Labnill. agg.
   - *V. megalotropis* Ledeb.
6. *Pedunculatae*
   - *V. altissima* Desf.
   - *V. cedretorum* Font Quer
   - *V. onobrychoides* L.
7. *Americanae*
   - *V. americana* Muhl. ex Willd.

Table 6.3. (*cont.*)

8. *Subvillosea*	
V. subvillosa (Ledeb.) Trautv.	
9. *Volutae*	
V. biennis L.	
10. *Panduratae*	
V. cappadocica Boiss. & Bal.	*V. cretica* Boiss. & Heldr.
V. cassia Boiss.	
11. *Ervum*	
V. laxiflora Brot.	*V. tetrasperma* (L.) Schreber
V. pubescens (DC.) Link	
12. *Ervoides*	
V. articulata Hornem.	
13. *Ervilia*	
V. ervilia (L.) Willd.	
14. *Lentopsis*	
V. caesarea Boiss. & Bal.	
15. *Trigonellopsis*	
V. cypria Kotschy. ex Unger & Kotschy.	*V. singarensis* Boiss. & Hausskn.
V. lunata (Boiss. & Bal.) Boiss.	
16. *Australes*	
V. andicola Kunth	*V. nana* Vogel
V. bijuga Gillies ex Hook.	*V. pampicola* Burkart
V. epetiolaris Burkart	*V. peruviana* Vichez
V. graminea Smith	*V. platensis* Speg.
V. linearifolia Hook. & Arn.	*V. setifolia* Kunth.
V. macrograminea Burkart	*V. stenophylla* Vogel
V. montevidensis Vogel	
17. *Mediocinctae*	
V. leucophaea Greene	
II. Subgenus VICIA	
18. *Atossa*	
V. oroboides Wulfen	*V. truncata* Fischer ex. M. Bieb.
V. sepium L.	*V. balansae* Boiss.
19. *Vicia*	
V. pyrenaica Pourret	*V. barbazitae* Ten. & Guss.
V. sativa L. agg.	*V. lathyroides* L.
V. grandiflora Scop.	*V. cuspidata* Boiss.
20. *Faba*	
V. faba L.	*V. hyaeniscyamus* Mouterde
V. narbonensis L.	*V. johannis* Tamamschian
V. galilaea Plitm. & Zohary	*V. bithynica* (L.) L.
21. *Hypechusa*	
V. anatolica Turrill	*V. hyrcanica* Fischer & C. A. Meyer
V. assyriaca Boiss.	*V. lutea* L.
V. ciliatula Lipsky	*V. melanops* Sibth. & Smith
V. esdraelonensis O. Warb. & Eig.	*V. noeana* Reuter ex Boiss.
V. galeata Boiss	*V. pannonica* Crantz
V. hybrida L.	*V. seriocarpa* Fenzl.
22. *Peregrinae*	
V. aintobensis Boiss. & Hausskn.	*V. michauxii* Sprengel
V. peregrina L.	*V. mollis* Boiss. & Hausskn. ex Boiss.

From Kupicha (1976).

Chemotaxonomy

In addition to comparative karyotype studies, Ladizinsky (1975*c*) has also compared extracted seed protein (albumin) profiles electrophoretically. *Vicia faba* produced a very distinctive profile from all the other three species. The profiles of *V. galilaea* and *V. hyaeniscyamus* were the most similar, both of which differed appreciably from that of *V. narbonensis*.

Experimental hybridisation and genetic resources

Although all forms commonly included in *V. faba* are readily cross-compatible, attempts at inter-specific hybridisation with other members of the genus have been uniformly unsuccessful. The causes of this failure have been studied by Pickersgill *et al.* (1983) in terms of failure of pollen germination, pollen tube growth, fertilisation, and embryo and endosperm development. Curiously enough, the inter-generic cross *Pisum sativum* × *V. faba* proceeds as far developmentally as intra-generic crosses involving *V. faba* (Gritton and Wierzbicka, 1975). The faba bean therefore appears to be virtually completely isolated genetically from any other member of the genus *Vicia* and the tribe *Vicieae*. This conclusion is supported strongly by both cytotaxonomic and chemotaxonomic data. We are therefore in a position where experimental data sheds little or no light on the origin and evolution of the faba bean.

Domestication and evolution

In sharp contrast to all other major grain legume crops, we are totally unable to identify a plausible prototype or wild progenitor, on morphological grounds, which under domestication could credibly have given rise to the faba bean. It seems probable that morphologically it could have been not unlike *V. narbonensis sens. lat.* Unfortunately, other lines of investigation have not been very productive. The earliest archaeological material cannot be identified with any certainty. Although the earliest beans found (Hopf, 1969) have been dated at *ca.* 6250 BC from Jericho, their identification as *V. faba* can only be tentative; they could equally well be *V. narbonensis*, which has been, and still is, cultivated in parts of the eastern Mediterranean.

In spite of the uncertainty and lack of compelling evidence, the opinion of most authorities (e.g. Zohary and Hopf, 1973) is that the faba bean was domesticated in the Neolithic. Extensive archaeological remains from the Bronze Age have been found. Zohary (1977) has argued that since materials dated between 2000 and 3000 BC have been found over the entire Mediterranean basin, domestication must have occurred at least 1500–2000 years previously. This argument was developed to counter that of Ladizinsky (1975*b*) who suggested an origin in Afghanistan largely on the basis of the distribution of the primitive ssp. *paucijuga*. Although archaeological remains cast little light on botanical origin they do give a strong indication that the probable centre of domestication was in the eastern Mediterranean and the Near East.

Domestication

Assuming an eastern Mediterranean – Near Eastern origin of the faba bean, extant archaeological discoveries of seed and historical evidence enables an approximate time scale and pattern for its dispersion to be suggested. Starting from a possible Neolithic Near Eastern domestication (Hopf, 1969), dispersal to the Western Mediterranean had occurred in the Iberian late Neolithic (Schultze-Motel, 1972) by about the third millennium BC. Bronze Age remains have been recovered from an even wider geographic range as far north as Switzerland, Austria and Czechoslovakia (Renfrew, 1973) and even Jersey in the Channel Islands. The earliest British remains were found in Somerset and date from the Iron Age.

The eastward spread of the faba bean can be presumed to have occurred in ancient times, since ssp. *paucijuga* is found in north India and Afghanistan. No doubt subsequent introductions have also been made in this area (which is very suitable indeed for its cultivation) during colonial times; this perhaps complicates matters. The greatest interest centres on the crop's introduction to China. De Candolle (1886) apparently believed that it had arrived in China at about 2822 BC, introduced by the Emperor Chin-nong (Shen Nung). The historical status of this legendary emperor has, however, been questioned by sinologists (Hymowitz, 1970). Good evidence of cultivation from the time of the Yuan dynasty (Hanelt, 1972*b*) is available; the suggestion has also been advanced that introduction occurred concurrently with establishment of the silk trade. It would appear that the major spread of the crop in China occurred in the fifteenth century AD (Hawtin and Hebblethwaite, 1983). Faba beans are mentioned in Japanese writings of the seventeenth century AD.

Post-Columbian dissemination to thc Ncw World dates from the sixteenth century AD. They have been successful in cooler montane areas of Latin America, most notably those of Brazil, Mexico and Peru. These are areas in which the indigenous *Phaseolus* beans cannot be grown as satisfactorily.

Breeding system

Not the least interesting feature of the faba bean is its breeding system. Essentially it is a random mater. No barrier to self-pollination exists and no evidence has been found for any kind of self-incompatibility system. The surface of the stigma bears numerous papillae which produce an exudate that stimulates pollen germination either when damaged by insect abrasion or spontaneously in autofertile genotypes (Paul *et al.*, 1978; Bond and Poulsen, 1983). Faba beans can be classified as autofertile or autosterile. Autofertile plants will self-pollinate spontaneously without insect visitation; autosterile plants on the other hand require abrasion of the stigmatic surface. After abrasion apparently any pollen present on the stigma, either from self- or cross-pollination, can germinate. The question of autosterility and autofertility was studied in depth by Drayner (1956, 1959). She found that in ssp. *faba* (ssp. *paucijuga*

Fig. 6.13. *Vicia faba* autosterility and autofertility. (*a*) Autofertile: pods set without tripping; (*b*) autosterile: pods set only after tripping (D. A. Bond).

is autofertile) varying degrees of autofertility could be found in the bean populations she studied. This could be related broadly to levels of inbreeding. Plants which had arisen from cross-pollination tended to be more autofertile than those which has arisen from selfing. Cross-bred plants were observed to produce pollen more abundantly than those which were inbred; the greater bulk of pollen produced would ensure that the stigmatic surface was well covered with pollen. This system ensures that successful pollination is not totally dependent on insect visitation. In poor weather conditions when bees are not flying pods could still be set.

Evolution

Since the progenitor type of the faba bean is quite unknown and may even be extinct (although it would be premature to make such an assumption!) evolutionary studies must largely concern themselves with the diversification within the cultigen. As far as the genetic resources available are concerned the species *Vicia faba* can be considered as a closed gene pool which exists for all practical purposes in a state of total domestication. Even if the progenitor type did come to light it might well prove to be a rare form with a very restricted range which would contribute little to the total gene pool of the species.

Evolutionary potential

The greatest success of *V. faba* as a crop has been achieved in China, which produces nearly two thirds of the world's total crop. The genotypes which have been selected and established in China therefore have a special claim for particular study. China is a vast country with a considerable range of ecologically distinct habitats and presumably faba beans can be produced successfully in a range of these. With its large population and the cardinal necessity to produce adequate supplies of food reliably, crops which suffered from yield instability would not be favoured. The Chinese experience might thus seem to be at variance with that in Western Europe where the crop has fallen into disfavour on account of its notoriously unstable yields. This appears to be related to poor (i.e. cool and wet) conditions. It has been thought that poor weather conditions resulting in low levels of bee activity could result in poor pod setting and low yields. Such conditions would be likely to set in train a high selection pressure for autofertility. It seems probable also from Drayner's (1959) studies that a

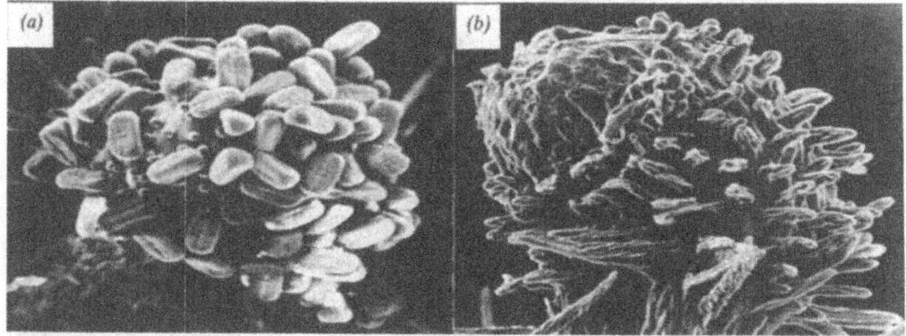

Fig. 6.14. Scanning electron micrographs of *Vicia faba* stigmas (line T2, autosterile): (*a*) after flower opening, stigma covered with pollen grains; (*b*) emasculated before anthesis, showing exudate on stigma 18 h after treatment.

significant level of self-pollination can occur in many faba bean populations. Failure of adequate pollination may therefore not be a major cause of seasonally variable yields. In the United Kingdom faba beans are regarded primarily as a crop of heavy land. In cool wet seasons it is possible that the efficiency of root nodule nitrogen fixation is impaired by a combination of waterlogging with low temperatures. This kind of climatic effect, which would seriously impair the capability of the faba bean to

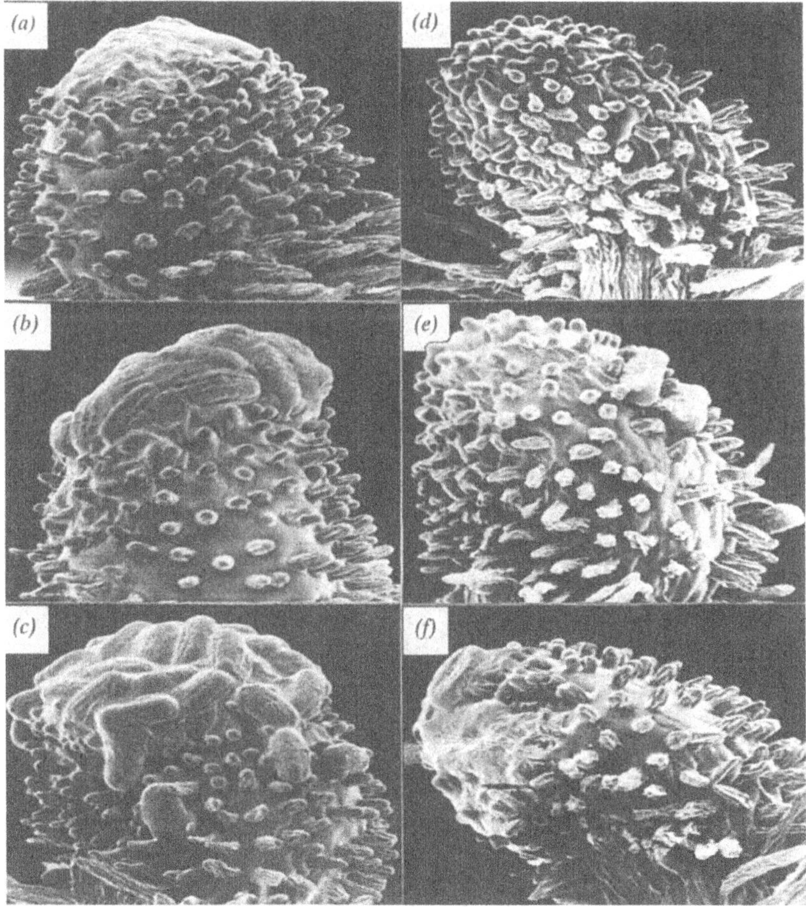

Fig. 6.15. Scanning electron micrographs of *Vicia faba* stigmas showing comparative development of stigma exudate in line T51 (autofertile) (*a–c*), and line TR2 (autosterile) (*d–f*). (*a, d*) Before anther dehiscence; (*c, f*) after flower opening. Note earlier appearance of exudate in the autofertile line (Paul *et al.*, 1978. Reprinted by permission from Nature, vol. 275, pp. 54–5. Copyright © 1978 Macmillan Magazines Ltd.)

Fig. 6.16. *Vicia faba*, determinate growth habit: (r) original Svalof mutant; (l) in adapted British background genotype (D. A. Bond).

Fig. 6.17. *Vicia faba* (winter bean): two inbred lines showing contrast in seed set (D. A. Bond).

mature a large crop, clearly merits further investigation. Of course this does not necessarily explain yield instability elsewhere.

The Chinese germplasm itself merits detailed study. This material has not been available for study generally; there can be little doubt but that this represents perhaps the greatest single geographically defined genetic resource in the species. Other significant resources are those of Latin America which again would merit very detailed agro-ecological and genecological study. The best collected and best understood segment of the world gene pool of the faba bean is the Mediterranean–Near Eastern–European block. More study is also indicated of the *paucijuga* forms and more extensive collection also. As with many other species of legumes, germplasm collection of the faba bean is beset with political problems. Conflicts in the Middle East, the Persian Gulf and Afghanistan make free exploration for the wild progenitor and landrace material very difficult if not impossible at the present time. Unfortunately, the recent political upheavals in China raise considerable potential difficulties for collaborative collection, conservation, evaluation and exploitation of the vast Chinese gene pool.

Although the faba bean does not appear to have any readily exploitable genetic resources outside the biological species *V. faba*, those within the species have very obviously not been exploited to the full. The role of the faba bean appears to be complementary to that of many other pulse species, notably the *Phaseolus* beans. In the United Kingdom *Phaseolus* beans and faba beans have complementary soil preferences, whereas in Peru (a centre of diversity for *Phaseolus* beans) faba beans are preferred for cultivation at the highest elevations. It is difficult to accept that there can be no role for faba beans in some, at least, of the geographic areas in which they are scarcely to be found at the present time in significant quantity, most notably North America and the Antipodes.

Genetic resources

According to the gene pool hierarchy of Harlan and de Wet (1971) *Vicia faba* has a primary gene pool (GP1A) consisting solely of the domesticated sub-species *faba* and *paucijuga*. No GP1B (conspecific wild population) is known, nor has any GP2 been established. The tertiary gene pool (GP3), such as it is, comprises those species with which (inviable) hybrids can be formed. These invariably abort during embryogenesis. Inter-specific hybridisation has been studied between *V. faba* and other members of section *Faba* (Ladizinsky, 1975*b*) and between *V. faba* and *Pisum sativum* (Gritton and Wierzbicka, 1975). Inter-specific hybridisation has also been studied between other species of section *Faba*, with interesting results (Ladizinsky, 1975*b*). The crosses Ladizinsky reported are between members of the *V. narbonensis* complex, i.e. *V. narbonensis*

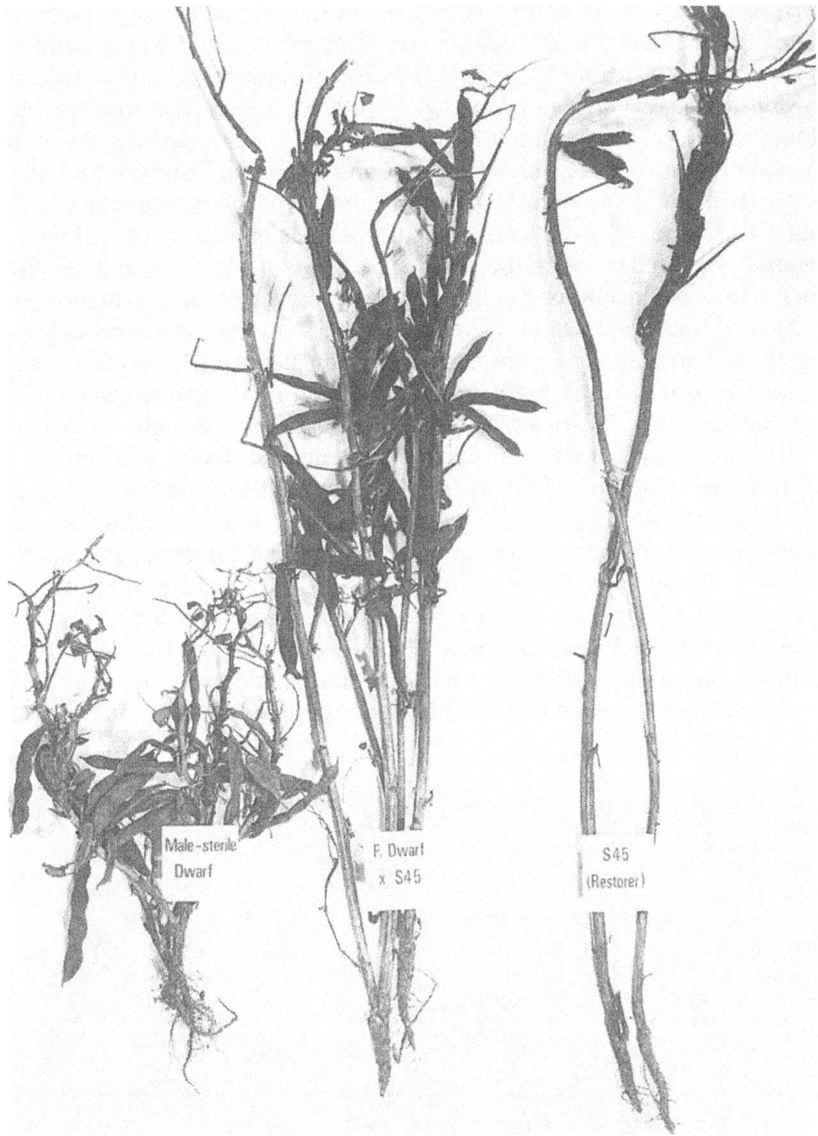

Fig. 6.18. *Vicia faba*. Single cross hybrid (centre); male sterile dwarf (l); inbred bean, restorer (r) (D. A. Bond).

sens. strict., *V. galilaea* and *V. hyaeniscyamus*. Germinable seed was produced in all combinations but viability after germination was generally poor. The combinations *V. galilaea* × *V. hyaeniscyamus*, *V. galilaea* × *V. narbonensis* and *V. hyaeniscyamus* × *V. narbonensis* all produced mostly albino seedlings. One example of *V. galilaea* × *V. narbonensis* grew into a small abnormal plant which flowered and whose meiosis was studied; five individuals of the cross *V. hyaeniscyamus* × *V. narbonensis* reached maturity. These individuals were small with pale green foliage and their meiosis was studied. On the whole Ladizinsky observed apparently normal bivalent formation but he did observe univalents and formation of a chain of four chromosomes, indicating chromosome structural differentiation involving a reciprocal translocation between *V. galilaea* and *V. narbonensis*. Similar figures were observed in *V. hyaeniscyamus* × *V. narbonensis* hybrids. These observations strongly support Ladizinsky's suggestion that chromosome structural change is an important component of the overall divergence between these species. A single F2 individual was also produced from the *V. hyaeniscyamus* × *V. narbonensis* hybrid but this was small, abnormal and chlorotic, and failed to flower. Pollen stainability counts were also recorded and these were very low: 3.1% for *V. galilaea* × *V. narbonensis* and 2.7% for *V. hyaeniscyamus* × *V. narbonensis*. These results support the view that considerable genetic divergence has occurred between members of the *V. narbonensis* species complex. Further experimental study of a wider

Fig. 6.19. *Vicia faba*. Expression of hybrid vigour in single cross hybrid (F349-545) centre; line 349 parent (l); line 545 parent (r) (D. A. Bond).

sampling from the species concerned is very clearly indicated. The results of such studies could materially help taxonomists to distinguish the component species of the complex with greater confidence. Present evidence indicates that effective isolating mechanisms may well have developed between divergent forms which still retain a high level of morphological similarity. This of course makes for considerable difficulty in identifying, with acceptable reliability, the species of the complex. Perhaps particular study of morphological intermediates between the typical forms of the species would be especially useful.

The difficulty of producing viable inter-specific hybrids between

Fig. 6.20. *Vicia faba*: performance of 4 cultivars, Blaze, Alfred, Ticol, Sti (left to right); 4 plants on right irrigated; 4 plants on left unirrigated. Note consistent performance of determinate cultivars under both régimes (D. A. Bond).

members of section *Faba*, even those which show the strongest karyotype and biochemical resemblances, enables us to appreciate and understand why it has so far proven impossible to produce any viable inter-specific hybrids with *V. faba*. The latter shows very marked karyotype and biochemical differences from other sect. *Faba* species. This also indicates that the prospects for using successfully conventional hybridisation techniques in the attempt to mobilise the genetic resources of the section to improve *V. faba* may well prove to be very bleak indeed. The possible use of genetic engineering techiques is also fraught with difficulty. The question then becomes, is the game really worth the candle? When extant genetic resources have not been adequately collected, conserved, characterised and evaluated, and a monumental task has to be done in this regard, biotechnology perhaps looks like a short cut. This it is not; our knowledge of the genetic map of *V. faba* itself is inadequate let alone that of other species in sect. *Faba*. The prospects of carrying out effective DNA manipulations are at present vanishingly small. We are still very far from doing with flowering plants what is now routine with prokaryotes.

Fig. 6.21. *Vicia faba*: early maturing winter bean, cv. Punch, and standard cv. Banner (D. A. Bond).

Fig. 6.22. *Vicia faba*: range of growth and fruiting habits shown by four spring bean cultivars (D. A. Bond).

6.4. The lentil (*Lens culinaris* Med.)

Introduction

The position of the lentil in agriculture is rather curious. It is a crop which can be produced in environments which are, agriculturally speaking, marginal; nevertheless it produces a pulse which is acceptable far beyond the areas in which it is actually grown. It is perhaps the combination of the high regard in which it is held by the consumer coupled with the fact that it can be grown satisfactorily in areas in which other hardy crops such as the chickpea cannot be grown (Cubero, 1981) that has maintained this crop in cultivation. The nutritional value of the crop is high, and levels of toxic and anti-metabolic materials are low, so therefore it is highly desirable that this very palatable pulse is retained in cultivation. The optimal climate for the crop is basically temperate. It is essentially a winter crop of the Mediterranean area and matures in the early summer when rainfall is low. Its optimal maturation under dry (even arid) conditions explains some features of the distribution of its cultivation. It is grown in the Indian sub-continent on residual soil moisture but it is not grown in northern Europe where wet conditions at maturity would be very detrimental.

In terms of production and trade, most of the crop is consumed in the major areas of production with over two thirds of the crop being grown in Asia and with comparatively little being sold commercially. The international market, though small, is comparatively stable. The production of growers in North America, Europe and the USSR is commercially orientated rather than to subsistence. Recent developments in dietary fashions favouring high-fibre foods and plant foods rich in protein can be expected to increase demand for lentils.

Biosystematics

Taxonomy

The genus *Lens* is one of the five constituent genera of the tribe Vicieae (the others being *Pisum*, *Lathyrus*, *Vavilovia* and *Vicia*). The vegetative morphology is decidedly vetch-like with pinnate leaves commonly bearing 8–10 leaflets and with a terminal tendril. The past confusion over the status of the genus and its constituent species has been discussed by Cubero (1981) and the following is a summary of his conclusions. Miller is accepted as the author of the genus *Lens*, which has survived attempts to merge it with *Vicia*. It is a small genus comprising only 6 recognised species; one of these is a recent addition to the list, *L. odomensis* Ladizinsky (1986).

The relationship between *L. orientalis* and *L. culinaris* is particularly close; Barulina as long ago as 1930 suggested that *L. orientalis* rep-

Fig. 6.23. The lentil, *Lens culinaris*, cv. Copticum (from Westphal, 1974).

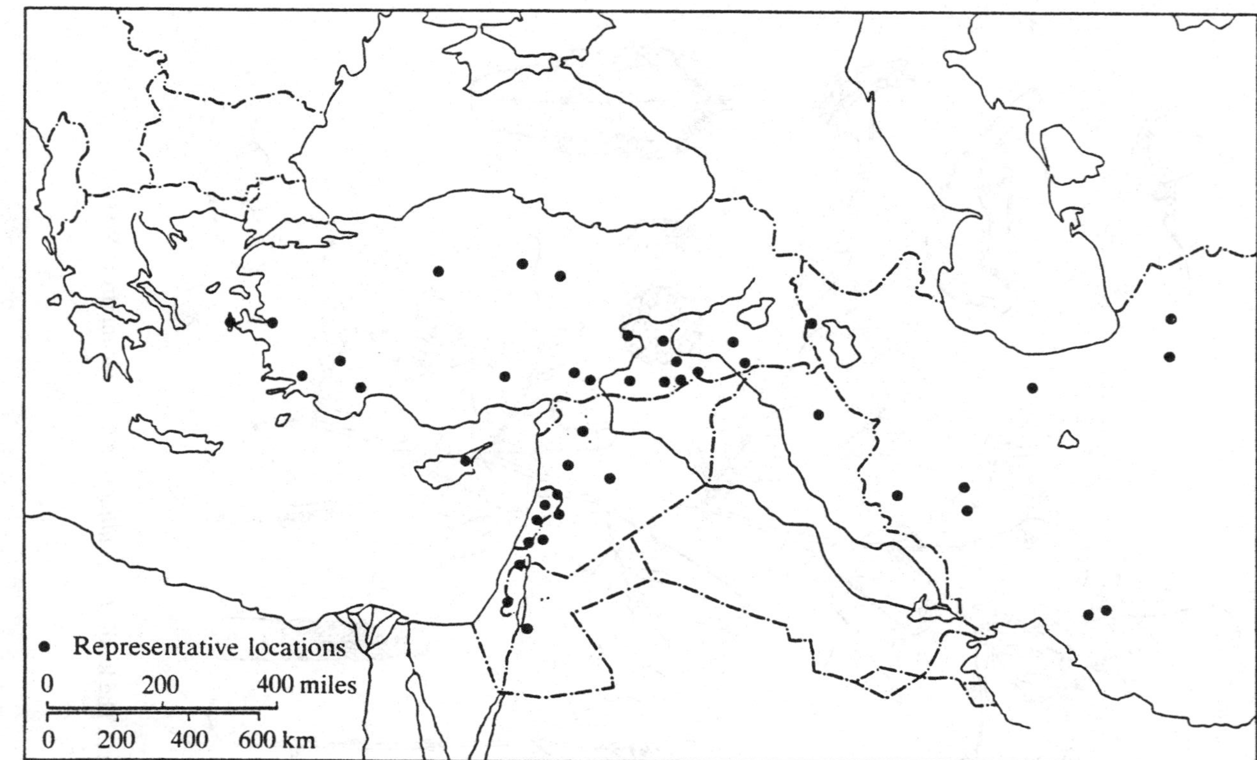

Fig. 6.24. Distribution of wild lentil, *Lens culinaris* ssp. *orientalis* = *L. orientalis* (from Zohary and Hopf, 1973.)

Table 6.4. Lens *taxonomy*

genus Lens Miller
1. *L. montbretii* (Fisch. & Mey.) Davis & Plitmann
2. *L. ervoides* (Brign.) Grande
3. *L. nigricans* (Bieb.) Godron
4. *L. orientalis* (Boiss.) Handel-Mazzeti
5. *L. culinaris* Medikus
6. *L. odomensis* Ladizinsky

resented the wild ancestral type of *L. culinaris*. This has led some authors (Williams *et al.*, 1974; Smartt, 1986) to suggest that these wild and domesticated forms be included in *L. culinaris* with sub-specific rank. Barulina's view of this relationship has been confirmed experimentally by Ladizinsky (1979*a*).

The distribution of the genus is basically Mediterranean but there are significant differences in the distribution of individual species. As its name implies *L. orientalis* has an eastern distribution, from Turkey and Israel eastward to Uzbekistan in the southern USSR. *L. nigricans* has a distribution largely on the northern shores of the Mediterranean from Israel to Spain and also in Algeria, Morocco and the Canary Islands. *L. ervoides* has a somewhat similar but on the whole more restricted distribution with interesting outlier populations in Ethiopia and Uganda. *L. montbretii* has the most restricted distribution, in the eastern Mediterranean around the headwaters of the Tigris and Euphrates. This distribution suggests that the eastern Mediterranean is the probable area in which the cultigen was originally domesticated.

Cytotaxonomy and hybridisation

Ladizinsky (1979*a*) has carried out some very significant experimental hybridisation studies between the domesticated *L. culinaris* and the two wild species *L. orientalis* and *L. nigricans* in all possible combinations, *L. culinaris* × *L. orientalis*, *L. culinaris* × *L. nigricans* and *L. orientalis* × *L. nigricans*. *L. orientalis* and *L. nigricans* are the species morphologically closest to the cultigen and while most authorities prior to Ladizinsky's study (e.g. Williams *et al.*, 1974) favoured *L. orientalis* as the ancestral form of the cultigen, Renfrew (1973) favoured *L. nigricans*. The establishment of *L. nigricans* as the progenitor type would have tended to favour a more westerly origin of the cultigen. Ladizinsky (1979*a*), however, has resolved this question beyond reasonable doubt in support of Barulina's (1930) suggestion and hence a probable eastern Mediterranean or West Asian origin for the cultivated lentil. In his study he has demonstrated that development of isolating mechanisms has occurred between *L. culinaris* and *L. nigricans* but not with *L. orientalis*. Hybrids *L. culinaris* × *L. orientalis* may be of very high fertility, whereas those of *L. culinaris* × *L. nigricans*, although not completely sterile, are much less fertile.

Chromosome number is $2n = 2x = 14$. Karyotype differences have been established between *L. culinaris and L. nigricans* (4 pairs metacentric and 3 pairs acrocentric) while karyotypes of *L. culinaris* and *L. orientalis* are very similar with 2 large sub-metacentric, 2 metacentric and 3 acrocentric pairs. Certainly a good case can be made for regarding the two forms *L. culinaris* and *L. orientalis* as essentially conspecific and recognising them as sub-species of *L. culinaris*. There is also a suggestion from Cubero (1981) that not only should *L. orientalis* be merged with *L. culinaris* but also *L. nigricans* and *L. ervoides*. This seems to be a rather extreme expedient. Cubero suggests that all belong to the primary gene pool of *L. culinaris* because somewhat fertile hybrids can be produced. In effect what is being suggested is to abolish the distinction which

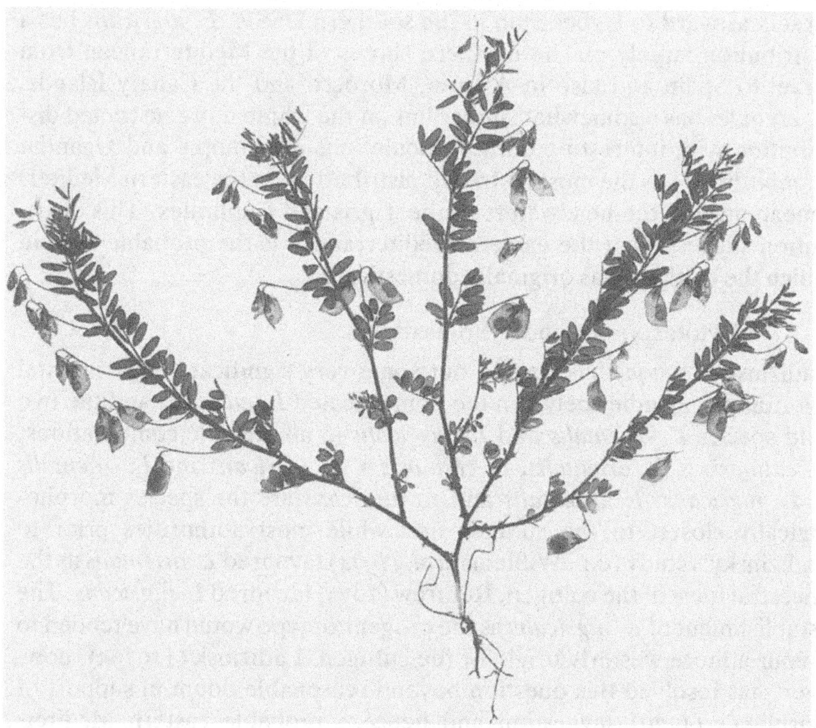

Fig. 6.25. Wild lentil (*Lens culinaris* ssp. *orientalis*) (D. Zohary).

Harlan and de Wet (1971) make between primary (GP1) and secondary (GP2) gene pools. The reduction by ±70% in pollen fertility observed by Ladizinsky (1979*a*) in hybrids *L. culinaris* × *L. nigricans* coupled with differentiation of the karyotypes suggests that gene flow would not be free between the two forms and would probably be accomplished with some difficulty. The situation is quite comparable with that of the *Phaseolus* species *P. vulgaris* and *P. coccineus*, which produce somewhat fertile F1 inter-specific hybrids and between which limited gene flow is possible (*vide* Chapter 4). A further point is that the pollen fertility of *L. culinaris* × *L. nigricans* F1 hybrids is most probably at or close to the lower limit at which conventional pollination techniques can be used to achieve reciprocal backcrosses to parents.

Genetic resources

The lentil possesses a primary gene pool with both wild and domesticated components. The species *L. nigricans* constitutes a secondary gene pool, since Ladizinsky (1979*a*) demonstrated a degree of fertility in F1 inter-specific hybrids between it and the cultivated lentil. The remaining species probably constitute a tertiary gene pool.

Fig. 6.26. Wild lentil (*Lens nigricans*) in natural habitat (G. Ladizinsky).

Chemotaxonomy

Although some biochemical studies of *Lens* species have been undertaken (Harborne *et al.*, 1971; Ladizinsky, 1979*c*) these have been largely concerned with elucidating relationships between genera rather than within *Lens* itself. Further comparative studies of the seed proteins of *Lens* species have been a useful extension of the experimental hybridisation and cytogenetic studies already carried out and deepen our appreciation of the nature of the species differences that have evolved (Pinkas *et al.*, 1985). These have also provided support for conclusions derived from studies of experimental hybridisation.

Domestication

The area of domestication indicated is in the eastern Mediterranean basin, in the Fertile Crescent, probably contemporaneously with that of the cereals.

Archaeology

This crop has an impressive archaeological record, well reviewed and summarised by Cubero (1981). Some of the oldest remains of food plants are of lentils dated 7500–8500 BC. The identification of this material as

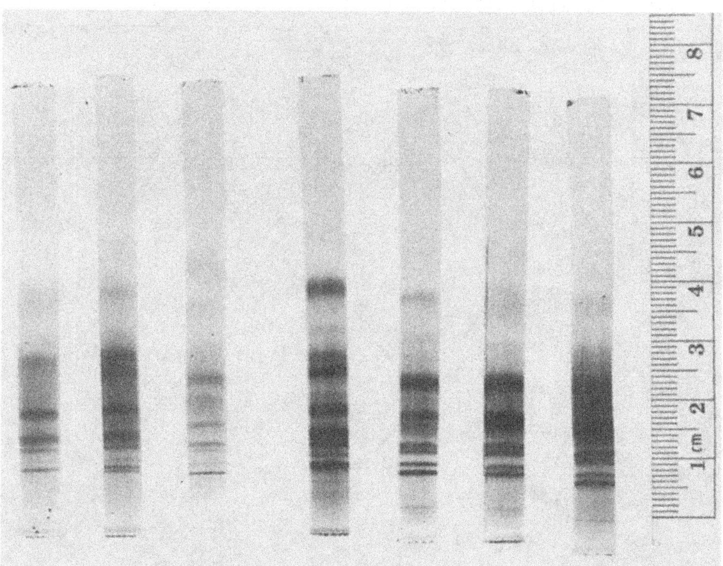

Fig. 6.27. PAGE seed protein profiles of *Lens* species. (Reproduced from Ladizinsky, 1979*c*. Copyright University of Chicago Press.)

wild or cultivated is not certain. This is an apparently insuperable problem, particularly as regards cultivated small-seeded *microspermus* material (distinct large-seeded *macrospermus* and small-seeded *microspermus* types are known in cultivation). Only if seed size is appreciably larger than that of wild species can domestication reasonably be inferred. Even so, domestication has been suggested for material dated at 7000–8000 BC. A good picture also emerges from the archaeological record of the probable post-domestication spread of the crop from the Near East (where cultivation is thought to have started) to North Africa, India, central and western Europe. It is possible that lentils had achieved virtually their present range in the Old World by about three thousand years ago. Earlier accounts of archaeological material include those of van Zeist (1976), Zohary (1972), Zohary and Hopf (1973) and Renfrew (1973).

Evolution

The nature of the evolutionary changes which have come about in the lentil are closely similar to those observed in other pulses. A lax prostrate growth form has given rise to a form with an ascending growth habit, much more amenable to cultivation. Changes paralleling those of other pulses are found: loss of seed dormancy and loss of pigmentation in flowers and stems are typical, as is the loss of patterning and deep pigmentation of the seed coat. It is a little surprising that the genetic control of these differences is as well and perhaps better understood in the lentil than most other pulses.

Studies reported by Ladizinsky (1979*b*, 1985) on genetic analysis of the differences between wild and cultivated populations of the lentil give a very clear picture of the nature of genetic divergence which has occurred. With regard to growth habit erect versus prostrate, a single locus with incomplete dominance seems to control this difference, giving 1:2:1 segregation in F2. The control of pod dehiscence was found to be a single locus effect with dehiscent pods dominant; a 3:1 ratio of dehiscent: indehiscent was found in F2. Hard-seededness was found to be controlled by a single recessive gene in the ssp. *culinaris* × ssp. *orientalis* cross. Interestingly, in the inter-specific cross *L. nigricans* × *L. culinaris*, hard-seededness was determined by a dominant allele. Genetic control of epicotyl and flower pigmentation is similar: pigmentation is produced by dominant alleles giving purplish epicotyls and blue flowers, the corresponding recessive states being green epicotyls and white flowers. A single-factor genetic difference was also shown to produce the difference between the brown-spotted wild type testa colour and pattern and the self-coloured grey of the cultigen parent.

These studies by Ladizinsky have been very informative indeed and indicate not only the genetic nature of the evolutionary changes which have been established in the lentil but in all probability that of other grain legumes as well.

Further evolutionary potential

The lentil is unique in that it is a legume crop for which there is a steady demand yet its future in agriculture is problematical. This uncertainty arises from the lack of spectacular success in improving its productivity. It has been relegated to the status of a crop of marginal environments; it is therefore worth considering what can be done to improve its productivity in just these environments. Measurement of productivity almost invariably concentrates on production of seed in the case of grain legumes. The quest for ever higher harvest indices is not necessarily justified for all such crops. In the case of the soyabean it is easy to justify high harvest index selection since the leaves are shed at maturity and the residual value of the straw is not great. In the case of healthy groundnut crops, which retain much leaf at maturity, the haulm has a high nutritive value for livestock. At present this material is lost in combining but not in traditional harvesting techniques. The economic value of such by-products to the farmer should not be ignored. In marginal agriculture environments where lentils are grown the value of the haulm to the farmer should not be discounted. Agricultural exploitation of marginal environments is something we have to live with and it really behoves us to make it as effective and non-destructive as possible. All too frequently there is a tendency to write off such environments and the crops which have been relegated to them. It is unrealistic to entertain the same expectations of productivity as are appropriate to mesic environments. A different set of criteria are clearly needed. These should recognise the fact that potential biological productivity is lower and attempt to set realistic levels for productivity targets and devise ways of achieving them.

Attempts to bring about improvements in crops of marginal environments are often misconceived and inadequately executed. Selections may well be made for yield, quality and other attributes in environments in which the crop would not actually be grown. There might well be much more attractive crops for these areas. Such selection work should be carried out under actual conditions of production and concurrent agronomic investigations also cairried out. It is important that the *forte* of a crop should be recognised and as far as possible exploited. In the case of the lentil it would seem that this is the exploitation of marginal environments in the production of modest yields of grain and fodder. Framing of research objectives in this revised context is clearly indicated. It is to be expected that the full range of available genetic resources would need to be carefully evaluated prior to effective utilisation.

6.5. The chickpea (*Cicer arietinum* L.)

Introduction

The chickpea is one of the world's major pulse crops, which apparently was domesticated in the Fertile Crescent some 7000 years ago (Ramanujam, 1976). It is cultivated from the Mediterranean basin to the Indian sub-continent and southward to Ethiopia and the East African Highlands. It has been introduced to the Americas where it has gained particular significance in Mexico. However, it is in the Indian sub-continent that the crop achieves its greatest importance.

Biosystematics

Taxonomy

The chickpea is the sole cultigen in the genus *Cicer* of any economic significance. It belongs to the monogeneric tribe, the Cicereae, which was formerly included in the Vicieae, but which Kupicha (1981) regards as having closer affinities with the Trifolieae. In Kupicha's view the genus comprises some 40 species. Van der Maesen (1972) monographed the genus, but before the wild progenitor type of the cultigen had been collected and identified (Ladizinsky, 1975*a*). It is a moot point whether the discovery of the progenitor, named *Cicer reticulatum* Ladizinsky, actually adds to the number of species in the genus since it clearly belongs to the same biological species as the cultigen (Ladizinsky and Adler, 1976*a*). The taxonomic breakdown of the genus produced by Ladizinsky has been generally accepted and is summarised in Table 6.5 (Witcombe and Erskine, 1984).

Cytotaxonomy, hybridisation and genetic resources

Biosystematic relationships between the cultigen and its annual wild relatives have been studied experimentally by Ladizinsky and Adler (1976*b*). The species studied were *C. arietinum*, *C. reticulatum*, *C. echinospermum*, *C. judaicum*, *C. pinnatifidum*, *C. bijugum* and *C. cuneatum*. The chromosome number of all these species was found to be $2n = 2x = 16$. The possibility of gene transfer from wild species to cultigen was examined and Ladizinsky and Adler concluded that such transfer was only possible, by direct hybridisation, with *C. reticulatum* and *C. echinospermum*. Other combinations failed to produce viable hybrids. Interspecific pollination probably resulted in fertilisation but only small empty pods were produced by species other than *C. reticulatum* and *C. echinospermum*. On the basis of their results Ladizinsky and Adler (1976*b*) assigned species they studied to three crossability groups; Group I containing *C. arietinum*, *C. reticulatum* and *C. echinospermum*, Group II

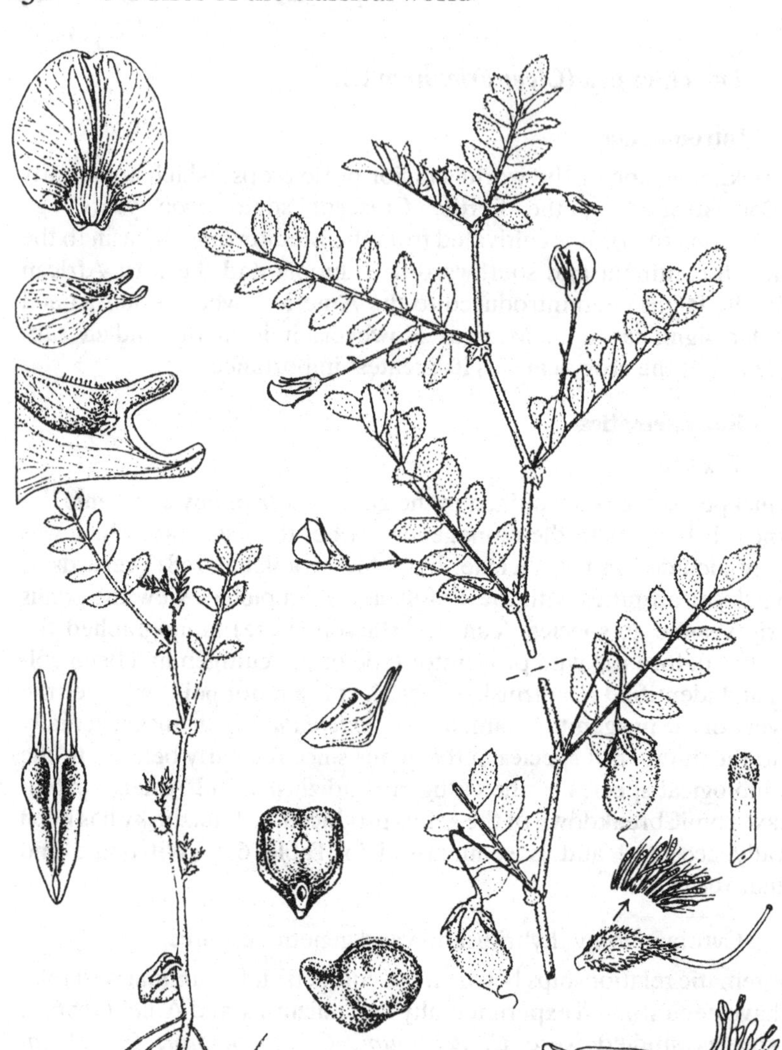

Fig. 6.28. The chickpea, *Cicer arietinum*, cv. Abyssinico-nigrum (from Westphal, 1974).

(a)

(b)

Fig. 6.29. Desi and kabuli types of chickpea (R. J. Summerfield).

Table 6.5. *Taxonomic breakdown of the genus* Cicer *L.*

Genus *Cicer* L.
- Sub-genus *Pseudononis* M. Pop.
  - Section 1 *Monocicer* M. Pop.
    - Series *Arietina* Lincz.
      - *C. arietinum* L.
      - (*C. reticulatum* Ladizinsky)
      - *C. bijugum* K. H. Rech.
      - *C. echinospermum* P. H. Davis
      - *C. judaicum* Boiss.
      - *C. pinnatifidum* Jaub. et Spach
    - Series *Cirrhifera* M.
      - *C. cuneatum* Hochst.
    - Series *Macro-aristae* van der Maesen
      - *C. yamashitae* Kitam.
  - Section 2 *Chamaecicer* M. Pop.
    - Series *Annua* van der Maesen
      - *C. chorassanicum* (Bge.) M. Pop.
    - Series *Perennia* Lincz.
      - *C. incisum* (Willd.) K. Maly
- Sub-genus *Viciastrum* M. Pop.
  - Section 3 *Polycicer* M. Pop.
    - Sub-section *Nano-polycicer* M. Pop.
      - *C. atlanticum* Coss. ex Maire
    - Sub-section *Macro-polycicer* M. Pop.
    - Series *Persica* M. Pop.
      - *C. kermanense* Bornm.
      - *C. oxyodon* Boiss. et Hoh.
      - *C. spiroceras* Jaub. et Spach.
      - *C. subaphyllum* Boiss.
    - Series *Anatolo-Persica* (M. Pop) Lincz.
      - *C. anatolicum* Alef.
      - *C. balaericum* Galushko
    - Series *Europaeo-Anatolica* M. Pop.
      - *C. floribundum* Fenzl.
      - *C. graecum* Orph.
      - *C. heterophyllum* Contandriopoulos
      - *C. isauricum* P. H. Davis
      - *C. montbretii* Jaub. et Spach
    - Series *Flexuosa* Lincz.
      - *C. baldshuanicum* (M. Pop) Lincz.
      - *C. flexuosum* Lipsky
      - *C. grande* (M. Pop) Korotk.
      - *C. incanum* Korotk.
      - *C. korshinsky* Lincs.
      - *C. mogoltavicum* (M. Pop) Karoleva
      - *C. nuristanicum* Kitamura

Table 6.5. (*cont.*)

Series *Songorica* Lincz.	
C. feldtschenkoi Lincz.	
C. multijugum van der Maesen	
C. paucijugum Nevski	
C. songaricum Steph. ex DC.	
Series *Microphylla* Lincz.	
C. microphyllum Benth. in Royle	
Section 4 *Acanthocicer* M. Pop.	
Series *Pungentia* Lincz.	
C. pungens Boiss.	
C. rechingeri Podlech.	
C. stapfianum K. H. Rech.	
Series *Macracantha* Lincs.	
C. macracanthum M. Pop.	
C. acanthophyllum Boriss.	
C. garanicum Boriss.	
Series *Tragacanthoidea* Lincz.	
C. tragacanthoides Jaub. et Spach.	
Species not assigned by van der Maesen (1984) to sections	*C. laetum* Rassulova and Sharipova *C. rassouloviae* Linczevski

After van der Maesen (1984).

C. judaicum, *C. pinnatifidum* and *C. bijugum* while Group III consisted of a single species, *C. cuneatum*. Within groups, hybrids could be obtained but fertility was variable; hybridisation was not successful between members of different groups. A conclusion which has been drawn from results of the crosses within Group I is that there are no apparent barriers to gene flow between *C. arietinum* and *C. reticulatum*; in fact crosses between these 'species' were no more difficult to make than hybrids between cultivars of *C. arietinum*. It was, however, much more difficult to produce hybrids within Group I which involved *C. echinospermum*. Although difficult to produce, viable hybrids were obtained, but were sterile. Meiotic configurations in the hybrid suggest that the parental species differ by a major reciprocal translocation. In addition these authors suggest that since a single interchange is not a sufficient explanation for hybrid sterility there is also cryptic structural hybridity in the F1. The F1 is not totally sterile; some F2 individuals have been produced but although these are also highly sterile they have nevertheless yielded backcross progeny with the cultigen. It may be possible to induce some gene flow from *C. echinospermum* to other species in Group I.

Although these studies of inter-specific hybridisation are not complete,

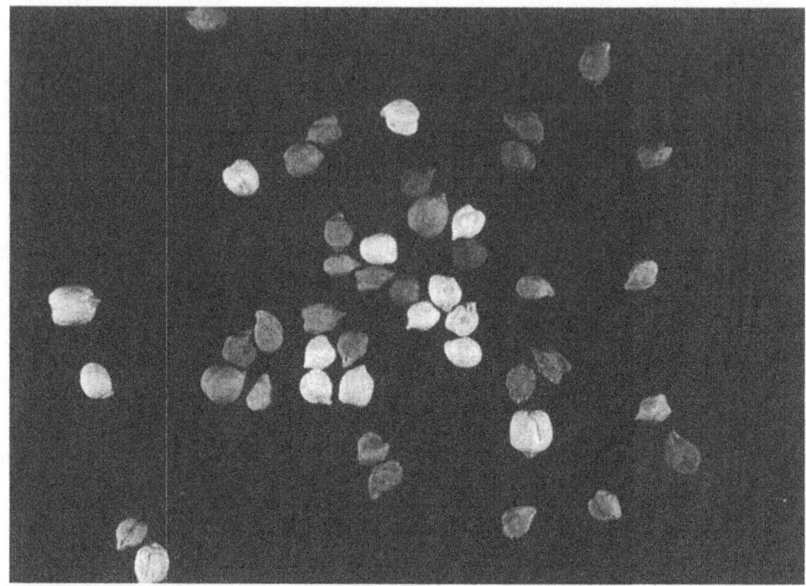

(a)

(b)

Fig. 6.30. Chickpea seed: (*a*) range of types; (*b*) comparison with garden pea (R. J. Summerfield).

it does seem very probable that the exploitable gene pool for the improvement of *C. arietinum* is confined to series *Arietina* in section *Monocicer*, sub-genus *Pseudononis* of the genus, and comprises only half of the species in the series. This situation is typical of the legumes where it has been pointed out that the range of related species genetically accessible to cultigens is usually very small. Isolating mechanisms are usually very well developed between species within genera of the Leguminosae (Smartt, 1979). In terms of Harlan and de Wet's (1971) gene pools, the primary gene pool (GP1) comprises *C. arietinum* (GP1A, the domesticated component) and *C. reticulatum* (GP1B, the wild component); the secondary gene pool (GP2) apparently consists of *C. echinospermum* while the remaining species tested can be assigned to the tertiary gene pool (GP3) which may well include many, if not all, of the remaining *Cicer* species. Van der Maesen (1984) appears to be somewhat reluctant to accept *C. reticulatum* as the progenitor type; Ladizinsky and Adler (1976,, *b*) have no such doubts. Although the changes which have occurred in the genetic makeup and morphology of the progenitor species are not known, the fact remains that *C. arietinum* and *C. reticulatum* belong to the same

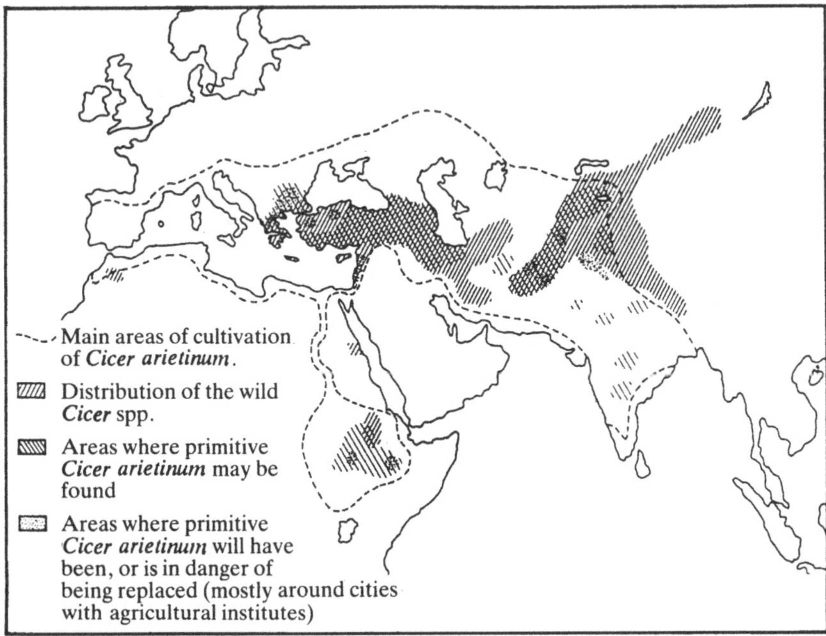

Fig. 6.31. Distribution of chickpeas (wild and cultivated) in the Mediterranean basin (from van der Maesen, 1984).

(a)

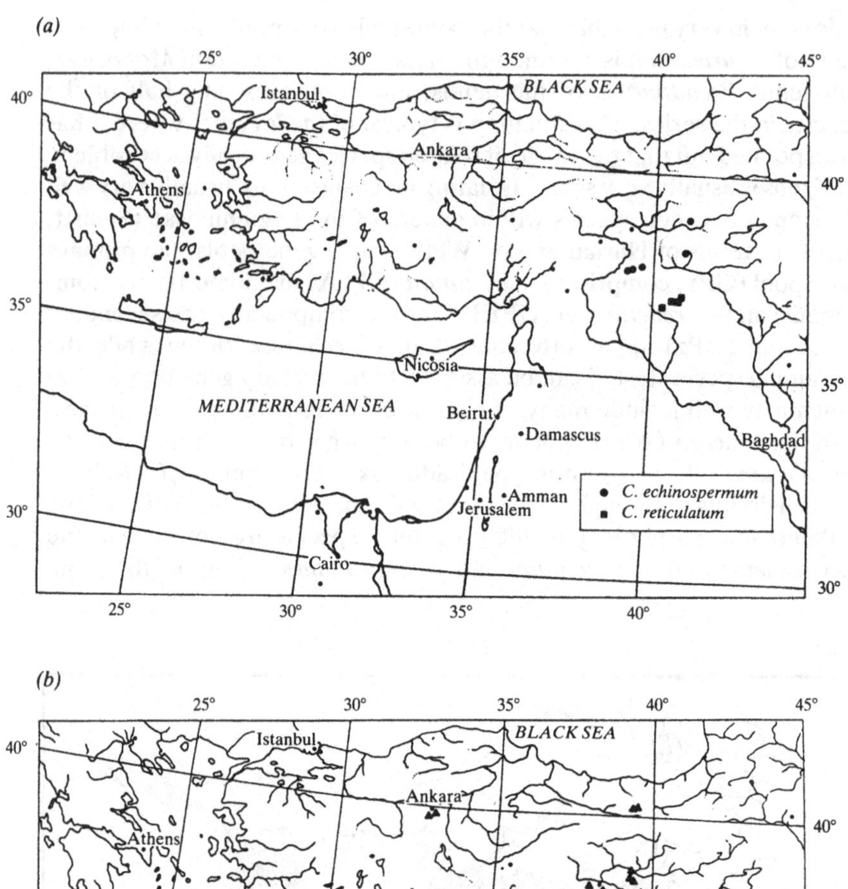

(b)

Fig. 6.32. Distribution of wild *Cicer* species. (*a*) *C. echinospermum* and *C. reticulatum*; (*b*) *C. judaicum*, *C. pinnatifidum* and *C. bijugum* (from Ladizinsky and Adler, 1976*b*).

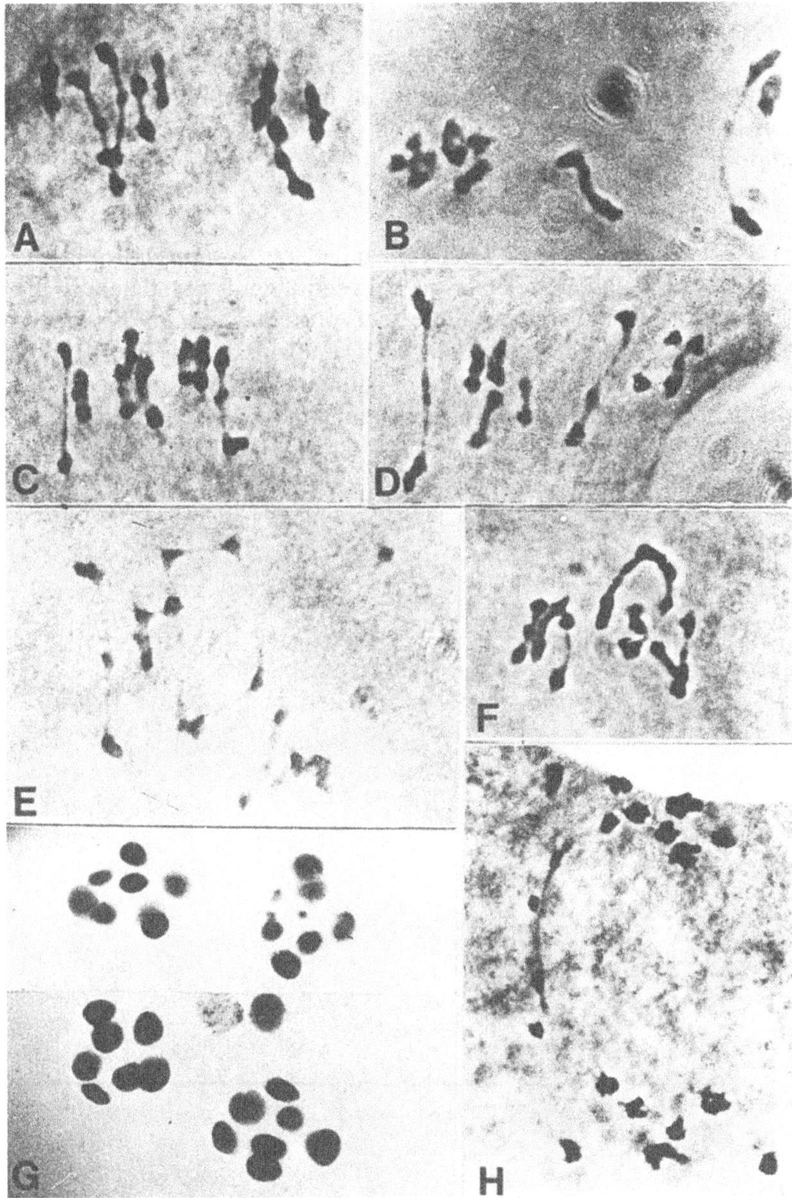

Fig. 6.33. Meiotic behaviour of *Cicer* species and hybrids. A, *C. arietinum*, 8 bivalents (8II); B, *C. reticulatum*, 8II; C, *C. echinospermum*, 8II; D, *C. arietinum* (no. 77) × *C. reticulatum*, 8II; E, *C. reticulatum* × *C. echinospermum*, 6II + IV; F, *C. arietinum* (58F) × *C. reticulatum*, 6II + IV; G, PMCs producing 7–9 meiotic cell products instead of the normal 4 in *C. reticulatum* × *C. echinospermum*; H, A bridge and fragment in *C. areitinum* (58F) × *C. echinospermum*, (From Ladizinsky and Adler, 1976*a*.)

biological species. In its general morphology, physiology and genetics *C. reticulatum* represents as good an approximation, in all probability, to the progenitor type as we are likely to see. One must also bear in mind the fact that ancestral populations could well have been polymorphic and patterns of domestication complex so only in broad general terms can we attempt to characterise crop plant ancestors.

Chemotaxonomy

The close affinity between *C. arietinum* and *C. reticulatum* shown by the lack of genetic isolating mechanisms between them is paralleled by the close similarity in the water-soluble seed protein fractions of the two forms, demonstrated electrophoretically by Ladizinsky and Adler (1975). The seed protein profiles examined of cultivars within *C. arietinum* are very uniform; this is interpreted as indicating a monophyletic origin of the cultigen.

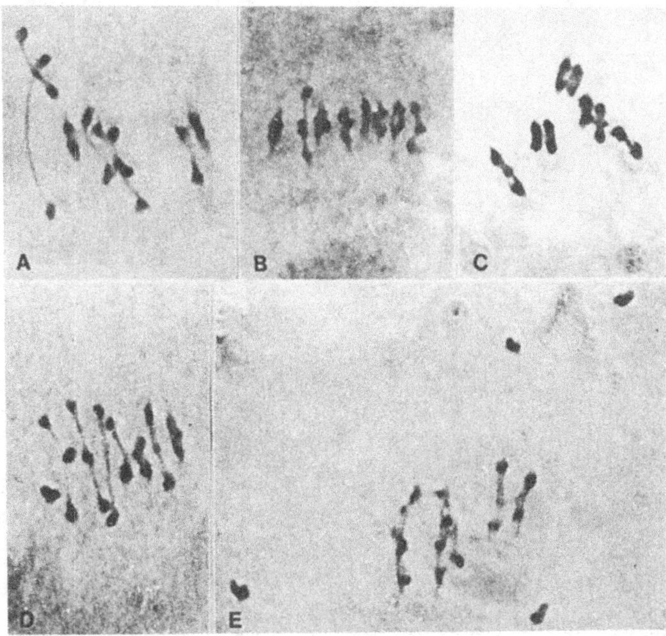

Fig. 6.34. Meiosis in some additional *Cicer* species and their inter-specific hybrids showing varying degrees of disturbance at metaphase I. A, B and C, 8II in parental species *C. pinnatifidum*, *C. judaicum* and *C. bijugum*; D, 7II + 2I in *C. pinnatifidum* × *C. pinnatifidum* F1 hybrid; E, 6II + 4I in *C. judaicum* × *C. pinnatifidum* F1 hybrid. (From Ladizinsky and Adler, 1976*b*.)

Domestication and evolution

If, as seems likely, *C. reticulatum* is regarded as the progenitor type of *C. arietinum*, then the location in which domestication occurred can be presumed to be in the general area of the present range of the wild form. This pre-supposes that the range of *C. reticulatum* has not changed dramatically within the past ten millennia. Anatolia in Turkey is the area in which the crop is thought to have originated (van der Maesen, 1984).

Seeds of the chickpea have only been occasionally found on prehistoric sites in the Near East (Renfrew, 1973). The shape of the seed with its prominent beak is conducive to damage, especially in the carbonised state (van der Maesen, 1972). The relatively poor archaeological record might be due to the difficulty of identifying damaged carbonised seed, which could be difficult to distinguish from that of the common pea. Ramanujam (1976), however, reports finds of chickpea radiocarbon dated at 5450 BC and that there is evidence for its cultivation in the Mediterranean basin 3000–4000 BC, but quotes no authority. Material from northern India has been dated at 2000 BC; from the south much later dates have been obtained (Chowdhury *et al.*, 1971): AD 200–150

Fig. 6.35. A novel leaf form in *Cicer arietinum*. This typifies the kind of genetic variant of potential breeding value which can be maintained in germplasm collections (R. J. Summerfield).

BC. Linguistic study of northern and southern Indian names for the chickpea has shown that these are not related and Ramanujam (1976) suggests that whereas the northern areas received their chickpeas overland, those of the south could have come by a sea route.

The biosystematic evidence already considered places *C. arietinum* closest to *C. reticulatum* although, as van der Maesen (1984) points out, on morphological grounds alone *C. bijugum* and *C. echinospermum* are as close to *C. arietinum* as is *C. reticulatum*. It is clearly apparent that

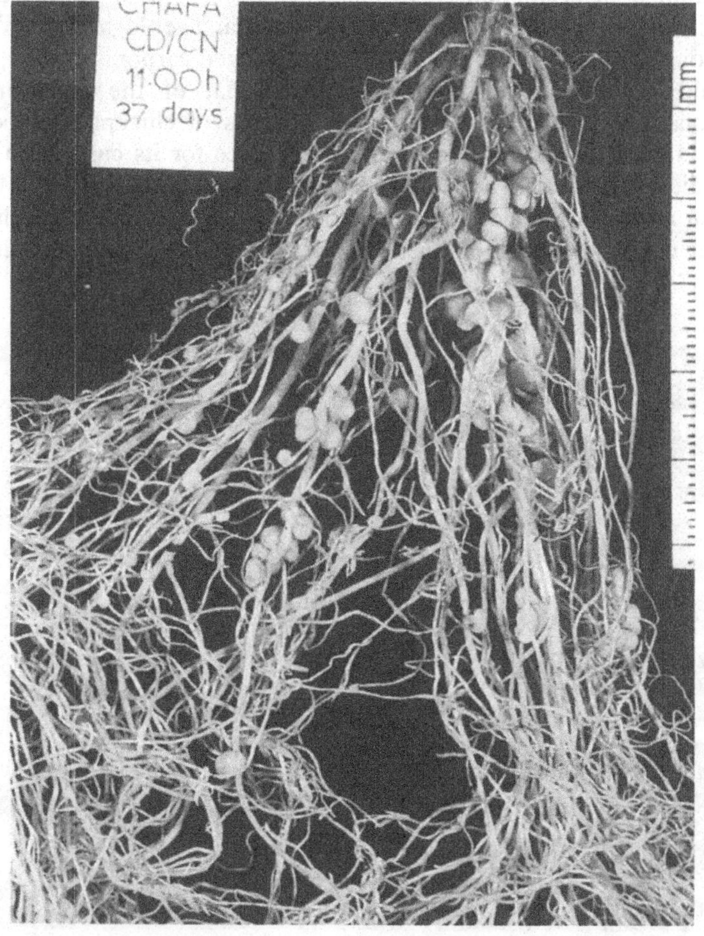

Fig. 6.36. A nodulated root system of the chickpea. High levels of nodulation and efficient symbiotic nitrogen fixation are increasingly important selection objectives (R. J. Summerfield).

isolating mechanisms must have evolved between *C. reticulatum* and the other wild species prior to domestication occurring. However, there is evidence to suggest that chromosome structural change, which is a significant component of the isolating mechanisms between *C. arietinum* and *C. echinospermum*, has also occurred within the cultigen (Smithson *et al.*, 1985), the accession 58 differing from *C. reticulatum* (and other accessions of *C. arietinum*) by a paracentric inversion and an interchange. This is not an uncommon situation within pulse species.

In common with other pulses the cultigen differs from its wild relatives principally by a more erect growth form and pods with reduced dehiscence. However, there is an interesting and peculiar feature of wild *Cicer* species which is suppressed in the cultigen. Seed dispersal in the wild is apparently a two-stage process: pods are shed from the plant and subsequently burst on the ground (Ladizinsky and Adler, 1976*b*). In the cultigen, pods are retained on the parent plant and dehiscence is reduced.

Under domestication two major forms have emerged, termed 'desi' and 'kabuli' types (the *microsperma* and *macrosperma* races of Moreno and Cubero, 1978). In addition, 'gulabi', pea-shaped forms are also recognised and are locally important. Desi chickpeas are small and angular with rough, brown to yellow testas, while kabuli types are relatively large, plump and with smooth, cream-coloured testas. Kabuli chickpeas can be regarded as the more advanced by virtue not only of their larger size but also of their reduced pigmentation, both characteristics produced by sustained selection. Although loss of pigmentation is thought to improve nutritional value of the crop it tends to increase susceptibility to pest and disease attack. Distribution of the two types may be related in part at least to distribution and severity of pest and pathogen attack.

The primary climatic adaptation of the chickpea is to a winter rainfall climate (Mediterranean) in which it may be winter–spring sown and mature its crop on residual soil moisture. It can be grown outside its area of origin where similar conditions prevail. It could very well be successful in the semi-arid tropics as a dry season crop in areas with a relatively high water table such as the flood plains of large rivers. It does not succeed as a crop either where heat and moisture stress develops or under conditions of heavy rainfall. One the puzzling features of the distribution of cultivation of this crop is its lack of northern penetration in Eurasia in marked contrast to the pea (*Pisum sativum*) and the faba bean. It is possible that success in northern Europe is too dependent on summer rainfall conditions; it may not succeed in the all too prevalent wet summers and autumns of the area.

Cultivation world wide is predominantly (85%) of the desi type in the Indian sub-continent, Ethiopia, Mexico and Iran. The kabuli type is found to a lesser extent in Indian cultivation but predominantly in the geographic area from Afghanistan through West Asia to North Africa and southern Europe, and in American cultivation apart from Mexico. It is obviously worth while to encourage and extend, if possible, the cultivation of this crop. However, as a result of Green Revolution technology in cereal cultivation, cultivation of chickpea and other pulses has declined. It is obviously desirable that balanced agriculture be the development goal rather than cereal monoculture; it is therefore desirable that high-yielding productive cultivars be developed which can economically maintain a position in improved farming systems. Indications from reported maximum yields of 5000 kg ha^{-1} (Duke, 1981) are that there is scope for significant improvement in yields by selection of more responsive genotypes to any agronomic improvement that can be brought about.

6.6. Comparative evolution, genetic resources and further potential

It appears that the five Mediterranean pulses originated in a fairly well-defined area of the eastern Mediterranean basin. This is clearly indicated on the basis of archaeological evidence. These five pulses have, however, developed two distinctly different patterns of distribution subsequent to their domestication. The common pea and the faba bean have shown the greatest northward spread and can be cultivated over virtually the whole of Europe. The grasspea has achieved only a limited northward spread, possibly due to a marked preference for the common pea. It has achieved a remarkable spread to the east due probably to its drought tolerance and in spite of the concomitant lathyrism problem. The distribution of the lentil's cultivation is broadly similar to that of the grasspea; little or no penetration of northern Europe has been achieved. The distribution of the chickpea in cultivation is not dissimilar to that of the lentil and grasspea; if anything its northward spread is even more restricted. The reasons for the failure of the lentil and chickpea to penetrate farther north in cultivation is probably related to the duration of growing season required. In northern latitudes during the summer good vegetative growth and flowering may be achieved but satisfactory maturation of pods and seeds does not occur reliably in cool, moist autumn conditions. In the most successful areas of cultivation maturation occurs under conditions of moderate temperatures; in northern areas, maturation tends to occur when temperatures and day lengths are declining rapidly. Under these conditions no worth while crop matures. Equally, under rain-fed conditions in the tropics, as in plateau Africa, satisfactory vegetative growth is achieved but maturation of the crop fails;

in addition, insect attack on developing pods may be severe (J. Smartt, unpublished). The case of the grasspea is rather different: growth and maturation of the crop could occur much farther to the north but there seems to be insufficient incentive to cultivate this crop. Somewhat similar species *L. odoratus* and *L. latifolius* flourish in north-west Europe as ornamentals.

With the exception of *Vicia faba*, the progenitor type of all the Mediterranean pulses has been identified and there seems to be a general consensus that wild and domesticated populations be recognised as subspecies within a single biological species. The grasspea however, although found in the wild and in cultivation, has either been insufficiently studied or has undergone too little differentiation under domestication for such a distinction to be made. The situation in *V. faba* is perplexing but further collections may, as Zohary (1977) suggests, provide an answer, when these are subjected to cytogenetic study.

The generally satisfactory state of biosystematics for Mediterranean pulse crops means that the general range and extent of germplasm resources can be gauged for the common pea, lentil and chickpea. According to Harlan and de Wet's gene pool system, these three species have both primary and secondary gene pools, with both wild and domesticated components of the primary pool extant. Tertiary gene pools doubtless exist but their extent is only tentatively known. The faba bean as far as is known has a domesticated primary gene pool, no known wild primary gene pool, no secondary gene pool and possibly an extensive tertiary gene pool. Understandably there is little incentive to investigate potential tertiary gene pools since these are very extensive, likely to be difficult to exploit and unlikely to be of much practical value in the immediate future.

Linguistic studies of vernacular pulse crop names have been carried out by de Candolle (1886) on lentils and Hanelt (1972*b*) on faba beans; curiously, in both cases the Greek names are distinct from Latin and other Indo-European names. This is intriguing but sheds little light on the domestication and dissemination of these pulses.

In common with pulses from other parts of the world, considerable parallel evolution has occurred. In general, seed size has responded to selection pressures and in *V. faba* very large seeds of ±1 g in weight have evolved. Suppression of seed dispersal mechanisms (by explosive pod dehiscence) has become general. Less rampant growth habits have become established in domestication, although the less rampant climbing or trailing growth habits may still be maintained in cultivation. Another common evolutionary trend is the reduction of pigmentation in flowers, stems and seeds. The more highly evolved cultivars tend to have white rather than violet flowers, pale (white or cream) seeds rather than dark (black or brown) and green rather than pigmented stems. The over-riding consideration here may be the seed character, with which the others are correlated to a greater or less degree. Testa pigments are often fungistatic (because of their phenolic nature) but can reduce palatability. Selection for improved palatability and tenderness could result in selection for reduced pigmentation. Selection against seed dormancy in domesticated populations has by and large been successful; hardseededness has virtually been eliminated.

Further improvement of pulses generally includes such features as improved yield, quality, pest and disease resistance; these individual aspects may of necessity have to be broken down further. Factors contributing to improved yield must be identified and for individual crops some determination made of the actual room for improvement which exists. For example there may or may not be scope for improving the light-intercepting efficiency of the leaf canopy; there equally may or may not be scope for improving the partitioning of assimilates between vegetative and reproductive sinks, etc. Quality improvement may be a relatively simple matter of selecting for improved 'cookability', the elimination or reduction in concentration of toxicants or anti-metabolites such as the incitants of lathyrism and favism. The ability of individual crops to respond to the new selection pressures that may be imposed upon them will determine their fugure in cultivation. This ultimately depends on the ability of each species to generate the right kind of variability and our own to conserve it, identify and incorporate it when required.

It is also appropriate to keep under review our selection and agronomic research strategies. We should not be afraid to re-think them when they appear to be failing. It is all too easy to abandon research on potentially valuable crops where failure to achieve progress has arisen from failure to appreciate the basic biology of the crop itself, resulting in the formulation of unrealistic and misconceived objectives.

7 The other legume oilseeds

At the present time there is a very lively interest in oilseed crops as a whole. This arises from a number of causes but principally the fact that the supply of animal fats and oils has been reduced as a consequence of conservation measures taken in the interests of whale populations. Perhaps more importantly, the role of saturated animal fats in human diets and nutrition has been called in question. This has led to an increased demand for polyunsaturated plant edible oils, which the supply of long established traditional oilseed crops has not been able to meet. This has meant that oils which twenty years ago were used almost entirely in industry are now extensively used in the food trade. The longer-established traditional oils, from the olive, groundnut and sesame, have been substantially replaced in part at least by a later generation of oilseeds: sunflower, soyabean, rapeseed, mustard and cottonseed, for example. Production of the traditional oils could not keep up with increased demand; there seems to be little prospect, for example, of a substantial increase in olive oil production. The very high culinary quality of oils such as olive and groundnut oils and their high cash value has led to problems of adulteration. Adulteration is not always easy to detect and control, with the increased tendency for oilseeds to be expressed in the country of origin.

Not all plant oils and fats are rich in polyunsaturates: palm oils, especially coconut oil and palm kernel oil, and also cocoa butter, are high in saturated fats. Some animal lipids such as pork fat may contain appreciable contents of polyunsaturates, particularly if diet had been rich in plant material containing these. The texture of such oily or soft fat is, however, not acceptable and production of carcases with a high polyunsaturate content is not a practical proposition at present.

The status of oil-rich legume crops depends to a large extent on the alternatives to use strictly as an oilseed. The groundnut is the legume oilseed *par excellence*, with a lipid content $\pm 50\%$ of total dry matter. However, in comparison with alternative outlets for this crop the edible oil market is not the most lucrative. In the USA the peanut butter market and that for roasted peanuts and use in confectionery greatly predominate. Only low-grade material has traditionally gone to the oil extraction market. The aflatoxin problem has seriously hit groundnut production in badly affected areas since very low tolerance levels for aflatoxin have been established. Because West Africa, and particularly Nigeria, a major source of groundnuts for oil expression, has suffered in this context to a considerable extent, the general availability of groundnuts for oil has been seriously affected. Central and Southern African production for export has been geared largely to the confectionery trade in the past.

The economic position of groundnuts as an oilseed explains in part at least the interest in other legumes which produce substantial oil contents in their seeds. The soyabean at best produces only half the oil content of the groundnut; the winged bean produces roughly comparable oil contents to the soyabean. The seeds of the lupins are slightly or substantially lower in seed oil contents (*L. mutabilis* ±18%, *L. albus* ±11%) (Gladstones, 1980). The economics of oilseed production demand markets for the oil itself and for the oilseed cake. Fortunately, the long-term market prospects for both are favourable. Not only are lipids of plant origin preferred currently to animal fats but the desirability of high levels of meat consumption has also been questioned on nutritional grounds. In addition, of course, vegetarian diets can be favoured on ethical grounds.

Of the crops here considered only the soyabean is an established and fully fledged oilseed crop. The winged bean, it has been suggested, can be utilised in a basically similar fashion to the soyabean (NRC, 1981) and could be regarded as a soyabean substitute. The more oil-rich lupins could be similarly exploited. The production and marketing of the winged bean obviously requires to be established on a sound economic footing. Acceptability to the processors needs to be determined. Whereas the white lupin (*L. albus*) can be regarded as an established (if minor!) crop, *L. mutabilis* would obviously require extensive selection before it reached a level of development as a crop plant for widespread production.

7.1. The soyabean (*Glycine max* (L.) Merr.)

Introduction

The soyabean is arguably the most important grain legume. It is first in volume of production (Probst and Judd, 1973) in the USA and probably in the world as a whole. It can be exploited both as a subsistence and as a commercial crop in many different ways. The diversity of present and

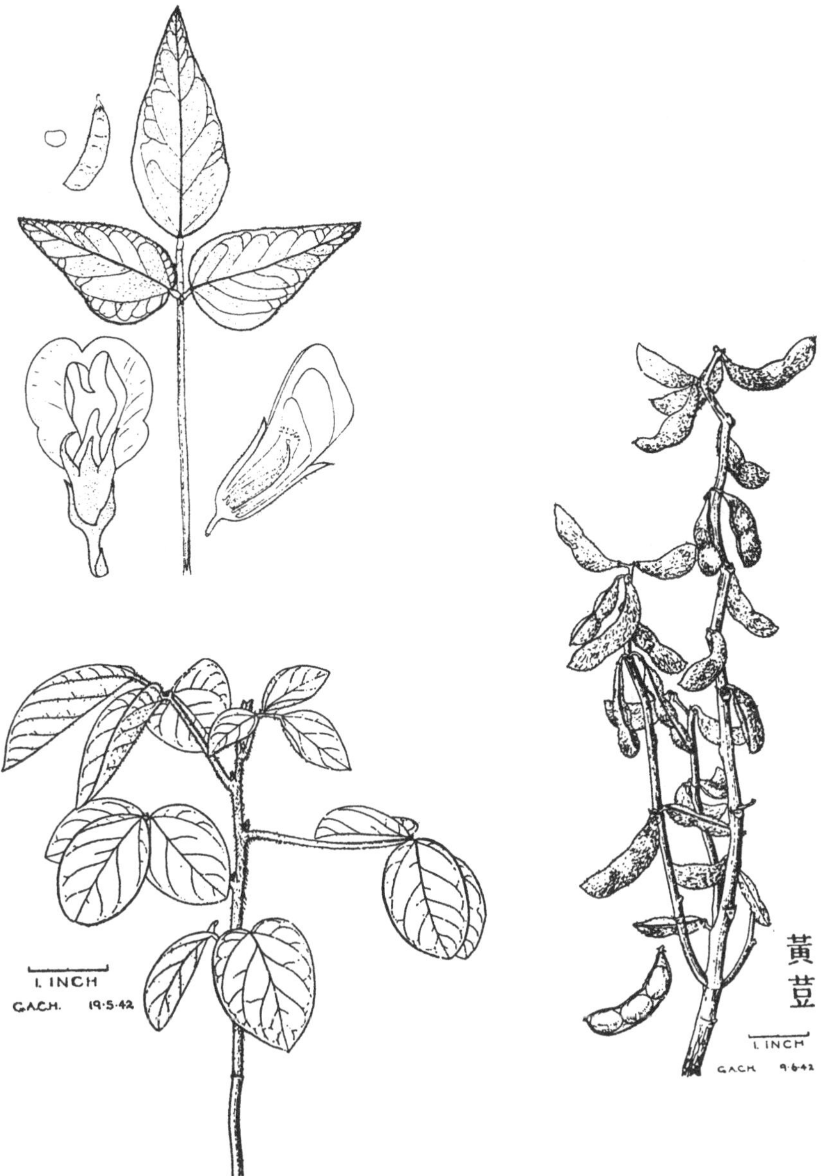

Fig. 7.1. The soyabean, *Glycine max* (from Purseglove, 1974 and Herklots, 1972).

potential uses would seem to ensure its present position in the agricultural economy indefinitely.

Biosystematics and classification

Taxonomic relationships in the genus *Glycine*

The soyabean is assigned to the genus *Glycine*, which has had a chequered career since it was established by Linnaeus (1754). This is reviewed in Hymowitz and Newell (1981) (Table 7.1). The present position has been developed by Hermann (1962), Verdcourt (1971) and Lackey (1981). The genus is assigned to the sub-tribe Glycininae of the tribe Phaseoleae. It is comparatively small and contains 9 species according to Lackey. The genus is divided (questionably in Lackey's view) into sub-genera *Glycine* and *Soja*. The latter includes the species *G. soja* and *G. max*, which breeding experiments show to comprise a single biological species. Piper and Morse (1910) regarded these as variant forms within a single taxonomic species. This view is well supported on genetical grounds (Hymowitz and Newell, 1980) and has been argued by Smartt (1986) also. Herman (1962) recognised the two species *G. soja* and *G. max*.

Cytotaxonomy and hybridisation

The relationships between *G. max* (*sens. lat.*) and other members of the genus have been studied experimentally by Newell and Hymowitz (1982).

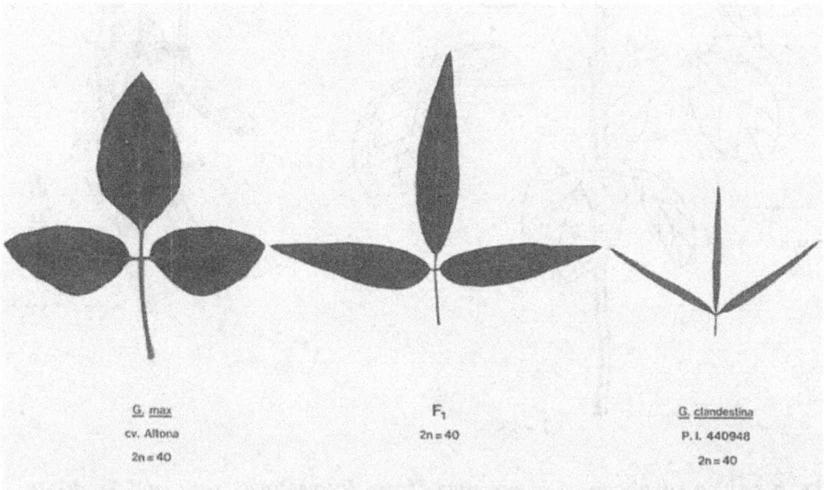

Fig. 7.2. Inter-specific hybridisation of *Glycine max* × *G. clandestina*. Herbarium specimens of leaves from parents and F1 inter-sub-generic hybrid (Singh *et al.*, 1987).

Table 7.1. *Synopsis of* Glycine *Willd.*

Genus *Glycine* Willd.
Sub-genus *Glycine*
1. *Glycine clandestina* Wendl.
var. *clandestina*
var. *sericea* Benth.
2. *Glycine falcata* Benth.
3. *Glycine latrobeana* (Meissn.) Benth.
4. *Glycine canescens* F. J. Herm.
5. *Glycine tabacina* (Labill.) Benth.
6. *Glycine tomentella* Hayata
Sub-genus *Soja* (Moench.) F. J. Herm.
7. *Glycine max* (L.) Merrill
ssp. *max*
ssp. *soja* (Sieb. & Zucc.) Ohashi

After Hymowitz and Newell (1980).

This essentially is a study of genetic relationships between the two sub-genera. These sub-genera have somewhat different geographical distributions: whereas sub-genus *Soja* (i.e. *G. max sens. lat.*) has an exclusively Asiatic distribution, that of sub-genus *Glycine* is predominantly Australian. *G. falcata*, *G. latrobeana* and *G. canescens* are exclusively Australian, while *G. clandestina* also extends to the South Pacific Islands; only *G. tabacina* and *G. tomentella* are found naturally on the Asiatic mainland (Hymowitz and Newell, 1980).

In crosses between wild (*G. soja*) and cultivated (*G. max*) soyabeans on the one hand and six wild *Glycine* species on the other (*G. canescens*, *G. clandestina*, *G. falcata*, *G. latifolia*, *G. tabacina* and *G. tomentella*) the only combination to produce viable (but sterile) progeny was *G. max* × *G. tomentella* (Newell and Hymowitz, 1982). Newell and Hymowitz (1982, 1983) have also studied intra- and inter-specific hybridisation in the sub-genus *Glycine*. Most intra-specific crosses were found to be fully fertile but within *G. clandestina*, *G. tabacina* and *G. tomentella* some genotype combinations were inviable. Inter-specific crosses gave varying results: the combination *G. canescens* × *G. clandestina* gave fertile F1 hybrids; *G. tomentella* × *G. tabacina* gave a sterile F1; *G. falcata* × *G. canescens* and *G. falcata* × *G. tomentella* gave inviable hybrids while *G. latifolia* × *G. tomentella* combinations showed seedling lethality. More recently the inter-sub-generic cross *G. max* × *G. clandestina* has been obtained (Singh *et al.*, 1987). It was produced by embryo rescue but was sterile, pods were set on backcrossing to *G. max* but all aborted.

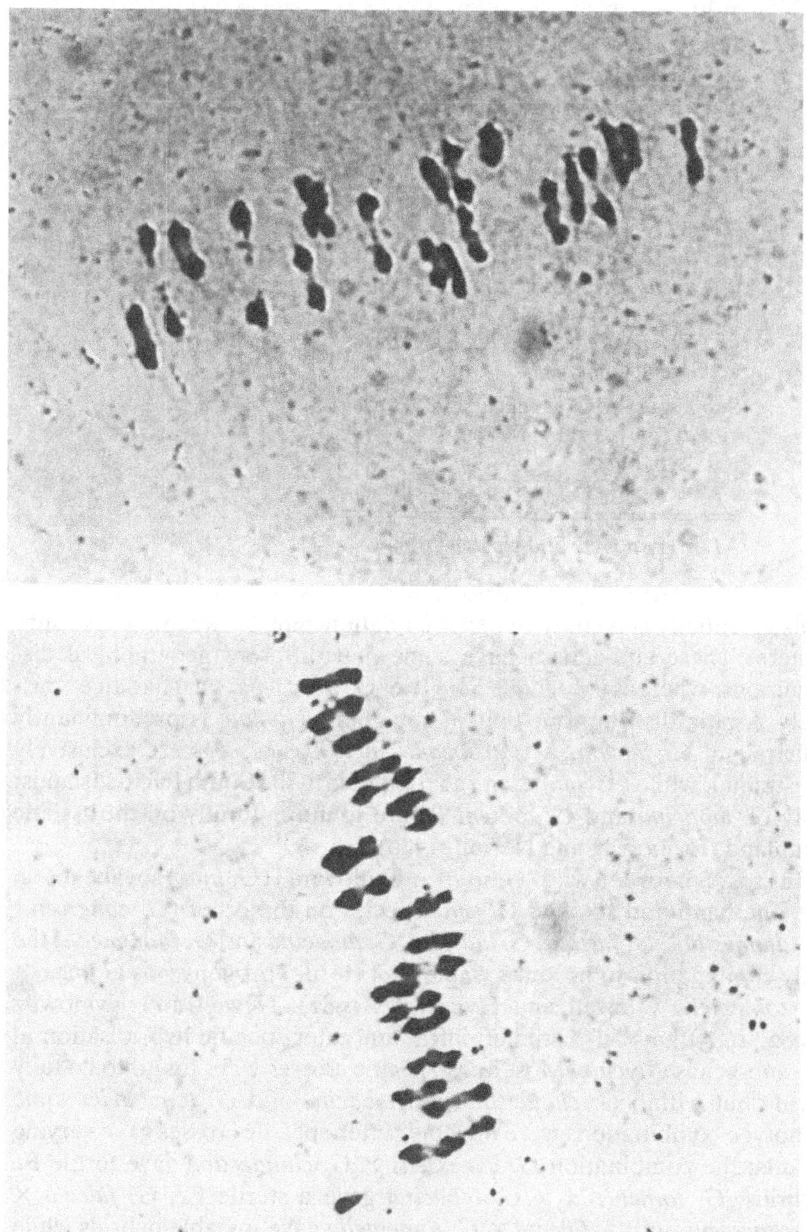

Fig. 7.3. *Glycine canescens*: two metaphase I plates. (The appearance of meiotic chromosomes in all $2n = 40$ *Glycine* species is similar.) (T. Hymowitz.)

The indications from these crosses are that some gene exchange between sections might be possible via *G. tomentella* but that this would not be easy to achieve. Genetic isolation between most species of the genus appears to be very well developed and gene transfers would require genetic engineering techniques. This suggests that the genus is ancient, with an Australian centre of distribution and origin from which it has spread to the Asian mainland via the islands of the South Pacific, and the Philippines. An interesting feature is the high chromosome number of *G. tabacina* and *G. tomentella*, $2n = 8x = 80$. These are presumed to be octoploids because the chromosome base numbers in the Phaseoleae are commonly $x = \pm 11$. As a general rule the Leguminosae are less prone to generate polyploid forms than the Gramineae and so *Glycine* must be regarded as somewhat exceptional in this respect.

The genetic resources available for the improvement of the soyabean consist of a GP1, with both wild and domesticated components, and a GP3. No GP2 has been shown to exist. This position is similar to that in the *Phaseolus* species *P. acutifolius* and *P. lunatus*.

The taxonomic status of the soyabean and its allies in section *Soja* is somewhat controversial. The problem arises from the ranking of the wild, weedy and cultivated forms which have been named *G. soja* Sieb. and Zucc. (*G. ussuriensis* Regal and Maack), *G. gracilis* Skvortz. and *G. max* (L.) Merr. respectively. This is precisely the same problem encountered by Maréchal *et al.* (1978) in naming the cowpea and its conspecific allies. Wild, weedy and cultivated forms are most appropriately recognised at the sub-specific level. Consistent practice in this has been very strongly urged (Smartt, 1986) in order to avoid confusion. The only evidence which has been put forward in support of giving these forms specific status is the reduced F1 fertility of hybrids between some domesticated and non-domesticated genotypes reported by Ahmad *et al.* (1979). However, chromosome polymorphisms within a species are not of themselves strong grounds for making taxonomic divisions.

Chemotaxonomy and biochemistry

Current levels of research interest in the chemistry and biochemistry of the soyabean seed are surprisingly low. Bernard and Weiss (1980) conclude that, by and large, protein and oil content are under polygenic control. However, monogenically controlled differences in seed protein properties have been observed (Larsen, 1967; Larsen and Caldwell, 1968, 1969). This suggests that there is scope for potentially informative electrophoretic studies of seed proteins should the need arise. In this connection the production of the Kunitz trypsin inhibitor studied by Orf *et al.* (1978) and Orf and Hymowitz (1979*a*, *b*) is under simple genetic control. It should be a simple matter to select against this anti-metabolite (and any others with similar genetic control) and improve the ease of utilisation of soyabean seed protein. Such an improvement would certainly be worth while if no yield penalty was entailed.

Overall contents of protein and oil are inversely related (Brim, 1980). Since in the past protein content has been considered the more important, no selection has been embarked on to improve oil content. There is scope perhaps for some improvement in quality by reduction in content of linolenic acid and a compensating increase in the content of other unsaturated fatty acids. To carry out this manipulation would obviously require further basic biochemical information to extend the scope of work that has already been done (Hume *et al.*, 1985). There does seem to be some prospect of improving oil quality in another way: Hildebrand and

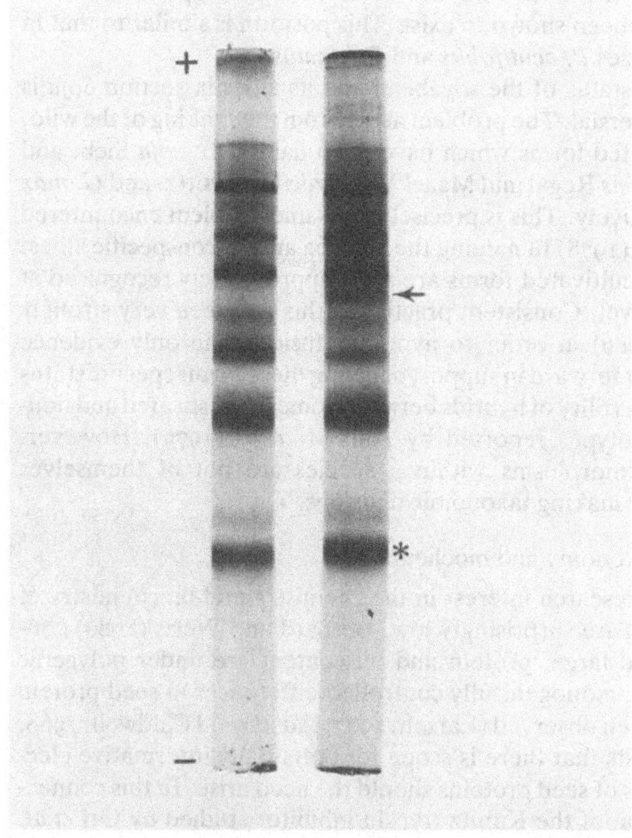

Fig. 7.4. *Glycine max* seed protein extract electrophoetogram (PAGE) demonstrating presence (r) and absence (l) of a lectin (from Orf *et al.*, 1978).

Hymowitz (1981) have identified genotypes lacking lipoxygenase-1. Stability of the oil is improved in such genotypes, and incidence of rancidity is diminished. The incentives to effect changes are higher now than they have been in the past. Improvement in protein content and quality are logical breeding objectives in most edible seed crops; however, the soyabean would seem to be something of a special case. Protein content and quality (in terms of amino acid profile) are both high in comparison with those of other grain legumes. It is pertinent to ask whether much improvement in either or both of these is likely to be possible or worth the effort to achieve it. The objective of selecting for improved methionine content in soyabeans (and other grain legumes), as well as that of select-

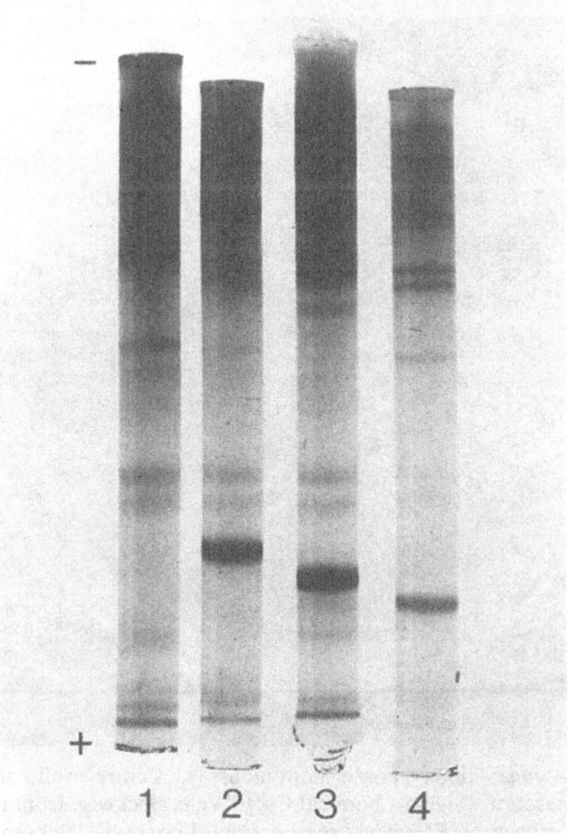

Fig. 7.5. *Glycine max*: PAGE gels of seed extracts showing, left to right: 1, absence of Kunitz trypsin inhibitor; 2, 3, 4, forms of the trypsin inhibitor (strong dark band in lower portion of gel) with different electrophoretic mobilities. (From Orf and Hymowitz, 1979.)

ing for higher lysine contents of cereals, implies that it is desirable to do so. This may not necessarily be so since it could produce a disincentive to maintain a reasonable diversity of crops in farming systems. It could tend to favour equally undesirable monocultures of cereals or soyabeans. The effort might very well be better channelled in other directions.

Domestication and evolution

The present distribution of the wild soyabean (*G. soja*) covers parts of China and the eastern USSR, Taiwan, Japan and Korea. Hymowitz (1970) and Hymowitz and Newell (1981) argue that domestication of the wild soyabean probably occurred in China. Certainly it appears that the

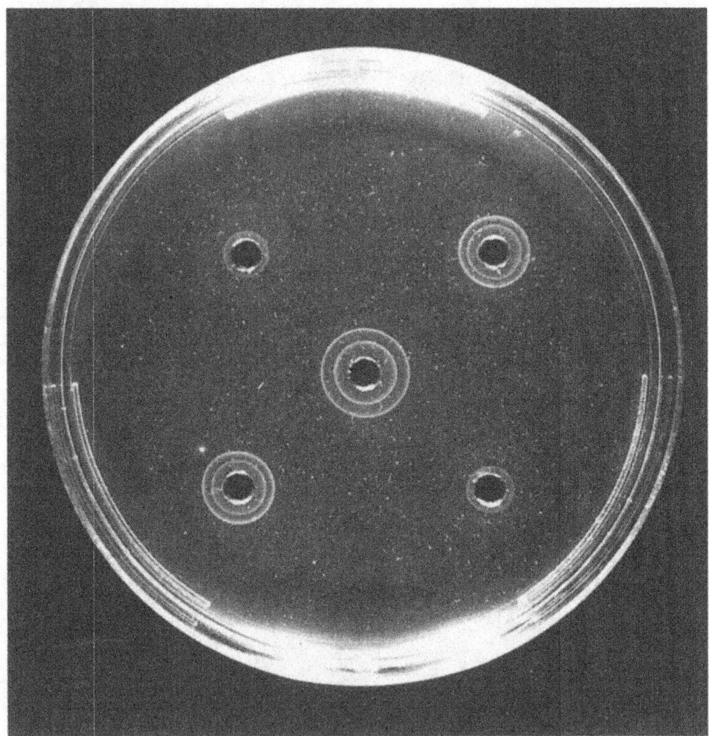

Fig. 7.6. *Glycine max*: lipoxygenase immunoassay. Centre well, standard lipoxygenase preparation (Sigma Chemical Co.). Wells clockwise from top left: 1, Williams seed extract; 2, PI 408251 seed extract, lipoxygenase absent; 3, PI 423800 seed extract; 4, PI 133226 seed extract, lipoxygenase absent. Lipoxygenase concentration affects the chemical stability of stored oils; high concentrations favour the development of off-flavours. (From Hildebrand and Hymowitz, 1981).

soyabean has been cultivated in north-east China for the past three millennia. The evidence which pinpoints the area of domestication is linguistic rather than archaeological. The evolution of the ancient character 'shu' traced by Hu (1963) suggests a date of at least the eleventh century BC. This indicates that the soyabean was established as an important crop in north-east China during the Chou Dynasty. From north-east China the soyabean spread south to central and southern China and also to Korea by the first century AD (Hymowitz and Newell, 1980). From this time to that of the voyages of discovery in the fifteenth century AD (and subsequently), soyabeans had become established in Japan, Indonesia, the Philippines, Indo-China, Thailand, the Malay peninsula, Burma, Nepal and north India. No significant archaeological finds of the soyabean, however, have been found in areas of ancient civilisation.

The introduction of the soyabean to the New World has been of the very greatest importance and this has been documented by Hymowitz and Harlan (1983). Apparently, the earliest introduction to North America was made by one Samuel Bowen in 1765. The explosion in the area of cultivation in the United States did not begin until one and a half centuries or more later. Between the two world wars, in the 1920s and 1930s, production expanded and has continued to do so until not only has soyabean production in the USA outstripped that of the Far East but it has also become the most important American cash crop (Hymowitz, 1976). Production elsewhere in the Americas has expanded and considerable production is now achieved in South America, particularly in Brazil. There is also great production potential in Africa.

The soyabean was known in Europe prior to its introduction to North America. Linnaeus (1737) described it in his *Hortus Cliffortianus*. It has so far made relatively little impact on European agriculture although a very real production capability exists in southern and central Europe. If agricultural conditions are suitable for maize production there is a strong likelihood that soyabean production would also be feasible. Although climatic conditions in northern and north-western Europe are marginal for the soyabean, success has in fact been achieved in some favourable seasons (Bowdidge, 1935). By and large the growing season is too short and harvest conditions at or about the autumnal equinox leave a very great deal to be desired in northern Europe at least. The establishment of the soyabean in Europe would be in competition with oilseed rape and sunflower, which are both well established.

The soyabean shows the typical transformation in growth form common to most grain legumes under domestication. Primitively it is a climbing or trailing vine. Mutations producing shorter internodes and erect, self-supporting growth have become established. In addition determinate and indeterminate patterns of branching have also evolved (cf. *Phaseolus* spp.). Variation also occurs in colour of both pods and seeds, which may both range from yellow through brown to black in colour. The paler colours are preferred for commercial production, but it is thought (Tindall, 1982) that the dark-seeded forms are more satisfactory performers in the tropics. This may be due to the phenolic pigments of testa and pod providing some measure of protection against fungal attack.

Perhaps the most remarkable polymorphism of the soyabean is in its photoperiod sensitivity. Virtually day-neutral genotypes are known (Shanmugasundarum *et al.*, 1980) as well as a wide range with differing photoperiod requirements. Individual genotypes range in flowering response from those that can flower in high latitudes such as Canada and northern Europe to those capable of flowering under considerably shorter days in the tropics. In North and Central America twelve maturity groups are recognised (Hymowitz, 1976); cultivars assigned to a particular maturity group tend to have a very narrow latitudinal range of successful cultivation. There are two components, nevertheless, of maturity: the first is photoperiod duration, the second is the time actually taken to mature under a given photoperiod régime. In practice photoperiod sensitivity can tend to synchronise flowering and fruiting of a genotype sown successively over a period of a month or even more. In effect this amounts to latitudinal adaptation and does have very important consequences in some conditions of ensuring the production of a crop in the monsoon tropics, for example, when the rains are late and the wet season is short. Under these conditions a crop, albeit lighter, can still be matured.

Further evolutionary potential

Hume *et al.* (1985) define four factors which determine success of soyabean cultivation. They are efficient production in the field, marketing, processing and utilisation. The soyabean is unique among grain legumes in that it is essentially an industrial crop: it requires processing before its seed can be exploited. The scale of industrial processing ranges from a cottage industry to that of a highly capital-intensive system. Processing is of two major kinds: fermentative and separative. In the former the whole seed is fermented by fungi (e.g. *Rhizopus oryzae*) or bacteria and products such as 'milk', 'cheese', tofu and soy sauce are produced with the neutralisation of the anti-metabolites and toxic materials. In the latter, basically, the lipid fraction is extracted leaving a protein-rich residue. This can be used more or less as it stands or further processed to produce textured vegetable protein and other protein products. The existence of these two distinct patterns of use must be taken into account in devising breeding policies to meet the needs of the developing and developed worlds. To illustrate this point one can consider the differences in chemical constituents required for different uses. The presence of protease inhibitors in soyabean seed is disadvantageous in the normal separative kind of utilisation. High residual active concentrations in the meal are disadvantageous. On the other hand, concentration of these and some other anti-metabolites (e.g. lectins) is virtually immaterial if a crop is to be fermented, as these materials are inactivated by microbial action. Presence or absence of anti-metabolites is a prime determinant (Orf and Hymowitz, 1979) of the amount of processing required to produce a safe cake for stock feed, say. The appropriate genetic manipulations should be straightforward and it may be possible to produce anti-metabolite-free genotypes without incurring a severe yield penalty.

Possibly the over-riding problem confronting the producer of soyabeans is yielding capacity; Hume *et al.* (1985) note that average yields in the USA are ±2.0 t ha^{-1} with a projected increase to 3.7 t ha^{-1} by the year 2002. To bring about this increase will require a further considerable effort in breeding and selection. Much of this is likely to be directed towards the production of a more efficient light-intercepting canopy. This is regarded as being much less efficient than that of the groundnut (Duncan *et al.*, 1978). High yields, once achieved, require protection from reductions exacted by pest and disease attacks. Appropriate resistances to meet local needs need to be incorporated.

At the present time, breeding objectives tend to be stated in terms appropriate to North American production. In the developing world the important objectives are good, stable yield levels combined with acceptable pest and disease resistance. The most appropriate seed quality selection objectives would depend on the production location. In the Far East the indigenous fermentation technology copes well with soyabean anti-metabolites. Unless this technology can be transferred successfully to other potential production areas, such as Africa, the introduction of the soyabean to subsistence and partly commercialised agriculture would be problematical. If it were not acceptable as a food its future could be only as a cash crop. Competition with existing large-scale producers in the Americas would be strong (except in times of shortage). Local use as food would be ruled out if local traditional cooking practices were followed, in Africa say, because to render soyabeans acceptable by boiling would necessitate unacceptably long cooking times and excessive use of fuel. Although there is considerable potential for production in the developing world, a market break-through in most areas has still to be achieved. It is regrettable that this valuable crop is even now used far below the feasible and desirable level of utilisation. The spectacular rise in American production during the present century indicates what can happen when economic and agricultural conditions are right.

7.2 The winged bean (*Psophocarpus tetragonolobus* (L.) (DC.)

Introduction

The winged bean is a crop with a great many positive attributes. It has been shown to nodulate freely and presumably fix nitrogen efficiently and to be a veritable pig of the plant world in that virtually every part of it that can be recovered is edible (NAS, 1975; NRC, 1981). First and foremost it produces a seed which is similar in composition to that of the soyabean (without some of its drawbacks). In addition it produces a pod acceptable as a green vegetable, the flowers are edible, the leaves can be plucked and used as a spinach, while young shoots can be used like asparagus. Furthermore, the root tuber it produces is not only edible and wholesome but is, for a tuber, usually rich in protein. Yet apart from cultivators in the hills of Burma and adjacent areas of India (Burkill, 1906), the New Guinea highlands (Khan, 1976) and more recently in Thailand, few seem willing to attempt field-scale cultivation. Its role appears to be that of a backyard crop. It is not entirely clear why this should be so. Available information suggests that much of its cultivation is being carried out close to its ecological tolerance limits. Optimal conditions may well be found in very few geographical areas. Moderately high elevations in the tropics with a good year-round rainfall distribution appear to meet the requirements of this crop. It would seem to be important that the nature of the factors limiting successful large-scale winged bean production be properly understood before this is attempted. At the present time there seems to be too much impatience to 'get on with the job' before it is known how the job should be tackled. Limiting factors need to be identified and problems adequately defined, and the boundaries of suitable eco-geographical areas drawn.

There may also be difficulties arising from palatability; the flavour of pods may not be generally acceptable on account of some bitterness. The seed also has a very tough testa and it is difficult to prepare dhal from mature seed as a result. Also the seed requires abrasion before sowing. Acceptability problems have perhaps been minimised; it never pays to overlook the consumer!

Biosystematics and classification

Taxonomy

The genus *Psophocarpus* has been monographed by Verdcourt and Halliday (1978). It is a small genus comprising eight African wild species plus the cultigen *P. tetragonolobus*, which is found in Asian native cultivation and in New Guinea. The relationship between the cultigen and the wild species has not been studied in depth experimentally, but on

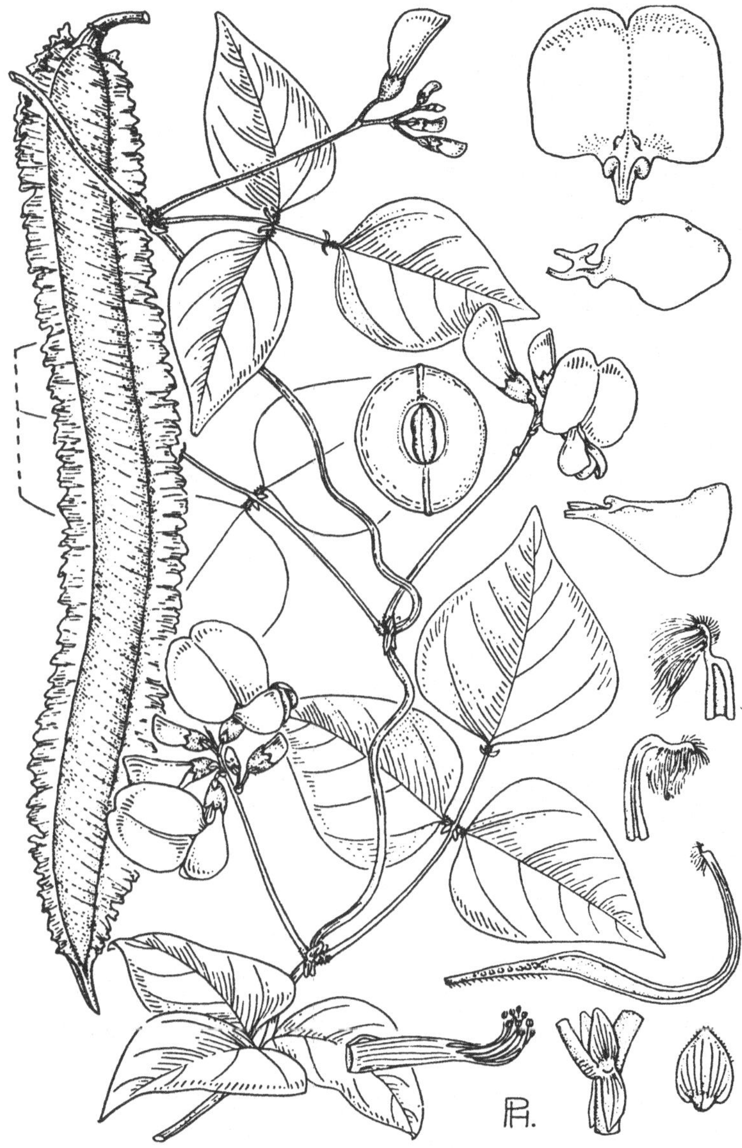

Fig. 7.7. The winged bean, *Psophocarpus tetragonolobus*. (From Verdcourt and Halliday, 1978. British Crown Copyright. Reproduced with permission of the Controller, Her (Britannic) Majesty's Stationery Office and the Trustees, Royal Botanic Gardens, Kew, © 1980.)

Fig. 7.8. *Psophocarpus tetragonolobus*: left, pods on the plant; right, range of size, shape and colour variation in winged bean pods from Papua New Guinea (T. Hymowitz and T. Khan).

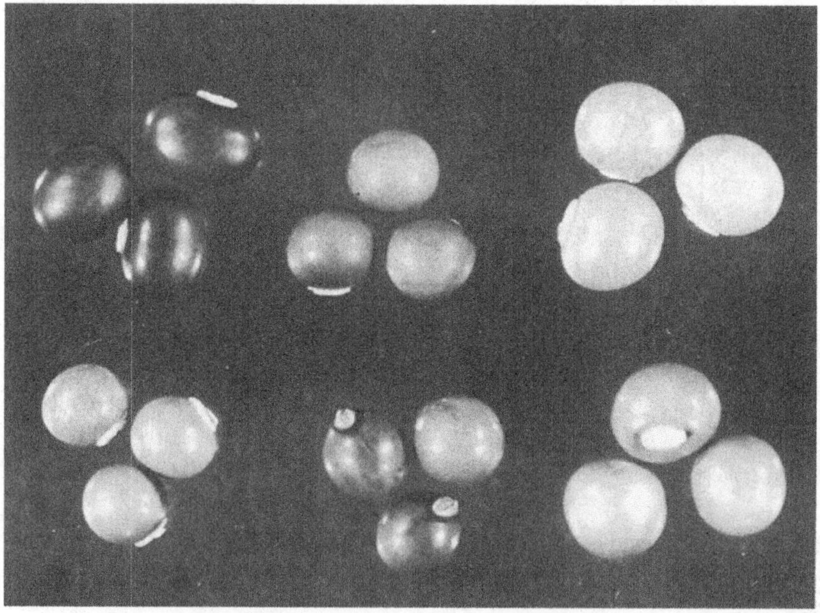

Fig. 7.9. *Psophocarpus tetragonolobus* seeds showing range of testa colours (T. Hymowitz).

morphological grounds the cultigen appears to be closest to the species *P. scandens*, *P. palustris* and *P. grandiflorus* in the section *Psophocarpus*. As Verdcourt and Halliday (1978) have observed, the genus presents little taxonomic difficulty but there has been some confusion and uncertainty in identification and naming of *Psophocarpus* species, particularly *P. scandens* and *P. palustris* and to a lesser extent *P. tetragonolobus* itself. However, in the wake of the taxonomic revision by Verdcourt and Halliday (1978), more consistency and accuracy is already apparent in identification (Table 7.2).

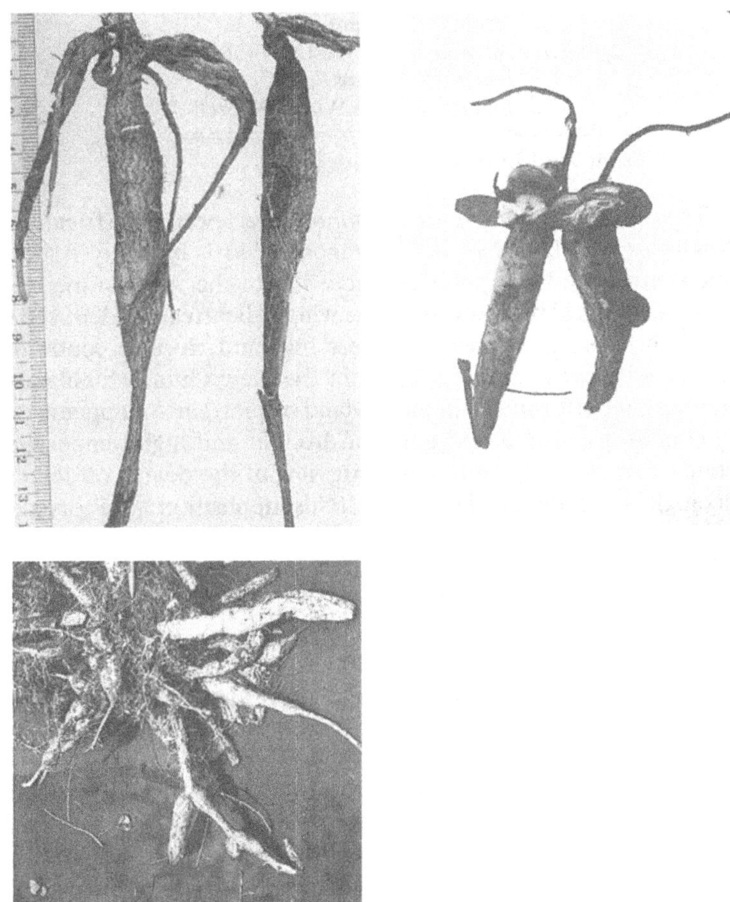

Fig. 7.10. *Psophocarpus tetragonolobus* root tubers. (From Hymowitz, 1980. British Crown Copyright. Reproduced with permission of the Controller, Her (Britannic) Majesty's Stationery Office and the Trustees, Royal Botanic Gardens, Kew, © 1980.)

Table 7.2. *Taxonomy of* Psophocarpus

Genus *Psophocarpus* Neck. ex DC.
Sub-genus *Psophocarpus*
Section *Psophocarpus*
1. *P. grandiflorus* Wilczek
2. *P. tetragonolobus* (L.) DC.
3. *P. palustris* Desv.
4. *P. scandens* (Endl.) Verdc.
5. *P. obovalis* Tisserant
Section *Unifoliolatae* A. Chev. ex Verdc.
6. *P. monophyllus* Harms
7. *P. lecomtei* Tisserant
Sub-genus *Vignopsis* (De Wild.) Verdc.
8. *P. lancifolius* Harms
9. *P. lukafuensis* (De Wild.) Wilczek

From Verdcourt and Halliday (1978).

The distribution of the wild *Psophocarpus* species is African (including Madagascar) while that of *P. tetragonolobus* is basically Asian and has been introduced comparatively recently to the African mainland. The broad ecological preferences of the winged bean can be deduced from the fact that it succeeds best in moist highland tropical conditions. This explains its considerable success in the New Guinea highlands and its indifferent performance in the lowland tropics, since it appears to flourish in conditions without extremes of drought and high temperature. This conclusion can only be tentative in view of the dearth of detailed agro-ecological knowledge of the crop. If this supposition were correct it would have a twofold effect: it would on the one hand circumscribe the area in which potential existed for intensive production of the crop while on the other it would enable promotional efforts to be concentrated on those areas in which success was most likely. There is clearly a very critical dearth of ecological knowledge of the crop which at the present time could be leading to a misdirection of effort in unsuitable areas.

Cytotaxonomy and hybridisation

It is important to bear in mind the cautionary note of Verdcourt and Halliday (1978) that references in earlier literature to *P. palustris* usually relate in fact to *P. scandens*. No successful inter-specific hybridisation has yet been reported in the genus; the attempts made so far probably have involved only *P. scandens* and *P. tetragonolobus*. For a number of years there was uncertainty about the exact chromosome number of *P. tetragonolobus*. These ranged from $2n = 26$ (Ramirez, 1960) to $2n = 18$ (Tixier, 1965; Brock, in Khan, 1976) for *P. tetragonolobus* itself and from

$2n = 18$ (Haq and Smartt, 1977) through $2n = 20$ (Frahm-Leliveld, 1960) to $2n = 22$ (Miège, 1960) for *P. scandens* (*P. palustris* in some reports). However, Pickersgill (1980) clearly showed that both species in fact have the same chromosome number, $2n \pm 18$, and very similar karyotypes. It seemed possible at one time that a chromosome numerical polymorphism might exist in the genus and even in species; this possibility can now be discounted. The variation in chromosome counts obtained by competent cytologists might well have been due, in part at least, to incorrect identification of the material examined, since for example $2n = 22$ is a very common chromosome number in the Phaseoleae.

Studies of inter-specific hybridisation have been severely limited by an extreme dearth of living material of *Psophocarpus* species other than *P. tetragonolobus* and *P. scandens*. It should not be unduly difficult to correct this deficiency except perhaps for *P. grandiflorus*, which is found wild in Zaire, Uganda and Ethiopia. It is to be hoped that the difficulties of access to parts of Uganda and Ethiopia will not persist for much longer.

At the present time, therefore, exploitable genetic resources can be considered as comprising only those of the cultigen itself. No conspecific wild relative has, as yet, been identified and no cross-compatible wild species producing viable (let alone fertile) inter-specific hybrids are known. The genetic resource situation for the winged bean is closely similar to that of the faba bean.

Chemotaxonomy and biochemistry

Chemotaxonomic studies of the genus are hampered by the same shortages of wild species material that bedevil other experimental biosystematic studies. However, the seed of the winged bean itself has been the subject of quite intensive recent study (Salunkhe *et al.*, 1982; Mossé and Pernollet, 1982; Jolivet and Mossé, 1982; Liener, 1982). This interest has been stimulated in large measure by the broad similarity in the chemical composition of the winged bean seed to that of the soyabean. Protein contents and amino acid profiles are similar and broadly the same kinds of anti-metabolites occur in the seeds of both species, although at lower concentrations in the winged bean. Similar treatments (e.g. moist heat) are effective in neutralising these compounds. Protein content of the seed is high, up to 35% crude protein; the seeds have a lipid content of 15–20% and oil extraction is feasible. Interestingly, although the oil content is similar to that of the soyabean, in its fatty acid composition it resembles groundnut oil, which commercially is more valuable. It would be interesting to ascertain how effectively winged bean oil could be substituted for groundnut oil as a salad and frying oil. Its shelf-life would depend on contents of anti-oxidants, such as tocopherols, which reportedly are high (Hymowitz and Boyd, 1977), since it is quite high in polyunsaturated fatty acids. The residue from oil expression is a high-protein meal which has been used experimentally with success in infant feeding. The content of flatus factors is also low relative to soyabean (NRC, 1981). There is no doubt that, in terms of both protein and lipid production, winged bean seeds are potentially of great value; this might be further improved by judicious selection.

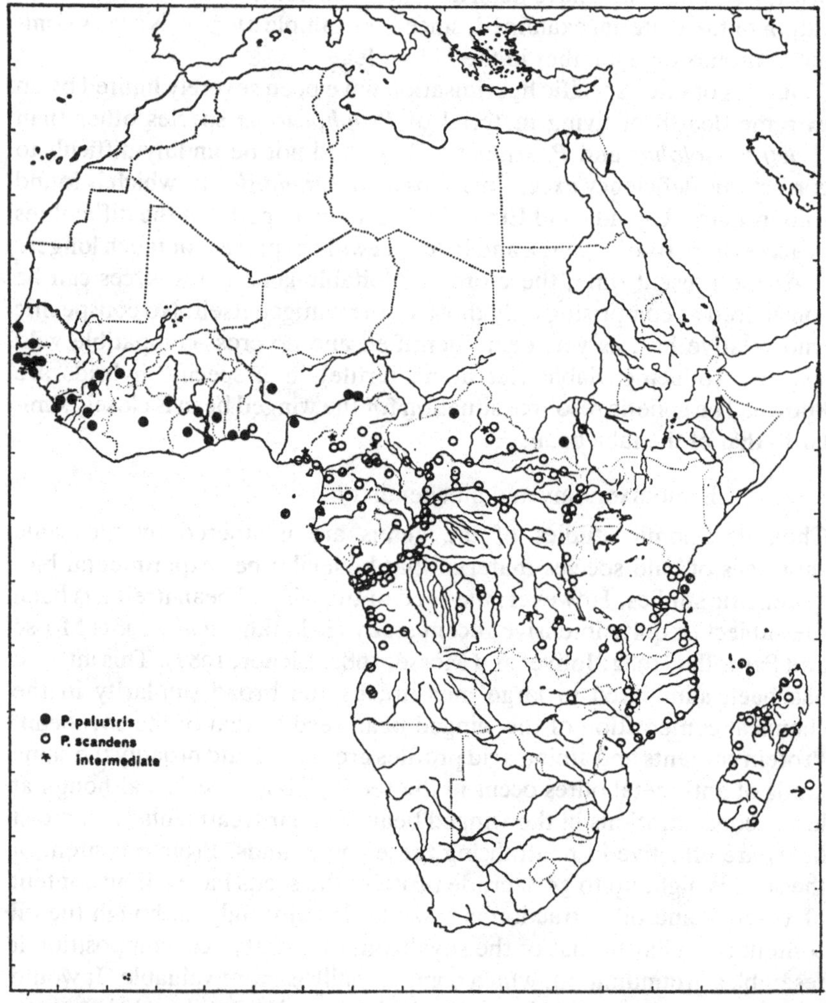

Fig. 7.11. Distribution maps of *Psophocarpus* species. (From Verdcourt and Halliday, 1978. British Crown Copyright. Reproduced with permission of the Controller, Her (Britannic) Majesty's Stationery Office and the Trustees, Royal Botanic Gardens, Kew, © 1980.)

Another feature of considerable nutritional significance is that the root tuber, when produced, has an exceptionally high protein content (8–20%) compared with common root crops such as cassava and potatoes (1–5%); the best widely cultivated tropical tubers, taro and yam (*Dioscorea*), may produce 6–8 and exceptionally 12% crude protein. An interesting feature of the amino acid profile of the tubers is their high

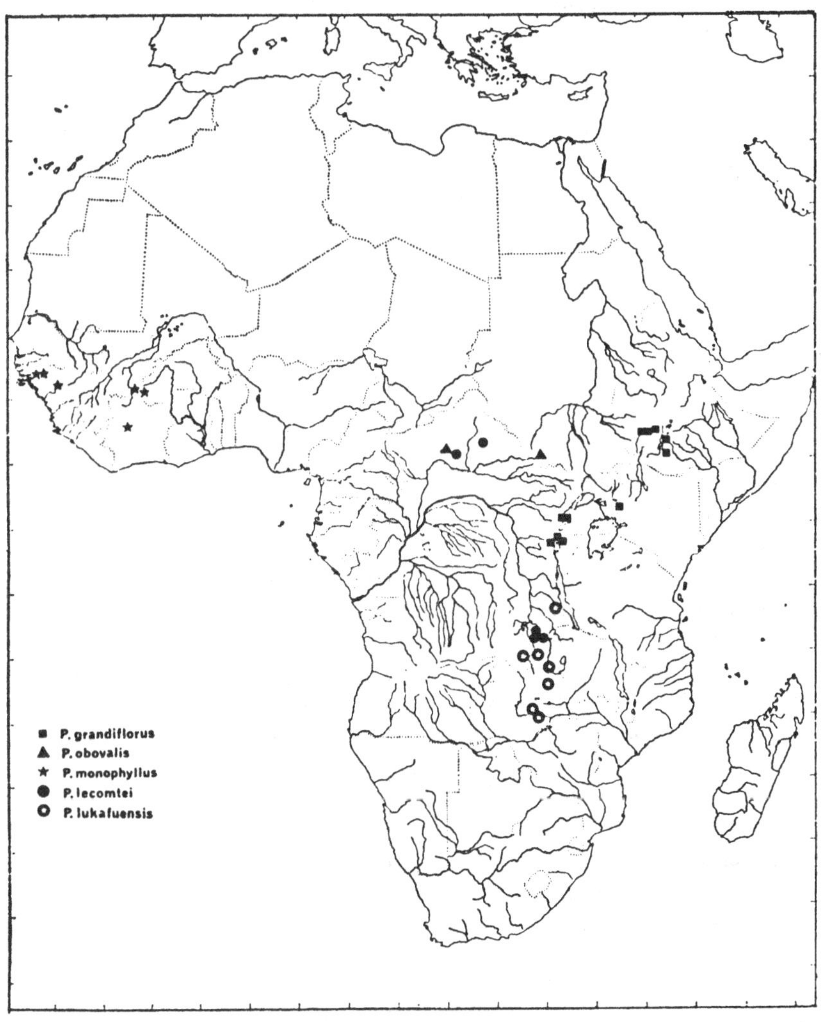

tryptophan content relative to that of the seeds (NRC, 1981). In the latter the first limiting essential amino acids are the sulphur amino acids as a group and the second tryptophan. There is thus some possibility of producing a better balanced amino acid profile when both tubers and seeds are available.

Domestication and evolution

Burkill (1935) and Hymowitz and Boyd (1977) have collected and published substantial lists of Asian vernacular names but these shed little light on winged bean evolutionary history, probably owing to the linguistic complexities of South-East Asia. However, Burkill (1935) suggested on the basis of some 'linguistic' evidence that the winged bean had an African origin. Fifty years on this still seems to be the most plausible hypothesis. Pickersgill (1980) has pointed out that there is a recognisable African element in the Asian flora, which can be correlated with the drift of the Indian sub-continent, on the break-up of Gondwanaland, to its present position. This could account for an Asian–African disjunction. However, since this disjunction is between an Asiatic cultigen on the one hand and an African wild species on the other, it cannot be considered as an entirely satisfactory hypothesis unless Asiatic wild species of *Psophocarpus* come to light. The closest affinity of the winged bean is with the species *P. scandens*, *P. palustris* and *P. grandiflorus*, according to Verdcourt and Halliday (1978). The last species is found wild in Ethiopia. It is possible that this or similar material could have been transported to Asia in a similar fashion to guar (Hymowitz, 1972) and the cowpea. It is possible that the winged bean is a transdomesticate (Smartt, 1980*b*), an hypothesis which is tenable in the absence of any evidence that truly wild Asiatic *Psophocarpus* species exist or have existed. Quite obviously more extensive collection of *Psophocarpus* in Africa is desirable and also further search in Asia. It would be particularly useful to resolve the status of *P. tetragonolobus* in Madagascar, which could have been a critical locality in the evolution of the cultigen.

Archaeology

The winged bean has no known archaeological record. The contribution of non-biological disciplines to our knowledge of its history is small, with the possible exception of comparative linguistics.

Further evolutionary potential

Although the winged bean produces a number of very acceptable products, the expansion of its production and commercial development is inhibited at present by the lack of the dwarf morphological variants common in other pulses, and the ecological uncertainties already mentioned.

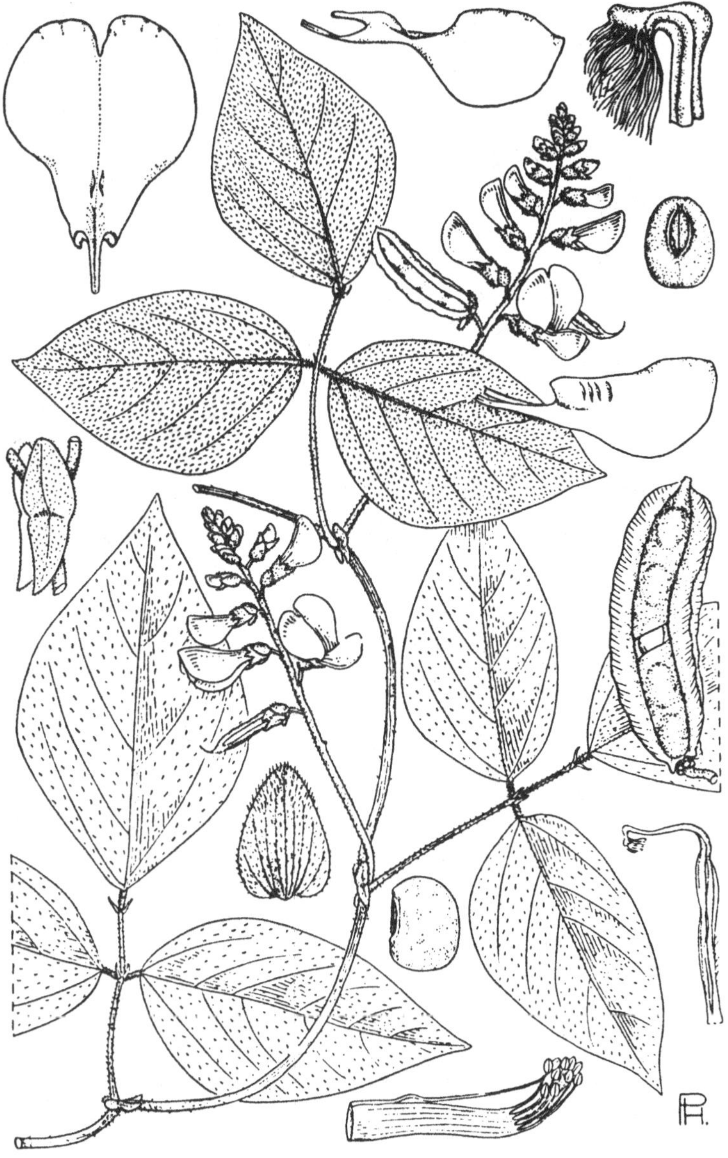

Fig. 7.12. *Psophocarpus palustris*. (From Verdcourt and Halliday, 1978. British Crown Copyright. Reproduced with permission of the Controller, Her (Britannic) Majesty's Stationery Office and the Trustees, Royal Botanic Gardens, Kew, © 1980.)

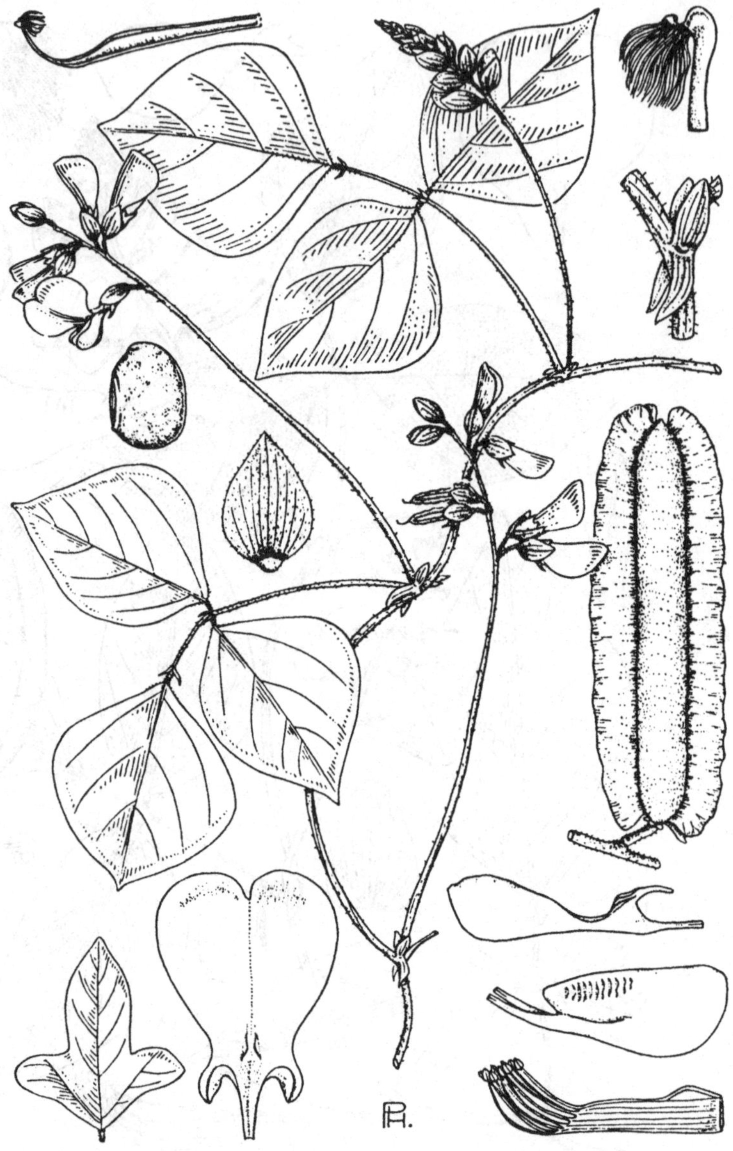

Fig. 7.13. *Psophocarpus scandens*. (From Verdcourt and Halliday, 1978. British Crown Copyright. Reproduced with permission of the Controller, Her (Britannic) Majesty's Stationery Office and the Trustees, Royal Botanic Gardens, Kew, © 1980.)

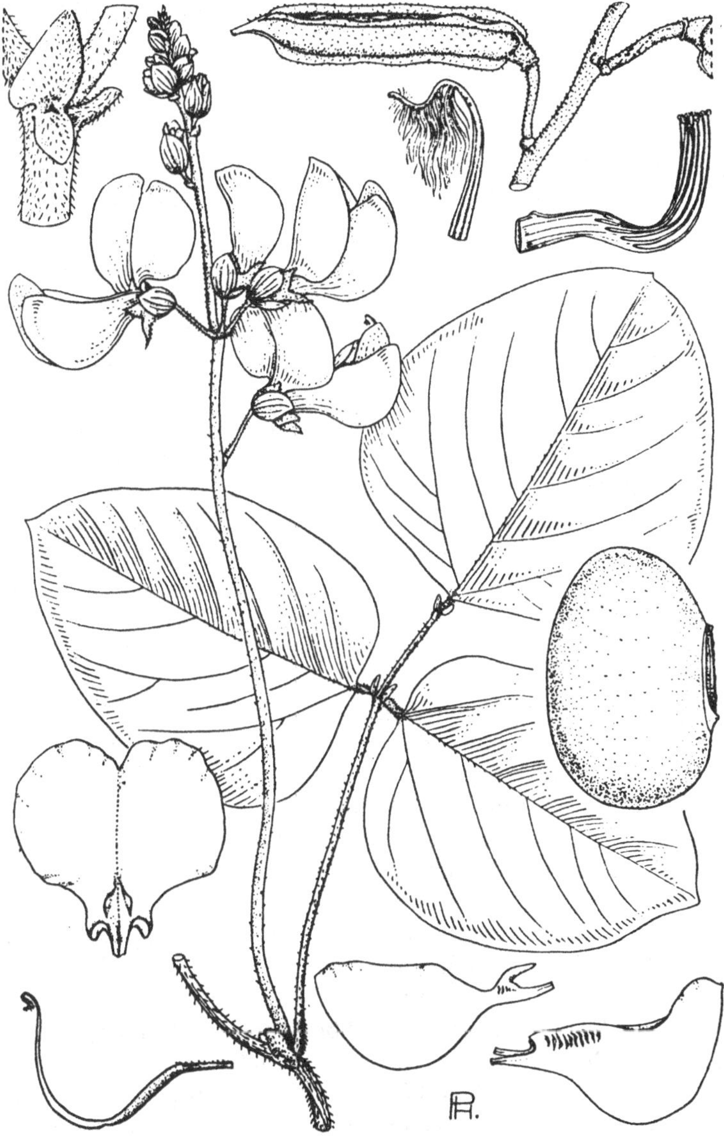

Fig. 7.14. *Psophocarpus grandiflorus*. (From Verdcourt and Halliday, 1978. British Crown Copyright. Reproduced with permission of the Controller, Her (Britannic) Majesty's Stationery Office and the Trustees, Royal Botanic Gardens, Kew, © 1980.)

It is not unreasonable to expect that further research and investigation may well resolve these difficulties. It sometimes is not appreciated that there may be a wide difference between the ability of a crop to grow and reproduce in an area and its ability to produce economically viable crops. A parallel with another legume crop, the groundnut, is very instructive in this context. These can in the United States be grown as a garden crop in states such as Maryland and Pennsylvania but intensive production of the commercial crop is based farther south. Caution is therefore necessary in drawing inferences concerning potentialities for intensive production of the winged bean from garden or backyard cultivation of the crop. The ecological limits of commercial production need to be established, otherwise unsuccessful attempts to launch intensive production are likely in areas which are marginal for the crop.

The morphological constraints are less straightforward than they appear on the surface. The winged bean, as already noted, is an indeterminate climbing plant which develops characteristic four-angled, long pods (up to 25 cm long) containing up to 20–21 seeds. Because of the considerable length and weight of these pods it is not practicable to grow winged beans without support, in the manner of the common cowpea. There is an interesting parallel with the asparagus or yard-long bean (a close relative of the cowpea) in which the climbing growth habit has been established. For successful maturation of long pods it is important that they mature out of contact with the soil, otherwise developing pods are soon destroyed by soil-inhabiting, facultatively parasitic fungi. The development of dwarf cultivars would necessitate devising an ideotype. A dwarf growth habit could be achieved in either of two ways, by a mutation producing reduced internode length and/or a mutation producing reduction of the number of nodes produced on the main axis and lateral branches. Since these mutations have not been recorded as having arisen naturally, it would probably be worth while to attempt to produce them artificially by ionising radiation or chemical mutagens. If a determinate mutant is produced, the number of nodes produced before axes are terminated by inflorescences could be important. Drawing parallels with *Phaseolus*, the indication is that a node number of 7–10 would be satisfactory (as in *P. vulgaris* and *P. lunatus*); substantially fewer nodes (as in *P. coccineus* with 4–5 nodes in determinate genotypes) would probably not be satisfactory. It would also be imperative to select for short-podded genotypes with fewer seeds per pod and it might also be necessary to select for higher intensities of flower production. This might entail selection for changes in the number of flowers in the inflorescence, the rate of inflorescence production and the duration of the flowering period. Drawing analogies with the changes produced in other pulses in which dwarf determinate forms have been selected, other important changes have occurred. The life form in dwarf determinate *Phaseolus coccineus*, for example, has changed from a short-lived, tuber-producing perennial to a non-tuberous annual. Such a change could have very important consequences in the winged bean as far as tuber production is concerned.

There are further implications in the production of dwarf cultivars. If a long growing season is available it may be necessary to sow them twice in the season whereas a single sowing of indeterminate climbers may be adequate. The advantages are thus by no means all with the determinate growth form. If these are ever produced, efficient utilisation could demand new systems of production and the nature of these can, at the present time, only be speculative. A further problem arises because most, if not all, winged beans are photoperiod-sensitive and will not flower readily in long days. The nature of this photoperiod-sensitivity is worth investigating; if it is similar in nature to that of the soyabean it could be used very effectively to determine the time at which crops mature. It is also possible that a photoperiod requirement satisfactory for an indeterminate climber may not be so for a determinate dwarf. Day-neutral genotypes could also be of value and a search for these is indicated. If none are found in germplasm collections, artificial mutagenesis could be helpful in inducing suitable mutants.

The winged bean, as a consequence of its wide Asiatic distribution, has been shown to have a wide range of genetic diversity. This seems to be most clearly expressed in variability of the reproductive parts: flowers, fruits and seeds. The range of morphological diversity at the present time appears to be small and the study of physiological variability has only barely begun. A great deal more fundamental biological knowledge will be necessary before substantial increases in volume of production are likely to be achieved. At the present time exploitable genetic resources are restricted as noted to those of the cultigen itself, the GP1A of Harlan and de Wet (1971). Further exploration of wild species, particularly those not previously used in interspecific crossing programmes, is clearly indicated.

The winged bean is a crop with great potential, but it would be a great mistake to over-promote it and minimise the problems which stand in the way of its realisation. Problems of acceptability and difficulties in utilisation should not be ignored or minimised.

7.3. The lupins (*Lupinus* spp.)

Introduction

Although the lupins are known to have been in cultivation for close on 4000 years they cannot be regarded as highly evolved crop plants

(Gladstones, 1970). In this respect they show some parallels with the grasspea (*Lathyrus sativus*). There is a very wide range of *Lupinus* species which appear to have potential in cultivation: the large-seeded lupins of Gladstones. These are mostly Old World species from the Mediterranean region and Africa; there is, however, one species, *L. mutabilis*, which has a long history of cultivation in South America.

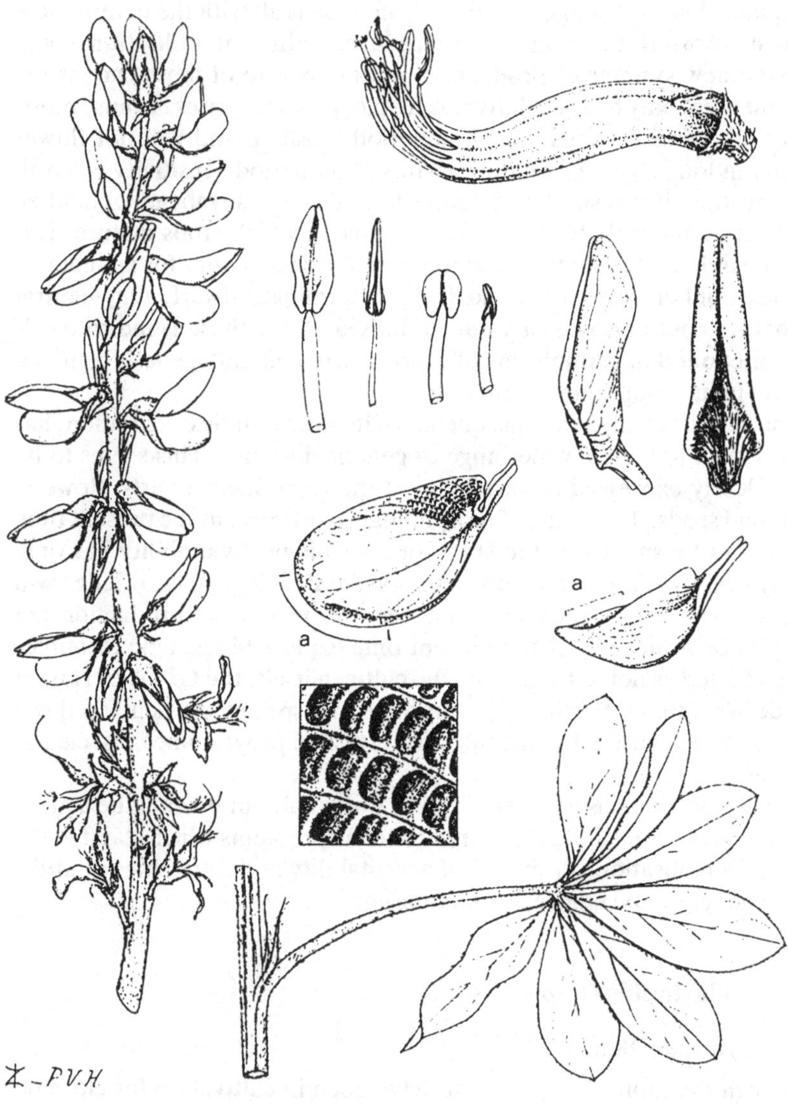

Fig. 7.15. *Lupinus albus* (from Westphal, 1974).

Table 7.3. *Inventory of* Lupinus *species with agricultural potential*

	Distribution	Chromosome number (2*n*)
Lupinus albus L.		
(*L. graecus* Boiss. and Sprun.).	circum Mediterranean	30, 40, 48, 50
L. angustifolius L.	circum Mediterranean	40, 48
L. micranthus Guss.	circum Mediterranean	48, ? 50
L. luteus L.	mainly W. Mediterranean	46, 48, 50, 52, 104
L. hispanicus Boiss. et Reuteur	Iberian peninsula	52
L. cosentinii Guss.	scattered W. Mediterranean	32
L. digitatus Forskal	Nile Valley and Saharan zone	? 36, 40, 42, 52
L. princei Harms	Kenya, Tanzania	?
L. pilosus Murray	E. Mediterranean	42, 50
L. palaestinus Boiss.	Israel and Sinai	42
L. atlanticus Gladstones	Morocco	?
L. somaliensis Baker	Somalia	?
L. mutabilis Sweet	S. American Highlands (Andean)	42, 48

From Gladstones (1980).

The most significant recent work on lupins is that which has been carried out in Western Australia by Gladstones (1970, 1974, 1980). His reviews are comprehensive and very valuable indeed and the best available sources of information on this crop. Valuable work has also been carried out on these species in New Zealand by Hill (1977).

Biosystematics and classification

Taxonomy

The taxonomy of the genus *Lupinus* has, according to Gladstones (1974), long been in a state of confusion which has still not been completely resolved. Gladstones (1974) has produced a taxonomic revision of those species with agricultural potential, to the number of 12. Since there are about 200 species in the genus (Bisby, 1981) the overall taxonomic problem is still not resolved satisfactorily. However from a practical point of view this is probably of little significance. A listing of species revised by Gladstones is given in Table 7.3, together with phytogeographic and cytological information.

Cytotaxonomy and hybridisation

In contrast with other grain legumes the chromosome complements show a lack of numerical consistency. All cultigens in the genera *Phaseolus* and *Vigna* (and several others in the tribe Phaseoleae) have a common chromosome complement of $2n = 2x = 22$. Reported chromosome numbers in *Lupinus* are $2n = 32, 36, 40, 42, 48, 50$ and 52. Darlington and Wylie (1955) suggest a basic genome $x = 12$, which would imply that the evolution of the chromosome complement has entailed polyploidy, a considerable degree of chromosome structural change and perhaps even aneuploidy. Many chromosome counts reported (Darlington and Wylie, 1955; Bolkhovskikh *et al.*, 1969) are inconsistent for species; these results can be explained partly by the fact that many counts date back to the period 1920–1940. Even allowing for this, chromosome number is highly variable between species, which is unusual in legumes. A cytological survey of the genus as a whole is obviously overdue with particular attention being paid to karyotype evolution within it.

Gladstones (1974) notes that, whereas in the New World lupin species (from North and Central America) inter-specific hybridisation is commonplace and species limits are ill-defined, the Old World species for the most part are separated by very effective genetic barriers. This suggests that these latter species are not of very recent evolutionary origin. It seems that not only have these species' karyotypes diverged strongly, as indicated by the range of chromosome numbers within the genus, but that strong divergence at the genetic level has also occurred. This implies, as Gladstones has observed, that each species is essentially a genetic island. Genetic resources immediately available for the improvement of individual large-seeded species reside entirely in the primary gene pools of the species themselves. This is a very considerable constraint, especially for those species which are at best only semi-domesticated. Artificial augmentation of utilisable genetic resources is clearly indicated by physical and chemical mutagenic agents. The outcome of such programmes would clearly be of interest not only to lupin breeders but also in the broader context of crop plant evolution.

Domestication and evolution

Three lupin species, *L. albus*, *L. luteus* and *L. angustifolius*, are in commercial production at the present time and can be regarded as fully domesticated. *Lupinus albus* has been established in cultivation for approximately 4000 years (Zhukovsky, 1929). In spite of the presence of alkaloids in the seeds, these were used both for livestock feeding and as human food. Boiling and soaking were required to remove the alkaloids. Lupin seeds are still used in this way in the Latin Mediterranean countries.

In its evolution as a crop plant *L. albus* shows the closest parallels of any lupin species with that of other grain legumes. It has been inferred (Gladstones, 1970) that it is essentially the domesticated form of a wild species described as *L. graecus*. The most significant differences from the wild form are in the greater seed size and the non-shattering pods of the cultigen. The persistence of high seed alkaloid contents over the millennia is interesting in view of the recent responses to selection for 'sweet' genotypes in this and other lupin species in Germany and Russia since the 1930s. It is quite probable that the alkaloids conferred protection against seed predation and made storage easier. This feature considered in conjunction with the ability of lupins not only to grow but also to thrive on skeletal soils have no doubt contributed substantially to the survival of this crop in cultivation.

The other two Old World species, *L. luteus* and *L. angustifolius*, show similar but less marked response to domestication. Whereas the wild (var. *graecus*) and cultivated (var. *albus*) populations are thought to merit taxonomic recognition by Gladstones (1974), he does not consider that the wild and domesticated populations of *L. luteus* and *L. angustifolius* have diverged sufficiently to merit any such recognition. In domesticated legumes increased seed size *vis-à-vis* wild forms is considered an almost constant concomitant of domestication. However, in lupins, Gladstones (1974) considers it possible that it may be of advantage in the wild under some ecological conditions. It is probable that there is a range of size optima in wild and domesticated populations and these may very well overlap, as they appear to in *Phaseolus vulgaris*.

Two other important characters in which wild and domesticated *L. albus* differ is in testa pigmentation and seed dormancy or hardseededness. The wild var.*graecus* has a mottled dark-brown and impermeable testa, whereas var. *albus* has a white or off-white, permeable testa. The loss of pigmentation to produce white seeds is a commonplace of domestication. The wild type of pigmentation which serves to reduce predation of shed seeds is superfluous under domestication, while the pigments themselves serve to reduce palatability and increase cooking times. Loss of testa impermeability and hence of seed dormancy is another common consequence of domestication. A further character of agricultural significance is the more erect growth habit of var. *albus*.

Gladstones (1974) considers that, although generally similar differences are found between the wild and domesticated populations of *L. angustifolius*, there is no clear discontinuity between them. In view of this he prefers to consider it rather as a polymorphic species. There is no good evidence for cultivation prior to the eighteenth century, but there is evidence to suggest exploitation in times of need in Italy (Bauhin *et al.*, 1651; Savi, 1798). Its principal use has been in land reclamation, soil conservation and for fodder. Non-shattering cultivar development has been reported by Gladstones (1980); these cultivars also are not hard-seeded and have low alkaloid contents. These developments show that significant improvements can be produced by selection over a relatively short time scale.

The response of *L. luteus* to selection closely parallels that of *L. angustifolius*. The most significant advances achieved by selection were incorporated in the cultivar Weiko II, developed in Germany before and during World War II. This had non-shattering pods, non-dormant seeds and low alkaloid content. Although protein seed contents are high, low harvest index and poor seed yields have severely constrained the adoption of this species as a crop on a wider scale.

Further evolutionary potential

There are two aspects which can be considered in relation to evolutionary potential, that of the three domesticated species already in cultivation and that of semi-domesticates and potential domesticates. Perhaps the greatest obstacle to further improvement of lupins is the dearth of usable genetic variability. The problems range from the poor harvest index and yielding capacity of *L. luteus* to the late maturity of *L. albus* and disease susceptibility of *L. angustifolius*. Since each of these species has a closed gene pool, introgression of appropriate germplasm from other species can be ruled out. The case for mutation breeding in these species would appear to be quite strong. Equally, a similar case could be made for attempts to produce suitable crop genotypes from species such as *L. mutabilis* and *L. cosentinii*. The response of *L. luteus* in this connection is particularly encouraging.

The impetus for these developments will come from economic necessity. Unfulfilled demand for edible oils and/or proteins could provide the necessary incentives. The necessity to exploit marginal agricultural land could also provide further justification for the further development of lupins as crop plants. It would also probably provide one of the severest tests of the general efficacy of mutation breeding.

7.4. Some conclusions on the legume oilseeds

The oilseeds, groundnut, soyabean, winged bean and the lupins, have few common characteristics apart from the fact that they are all legumes and that they all produce seeds containing commercially extractable oils. Between them they represent three different tribes of the family: Aeschynomeneae, Phaseoleae and Genisteae. Even the two which are members of the Phaseoleae belong to different subtribes, the Glycinieae and Phaseolineae. The common feature of an oleiferous seed is thus not a reflection of any close biosystematic affinity.

The economic position of the four crops is also markedly different. The groundnut, while the most abundant producer of oil in its seeds, is perhaps tending to go up-market from the edible oil trade. Its value in confectionery and for its very palatable seed has rather tended to price it out of the bulk edible oil market. The economic base for the soyabean is rooted both in the middle–lower end of the edible oil market (with also industrial use) and the protein concentrate market. The latter market has of late increased considerably in complexity with progressively higher quantities being taken by the processed food industry, for feeding both the human population and livestock. This is a market which one can reasonably expect to expand. The soyabean market is essentially the one towards which increased production of the winged bean is aimed. However, the heady optimism of the 1970s is giving way now to a more hard-headed appreciation of reality. Nevertheless, the winged bean does have great potential value in subsistence agriculture if production and acceptability problems can be overcome.

The economic problems of the lupin crop are not dissimilar. There is a question of what might be called consumer education involved. The consumer, in a word, must be convinced that he can use the crop in the first place, then he needs to be further convinced that he requires the crop; then its economic future is on a firmer footing. Both the winged bean and the lupin share common production and consumer-orientated problems. The future for potential oilseed crops seems to be bright, in the context of a favourably balanced supply–demand situation and the agricultural economic imperative of crop diversification in the agriculture of both developed and developing worlds.

8 The pigeonpea (*Cajanus cajan* (L.) Millsp.)

The pigeonpea can justifiably be regarded as an under-exploited legume; the reasons why this should be so are not entirely clear. The most likely explanation is that there is an acceptability problem in some parts of the world. Curiously, in the United Kingdom the pigeon pea has been consumed in small quantity, but consistently, as yellow split peas for making pease-pudding. The problem of limited acceptance is one which could at the present time be overcome by judicious stimulation of demand. Present dietary recommendations favour increased consumption of pulses by way of an alternative to red meat as a protein source (avoiding excessive intake of saturated fat) and as a source of dietary fibre. There is also a possible role as a protein source for the manufacture of textured vegetable protein for use in meat substitutes, meat extenders and the like.

The major centre of world production is undoubtedly India, where it is the second most important pulse crop. Production is about 2 million tonnes annually world wide, of which a little less than 85% is produced in India. Potential yield levels in excess of 2 t ha^{-1} are indicated from trials at ICRISAT (Annual Report, 1982) while in Queensland some breeding lines have indicated yields of over 4 t ha^{-1}.

8.1. Classification and biosystematics

Taxonomy

The pigeonpea is a member of the sub-tribe Cajaninae of the Phaseoleae and it is the only member of its sub-tribe to have been domesticated. The taxonomy of *Cajanus* DC. and its allies has recently been revised by van der Maesen (1985) and many uncertainties and anomalies have been removed. Probably the greatest change has been the merging of the genera *Cajanus* and *Atylosia* W. & A. The separation of these two genera had been questioned as long ago as 1864 by Mueller (Lackey, 1981). Recently, considerable experimental evidence has been produced from studies of experimental inter-specific hybridisation, cytology, biochemi-

Fig. 8.1. Pigeonpea *Cajanus cajan*, cv. Duke of Harar (from Westphal, 1974, 1985; van der Maesen, 1985).

cal and morphological features which supported merging the two genera. The high fertility of many 'inter-generic' hybrids *Cajanus* × *Atylosia* caused De (1974) and McComb (1975) to question their status. Lackey (1981) considered a merger to be justified but thought that, although this was so, nomenclatural confusion could result, particularly among applied biologists. This difficulty is now unlikely to materialise because the revised genus is effectively an expanded *Cajanus* not an enlarged *Atylosia*; a change in the generic name of the cultigen might well have caused nomenclatural confusion. Applied biologists by and large appreciate both the utility and validity of taxonomic revision and nomenclatural changes when they arise from a better understanding of the biosystematic realities. A much more profound taxonomic revision in the Phaseoleae, that of the Phaseolinae (Maréchal *et al.*, 1978) has been very readily accepted by applied biologists who have appreciated the effective and rational arrangement produced.

The genus comprises 32 species; these are listed with the most common synonym for each (where appropriate) given in Table 8.1. A sectional breakdown has also been proposed by van der Maesen (Table 8.2). Six sections are recognised. Section *Cajanus* contains the pigeonpea itself and its closest known relative, *C. cajanifolius*. Sections *Atylia* Benth. and *Fructicosa* van der Maesen include the erect species; sections *Volubilis*

Fig. 8.2. Fruiting branches of *Cajanus cajan* (from van der Maesen, 1985).

van der Maesen and *Cantharospermum* (W. & A.) Benth. contain climbing forms. Section Rhynchosoides Benth. has three trailing species which strongly resemble species of *Rhynchosia* such as *R. aurea* (van der Maesen, 1985).

An interesting outcome of inter-specific hybridisation has been the clear demonstration that one of the diagnostic characters used to separate *Cajanus* and *Atylosia*, namely presence of a seed strophiole in *Atylosia* and its absence in *Cajanus*, was under quite simple genetic control. Reddy *et al.* (1981) demonstrated that the difference was controlled by only two loci. Van der Maesen (1985) also reported that some pigeonpea

Fig. 8.3. *Cajanus crassus* var. *burmanicus* (l) and var. *crassus* (r) (from van der Maesen, 1985).

Table 8.1. *Nomenclature of* Cajanus *with commonest synonyms*

Species	Synonym	Distribution
1. *Cajanus acutifolius* (F. von Muell.) van der Maesen	*Rhynchosia acutifolia* F. v. Muell.	Australia
2. *C. albicans* (W. & A.) van der Maesen	*Atylosia albicans* (W. & A.) Benth.	S. India, Sri Lanka
3. *C. aromaticus* van der Maesen		Australia
4. *C. cajan* (L.) Millsp.	*C. indicus* Spreng.	pantropical
5. *C. cajanifolius* (Haines) van der Maesen	*A. cajanifolia* Haines	S.E. India
6. *C. cinereus* (F. von Muell.) F. von Muell.	*A. cinerea* F. v. Muell. ex Benth.	Australia
7. *C. convertiflorus* F. von Muell.	*A. pluriflora* F. v. Muell. ex Benth.	Australia
8. *C. crassicaulis* van der Maesen	—	Australia
9. *C. elongatus* (Benth.) van der Maesen	*A. elongata* Benth.	N.E. India, Vietnam
10. *C. goensis* Dalz.	*A. barbata* (Benth.) Bak.	India, S.E. Asia
11. *C. grandiflorus* (Benth. ex Bak.) van der Maesen	*A. grandiflora* Benth. ex Bak.	N.E. India, S. China
12. *C. heynei* (W. & A.) van der Maesen	*Dunbaria heynei* W. & A.	S.W. India, Sri Lanka
13. *C. kerstingii* Harms	—	W. Africa
14. *C. lanceolatus* (W. V. Fitzg.) van der Maesen	*A. lanceolata* W. V. Fitzg.	Australia
15. *C. lanuginosus* van der Maesen	—	Australia
16. *C. latisepalus* (Reynolds & Pedley) van der Maesen	*A. latisepala* Reynolds & Pedley	Australia
17. *C. lineatus* (W. & A.) van der Maesen	*A. lineata* W. & A.	S. India, Sri Lanka
18. *C. mareebensis* (Reynolds & Pedley) van der Maesen	*A. mareebensis* Reynolds & Pedley	Australia
19. *C. marmoratus* (R. Br. ex Benth.) F. von Muell.	*A. marmorata* R. Br. ex Benth.	Australia
20. *C. membranifolius* van der Maesen		Philippines, Indonesia
21. *C. mollis* (Benth.) van der Maesen	*A. mollis* Benth.	Himalaya foothills

22. *C. niveus* (Benth.) van der Maesen	*A. nivea* Benth.	Burma. S. China
23. *C. platycarpus* (Benth.) van der Maesen	*A. platycarpa* Benth.	Indian sub-continent, Java
24. *C. pubescens* (Ewart & Morrison) van der Maesen	*A. pubescens* (Ewart & Morrison) Reynolds & Pedley	Australia
var. *mollis* Reynolds & Pedley	var. *mollis* Reynolds and Pedley	
var. *pubescens*		
25. *C. reticulatus* (Dryander) F. von Muell.	—	
var. *grandifolius* (F. von Muell.)	*A. grandifolia* (F. v. Muell.) Benth.	Australia, N. Guinea
var. *reticulatus*	*A. reticulata* (Dryander) Benth.	Australia
var. *maritimus* (Reynolds & Pedley) van der Maesen	—	Australia
26. *C. rugosus* (W. & A.) van der Maesen	*A. rugosa* W. & A.	S. India, Sri Lanka
27. *C. scarabaeoides* (L.) Grah. ex Wall.	—	
var. *pedunculatus* (Reynolds & Pedley) van der Maesen	var. *pedunculatus* Reynolds & Pedley	Australia
var. *scarabaeoides*	*A. scarabaeoides* (L.) Benth.	S., S.E. Asia, Pacific Coastal Africa
28. *C. sericeus* (Benth. ex Bak.) van der Maesen	*A. sericea* Benth. ex Bak.	S. India
29. *C. trinervius* (DC.) van der Maesen	*A. candollei* W. & A.	S. India, Sri Lanka
30. *C. villosus* (Benth. ex Bak.) van der Maesen	*A. villosa* Benth. ex Bak.	N.E. India
31. *C. viscidus* van der Maesen	—	Australia
32. *C. volubilis* (Blanco)	—	
var. *burmanicus* (Collett & Hemsley) van der Maesen	*A. burmanica* Collett & Hemsley	Burma
var. *volubilis*	*A. crassa* Prain ex King	India, S.E. Asia

From van der Maesen (1985).

Table 8.2. *Classification of* Cajanus *in sections*

Section	*Cajanus*
	C. cajan
	C. cajanifolius
Section	*Atylia* Benth.
	C. cinereus
	C. confertiflorus
	C. lineatus
	C. lanuginosus
	C. reticulatus
	C. sericeus
	C. trinervius
Section	*Fruticosa* van der Maesen
	C. acutifolius
	C. aromaticus
	C. crassicaulis
	C. kerstingii
	C. lanceolatus
	C. latisepalus
	C. niveus
	C. pubescens
	C. viscidus
Section	*Cantharospermum* (W. & A.) Benth.
	C. albicans
	C. elongatus
	C. goensis
	C. rugosus
	C. scarabaeoides
Section	*Volubilis* van der Maesen
	C. crassus
	C. grandiflorus
	C. heynei
	C. mollis
	C. villosus
	C. volubilis
Section	*Rhynchosoides* Benth.
	C. platycarpus
	C. mareebensis
	C. marmoratus

From van der Maesen (1985).

Table 8.3. *Sections of the genus* Cajanus *(*sensu lato*)*

Section	No. of species
1. *Atylia* Benth.	7
2. *Cajanus*	2
3. *Fruticosa* v. d. Maesen	9
4. *Cantharospermum* (W. & A.) Benth.	5
5. *Volubilis* v. d. Maesen	6
6. *Rhynchosoides* Benth.	3

From van der Maesen (1985).

Table 8.4. *Pigeon pea gene pools*

Primary gene pool:	cultivar collections
Secondary gene pool:	*C. acutifolius*, *C. albicans* *C. cajanifolius*, *C. lanceolatus* *C. latisepalus*, *C. lineatus*, *C. reticulatus* *C. scarabaeoides*, *C. sericeus*, *C. trinervius*
Tertiary gene pool:	*C. goensis*, *C. heynei*, *C. kerstingii* (?) *C. mollis*, *C. platycarpus*, *C. rugosus* *C. volubilis*, other *Cajanus* spp. ? other Cajaninae (e.g. *Rhynchosia*, *Dunbaria*, *Eriosema*)

accessions in the ICRISAT collection (±200 of ±10000) actually had a strophiole. This was perhaps the final demonstration that on morphological grounds the two former genera together constitute a single natural taxon.

The geographical distribution of the genus is basically Old World with a strong Australasian element. There are 18 Asiatic species, 15 Australian and 1 African. Most Australian species are endemic (all but two). There are 8 species endemic to the Indian sub-continent and Burma. Apart from the cultigen itself, only one species has a widespread Asiatic–Australasian distribution, namely *C. scarabaeoides*. Van der Maesen (1985) has observed that Burma, Yunnan (China) and northern Australia are the areas in which the greatest numbers of wild species occur. The habitat range of the species is from grassland to open tropical woodland. Most tend to occur at the forest fringes. Altitudinal range is also varied; *C. trinervius* in India is found at 2000 m+ above sea level (van der Maesen, 1985) (Tables 8.1–8.4).

Cytotaxonomy and hybridisation

Congenericity of *Cajanus* and *Atylosia* was to a very large degree established by the relative ease with which pigeonpeas produced hybrids successfully with wild *Atylosia* species (Deodikar and Thakar, 1956; Kumar *et al.*, 1958; Roy and De, 1965; Sikdar and De, 1967; Reddy, 1973; Reddy *et al.*, 1981; De, 1974; Pundir, 1981). De (1974), for example, reported success in making hybrids of *C. cajan* with *Atylosia lineata*, *A. sericea* and *A. scarabaeoides*; reciprocal crosses using *C. cajan* pollen were only successful on *A. sericea*. The chromosome number $2n = 2x = 22$ is common to both *Cajanus* and species formerly assigned to *Atylosia*. Karyotype studies by Deodikar and Thakar (1956), Kumar *et al.* (1958) and Sikdar and De (1967) all demonstrated strong resemblances between the chromosome complements of *C. cajan* and *Atylosia* species. Kumar *et al.* (1958), who studied meiosis of the F1 inter-specific hybrid *C. cajan* × *A. lineata*, reported nearly normal chromosome behaviour; Reddy (1973) showed almost complete pairing of all homologues at pachytene. Pollen fertility of the hybrid *C. cajan* × *A. lineata* is high (78%) and seed setting is comparable (74%); these represent a reduction of approximately 20% fertility in the hybrid. These observations, coupled with the very low success of the reciprocal cross *A. lineata* × *C. cajan*, suggest that although

Fig. 8.4. *Cajanus cajan* (l); *C. albicans* (r); F1 inter-specific hybrid (*C. cajan* × *C. albicans* (centre) (from van der Maesen, 1985).

A. lineata is very close genetically to *C. cajan* it is unlikely to be the progenitor type. De (1974) suggested that investigation of other erect *Atylosia* species such as *A. geminiflora*, *A. candollei* and *A. cajanifolia* might help to identify wild species with an even closer phylogenetic relationship with *C. cajan*. Reciprocal crosses *C. cajan* × *A. cajanifolia* have been reported by van der Maesen (1980); the cross *A. cajanifolia* × *C. cajan* has apparently occurred spontaneously in the ICRISAT nurseries. Further studies of this material would obviously be of the greatest interest.

Chemotaxonomy and biochemistry

Biochemical studies of *Cajanus* (*sens. nov.*) species have not been carried out in any systematic fashion, owing in part at least to the lack of the necessary materials. The compounds which have been studied are not necessarily those which are of the greatest value to the taxonomist. Harborne *et al.* (1971) have collated information on chemical constituents of legumes, much of which is of little taxonomic value. Lackey (1977) studied the pattern of canavanine distribution in the Phaseoleae and found that it was absent consistently from the seeds of the Cajaninae (including 7 *Cajanus* species). It is obviously important to determine the appropriate level in the taxonomic hierarchy in which to use biochemical information. Kloz (1971) included the pigeonpea in his serotaxonomic study of legumes and found little cross-reaction between its seed proteins and anti-serum against those of *Phaseolus vulgaris*. Such studies could well be of value in *Cajanus* taxonomy if they were based on a standard *C. cajan* seed protein anti-serum. Serotaxonomic studies are most effective at the lower levels of taxonomic hierarchies in all probability.

Probably the most productive chemotaxonomic studies carried out to date are those of Ladizinsky and Hamel (1980) and Singh *et al.* (1981) using electrophoresis. The seed protein separation patterns produced were generally similar in *Cajanus* and former *Atylosia* species, confirming congenericity. It was also observed that species which had been successfully crossed with the pigeonpea had more closely similar patterns than those which failed to cross. Variation within the pigeonpea was also found to be less than between it and *C. cajanifolius*, the most similar wild species.

The broad chemical composition of the pigeonpea is that of a typical pulse with a crude protein content of 24.1%, carbohydrate 62.9% and lipids 1.7% (Gupta, 1982). In the review of Toms and Western (1971) its seed is reported as being free of measurable haemagglutinating activity; Liener (1982), in a review of literature on legume seed toxins and anti-metabolites, mentions the pigeonpea solely on account of a very low cyanide content (approximately one thirtieth of that normally found in

Phaseolus lunatus). Its use as a pulse therefore seems to be quite straightforward. It is in India that it finds its greatest range of uses. The major usage is probably as dhal (split seed) and it can be used in Indian cuisine in the preparation of a great diversity of dishes. It can also be used as a fresh vegetable in a similar way to the common pea.

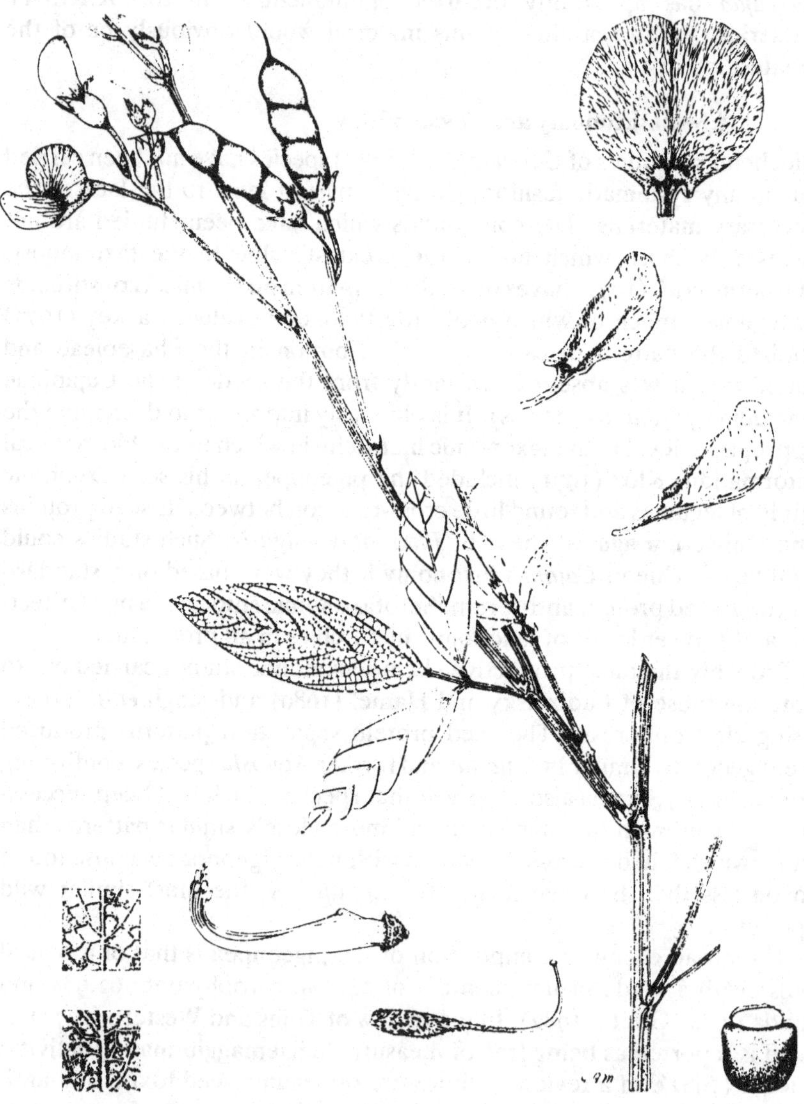

Fig. 8.5. *Cajanus cajanifolius* (from van der Maesen, 1985).

8.2. Domestication and evolution

Archaeology

Archaeological remains of the pigeonpea are scanty. An early purported find in an Egyptian twelfth dynasty burial vault should be re-examined by modern methods (van der Maesen, 1980). Remains dated between 200 BC and AD 300 have been found in India (Kajale, 1974), and De (1974) has found references to the pigeonpea in Indian literature dating from about AD 200–400. He has suggested that it was a well-established crop by the sixth century AD. He has also considered the derivation of modern Indian names of the pigeonpea from old Sanskrit names, *tur* from *tuvarica* and *arhar* from *adhaki*. *Tur* is the current name in south India; *arhar* is used in the north. The earliest Sanskrit name (*tuvari*) appears in writings of AD 300–400 but appears to be based on earlier Dravidian (south Indian) names. Further searches of ancient Indian writings may bring to light additional information on the origin and history of the pigeonpea.

Modern vernacular names of the pigeonpea reflect the complex migration and dispersal pattern of the crop. Van der Maesen (1985) has suggested that the Portuguese name *guandu* and its Spanish equivalent *gandul* are derived from the Indian Telugu *kandulu*. The name 'pigeonpea' itself was first recorded in Barbados (Plukenet, 1692) and literal translations are found in German, Dutch, Russian, French and Spanish as vernacular names in these languages. Names such as pois d'Angole, Congo pea, lentille du Soudan etc. reflect fairly recent provenance.

Geographic origin

There has recently been some controversy over the geographic origin of the pigeonpea. Westphal (1974) favoured an African origin, whereas De (1974) argued strongly that it had an Indian centre of origin. The weight of current opinion is overwhelmingly in favour of an Indian origin (van der Maesen, 1980, 1985) on the grounds of phytogeographical and experimental evidence. The original supposition of de Candolle (1886) that it had an African origin was probably based on a general misunderstanding of the real nature of the genera *Cajanus* and *Atylosia*.

The identity of the progenitor type has not as yet been determined. The closest we are apparently to it is *C. cajanifolius* but van der Maesen (1985) is strongly of the opinion that this is not the prototype. In his view there are good reasons not to regard *C. cajanifolius* as merely a sub-species of *C. cajan* on the grounds of quite well-developed barriers to crossing. It can with some justification at the present time be regarded as the best available approximation to the progenitor type. A similar conclusion has been drawn by Krishna and Reddy (1982) from isozyme studies. The detection of truly wild pigeonpea is complicated by the relative ease of establishment of feral populations of the cultigen (cf. *Gossypium*). Long-term survival in the wild of most herbaceous cultigens is rare. The germplasm of cultigens is most likely to survive in the wild in introgressed forms produced from hybrids between cultigens and progenitor types in areas in which both occur. Shrubby cultigens appear to be in a different situation and are better fitted for survival in the wild than are their herbaceous counterparts.

Dispersion of the pigeonpea, geographically, has been plotted by van der Maesen (1980) using information in papers by De (1974) and Royes (1976). Movement to the New World from Africa apparently occurred soon after its discovery in the sixteenth century. There is very little information on the movement of the crop in Africa. Acceptance has been greater in the Sudan zone than in plateau areas farther south. It is possible that the great southward migration of the Bantu was initiated from areas which had not yet received the crop. Certainly there appears to be very little long-standing cultivation of the crop in southern Africa.

Evolution of form

The main study of evolutionary changes in the pigeonpea has been carried out by van der Maesen (*in ed.*). A practically useful distinction can be drawn between *tur* and *arhar* cultivars and landraces. There is probably no sharp discontinuity between the two groups. Comparison with wild *Cajanus* species and other pulse crops suggests that the *arhar* type is the more primitive. Since the pigeonpea is a shrubby plant and most other grain legumes are herbaceous, parallels in the evolution of growth habits between the pigeonpea and the other are arguably less close. However, the restricted growth habits favoured by selection in herbaceous pulses have a parallel in the pigeonpea. *Tur* cultivars are less luxuriant in vegetative growth than those of the *arhar* type. Other parallel changes found in other pulses are also to be found in the pigeonpea. Pigmentation of flowers, pods, seeds and vegetative parts is reduced in *tur* cultivars as compared with those of the *arhar* group. Closely similar trends are thus apparent in the pigeonpea and other pulses. Van der Maesen (*in ed.*) has found it possible to follow the scheme of Schwanitz (1966) developed by Harlan (1975) and used specifically on legumes by Smartt and Hymowitz (1985). In this, evolutionary changes have been considered by van der Maesen in the following:

1 Plant habit
2 Crop duration

3 Photoperiod sensitivity
4 Flower number and inflorescence size
5 Fruit and seed size
6 Seed colour
7 Taste
8 Pod dehiscence
9 Seed dormancy
10 Seedling vigour
11 Habitat
12 Biochemical constituents.

This scheme represents sub-division of several broad classes of evolutionary change used previously.

1. *Plant habit* varies in two ways, height and lateral extension. Growth may be up to 4 m in height. The plant can be cut down annually and will rattoon (produce new shoots from cut stem bases). This is useful in management for fodder production; highest seed yields are achieved with annual sowings. The more compact growth forms are favoured in more intensive management systems. The annual life form has not yet become well established but some at least of the more dwarf forms have substantially reduced rattooning ability.

2. *Crop duration* is determined by the time of onset and duration of the flowering period in relation to the length of the period over which the crop actually matures. This is determined in part at least by photoperiod.

3. *Photoperiod sensitivity*. Pigeonpeas in general are short-day plants but some genotypes are less sensitive. There also seems to be an interaction with temperature. Duration of the vegetative growing season can be manipulated to some extent by varying time of sowing.

4. *Flower number and inflorescence size*. Although flower number per inflorescence may increase in some pulses under domestication (e.g. *Phaseolus coccineus*), in the pigeonpea and its allies flower production is high relative to fruit set. This prolific flowering over an extended period can be a useful strategy to exploit favourable conditions for pod setting as and when they occur. The size of both inflorescence and flower is by no means larger in pigeonpea than in wild *Cajanus* species; in the genus as a whole it is intermediate regarding the size of both.

5. *Fruit and seed size*. In common with most grain legumes, pigeonpeas have larger seeds than their wild relatives and apparently have been subjected to selection for larger seed. The larger seed sizes are favoured for use as a fresh vegetable; medium-sized genotypes are favoured for milling and dhal (split pea) production. Larger seed implies larger pods and these require more robust stems for their support, which appear to have coevolved in the pigeonpea. The weak stems of some wild species at least

would certainly not be robust enough to support adequately pods as heavy as those of the cultigen.

6. *Seed colour*. On the face of it this is a relatively trivial character, but it is a consistent feature of pulse evolution that lighter-coloured testas have consistently been favoured in selection. This is frequently the case even when the testa is removed as in the preparation of split peas. Highly pigmented testas tend to be more commonly found in the mixed populations favoured by the more primitive agriculturalists.

7. *Taste*. Taste and flavour are not unrelated to seed colour. Pigments often impart a bitter and astringent flavour, especially to fresh seed. Selection against constituents reducing palatability has undoubtedly occurred and been effective, as no doubt has selection in favour of constituents which improve taste and aroma.

8. *Pod dehiscence*. Delayed pod dehiscence is an essential requirement for efficient harvesting of any pulse crop. Van der Maesen (*in ed.*) notes that pod dehiscence in *C. cajanifolius* is quite late while in *C. cajan* pods will eventually dehisce if left on the plant at maturity. Delayed dehiscence in a wild legume could provide some stimulus and inducement to domesticate. The ultimate dehiscence of pigeonpeas left on the plant after maturity provides a seed source for establishment of feral individuals.

9. *Seed dormancy*. Loss of seed dormancy is a prime requirement for a cultivated crop plant, and most collections are non-dormant. Hardseededness, although exceptional, is not totally unknown. Wild species have characteristically dormant seeds. Dormancy may only persist for a matter of a few months, however.

10. *Seedling vigour*. This is not a pronounced feature of the pigeonpea; for its first 2–3 months of growth it is not a very effective competitor with weeds. It is interesting to note, as van der Maesen has, that larger seed confers a transient advantage only, which is lost 4–6 weeks post-emergence. This is a common feature of other pulses in which the larger-seeded genotypes do not appear to have any substantial competitive advantage (e.g. *Phaseolus vulgaris*), perhaps even the reverse.

11. *Habitat preferences*. A substantial segment of the Cajaninae has a strong habitat preference for forest margins. The favoured woodland types are those of the seasonally dry tropics. Distribution tends to be determined by habitat distribution, which tends on the whole to be discontinuous. The near-progenitor type *C. cajanifolius* is not a common plant, but knowledge of its distribution range has been extended as a result of recent collecting activities. Its comparative rarity may be due to habitat loss. The lack of aggressiveness of the pigeonpea in early competition with weeds may be a legacy of its ecological history. In its early growth phases it probablay tolerates shading well and emerges into full sunlight while approaching maturity.

12. *Biochemical constitution.* Ladizinsky and Hamel (1980) have demonstrated improved protein solubility in cultivated pigeon pea relative to that of the wild species; this may reflect improved nutritional quality (and digestibility?). Seed protease inhibitor activity (of trypsin and chymotrypsin) of wild *Cajanus* species is higher than that of the pigeonpea and *Cajanus cajanifolius*. Seed protein polymorphisms have been demonstrated in a range of pigeonpea genotypes. Some of the variants are characteristic of wild species and van der Maesen has suggested that this may indicate introgression from the wild species into the cultigen.

8.3. Unrealised evolutionary potential

The pigeonpea can justly be regarded outside the Indian sub-continent as an under-valued and under-exploited crop. Among pulse crops it is unique in being a shrub; among grain legumes as a whole, the only other shrubs are some lupin species. It is perhaps the fact that the pigeonpea requires an appreciably longer growing season than other pulses that has inhibited to some extent its more extensive cultivation. In shifting cultivation and semi-permanent systems (Ruthenberg, 1980) it can persist for several years in the fallow phase and be exploited as a source of seed and forage for livestock. What is cannot be used for is a catch crop, as selected cowpeas and common beans can. Although it is unlikely that a very short-season pigeonpea could be bred, it would be an advantage to have genotypes which matured much earlier than the average pigeonpea, perhaps a truly annual pigeonpea. The specification for an advanced pigeonpea could well be for an early-maturing, annual form producing a substantial yield of medium–large white seeds of good flavour and cooking quality. Selection could also be directed at the production of dhal or green vegetable types.

There is no doubt but that the pigeonpea could be usefully cultivated in vast areas, of Africa for example, where it is under-exploited at present. Its position in Indian agriculture clearly shows how valuablc it can bc. Care must be taken, in devising breeding programmes, not to lose sight of the more traditional types of pigeonpea; the objective should be to broaden the range of farming systems in which the pigeonpea can be exploited effectively. Improvement projects should not lose sight of the needs of growers following traditional systems.

9 Minor grain legumes

This chapter concerns legumes of minor current economic significance which have not been covered previously. These include those species which have some food use but which may have other more important uses. In practice it is difficult to draw the line as grain legumes between a crop such as the winged bean, which produces edible pods, seeds and tubers, and one like the yam bean (*Pachyrrhizus erosus* (L.) Urban), which produces edible pods and tubers but whose mature seeds are toxic (Purseglove, 1974). Guar (*Cyamopsis tetragonoloba* (L.) Taub.) is another case in point (Hymowitz, 1972). This produces a mucilage-rich seed useful not only for paper-making and textiles but also in food products. The velvet beans (*Mucuna* spp.) merit at least a brief mention. They are capable of producing prodigious yields of pods, seeds and forage. Their seed can be used as food in times of scarcity if sufficient care is taken during preparation and cooking to eliminate two toxic amino acids (stizolobic and stizolobinic acids) which are present in the seeds.

The crops of major concern in this chapter are the hyacinth bean, the horse gram (formerly included in the genus *Dolichos*), the Hausa (or Kersting's) groundnut (formerly *Kerstingiella*) and the sword and jack beans (*Canavalia* spp.).

9.1. The hyacinth bean (*Lablab purpureus* (L.) Sweet)

Introduction

The hyacinth bean can justifiably be regarded as an under-exploited legume. Herklots (1972) observed that it is useful for soil improvement, as a cover crop and for forage. Tindall (1983) gives useful information on its exploitation as human food. The immature pods can be used as a green vegetable and the leaves can be picked and used as a salad or spinach. The mature seed can also be used but, in common with many less intensely exploited grain legumes, should be well cooked before consumption. Perennial forms produce an edible root tuber also.

The crop is quite drought-tolerant once established, and crops well into the dry season on residual moisture in Central Africa. Under such a system it is unlikely that very high yields per unit area can be obtained, but the crop though small may be useful if it can be harvested during the dry season.

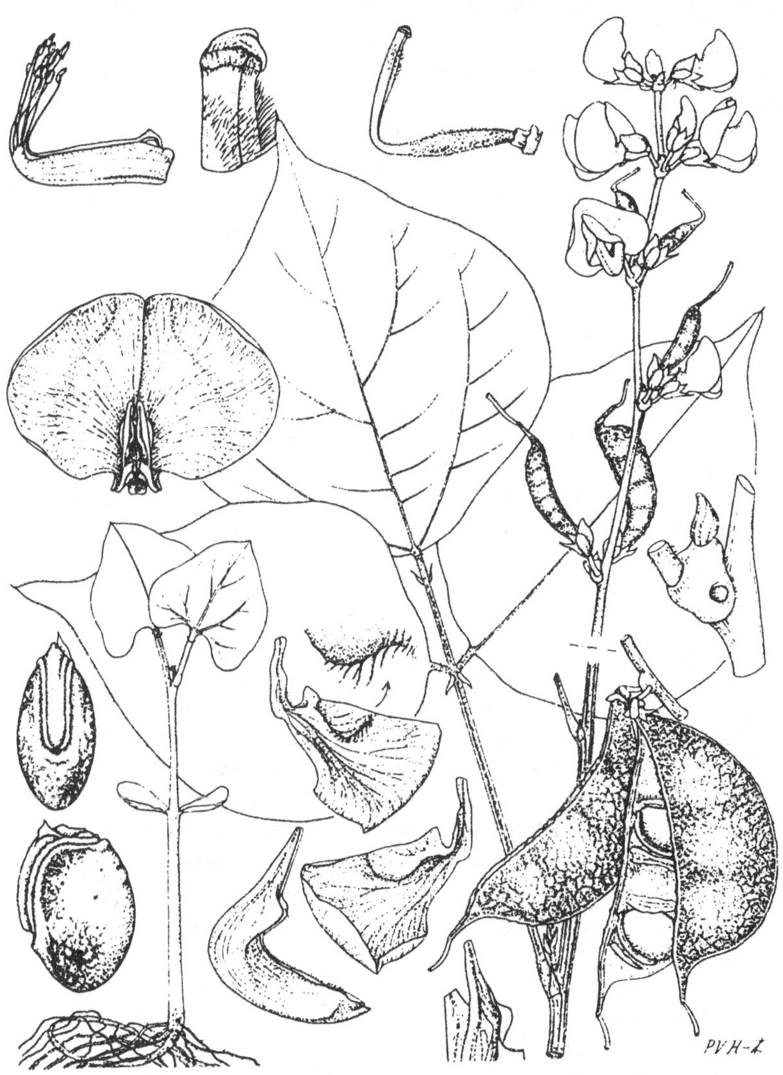

Fig. 9.1. The hyacinth bean, *Lablab purpureus* (from Westphal, 1974).

Classification and biosystematics

The biosystematic relationships and nomenclature of the hyacinth bean and its allies have recently been thoroughly reviewed and revised. Formerly the horse gram (*Macrotyloma uniflorum* (Lam.) Verdc.) and the hyacinth bean were both included in the genus *Dolichos* following Linnaeus. More recently both these cultivated species have been removed from the genus and the hyacinth bean has been assigned to the monotypic genus *Lablab* (Verdcourt, 1980, 1982). This appears to be justified on the distinctive floral morphology and palynological characters.

Westphal (1974) discusses the classification of the hyacinth bean and usefully reviews the history of its taxonomy. He disagrees with the taxonomic conclusions of Verdcourt (1970*a*, 1971), which, however, are now generally accepted and can be summarised. The intra-specific taxa recognised by Verdcourt are:

1. ssp. *uncinatus* Verdc., which is the wild form having pods approximately 40 mm × 15 mm. This plant appears to be taken into cultivation in East Africa.
2. ssp. *purpureus* includes cultivated plants with larger pods of similar shape to the foregoing, 100 mm × 40 mm in size.
3. ssp. *bengalensis* has characteristically longer pods than either of the other subspecies, proportionately twice as long, up to 140 mm × 10–25 mm.

Cytogenetics and hybridisation

Two different chromosome counts have been recorded for the hyacinth bean, $2n = 2x = 22$ and 24. According to Goldblatt (1981) the correct count is $2n = 22$ for *L. purpureus* and for *Macrotyloma* species, including *M. uniflorum* (the former *Dolichos uniflorus*), $2n = 20$. This suggests that there is a cytological basis for separating these two species formerly assigned to *Dolichos*. However, Stanton (1966) reported successful hybridisation of hyacinth bean and horse gram, which would imply a close genetic affinity. McComb (1975) has argued that successful inter-generic hybridisation in legumes suggests misplaced generic boundaries, since even inter-specific hybridisation within most legume genera fails more often than not.

Biochemistry and chemotaxonomy

Although few detailed biochemical and chemotaxonomic studies have been carried out on the hyacinth bean and its relatives, it has been considered (if rather peripherally) in studies of other genera such as

Phaseolus and *Vigna* (Kloz, 1971). In studies of graft compatibility *L. purpureus* showed rather poor compatibility with *Phaseolus vulgaris*. When its seed proteins were tested against antisera to selected seed proteins of *P. vulgaris* and *V. radiata* (*P. aureus*), the intensity of cross-reaction was moderate, about 30–40% of the intensity of reaction to the inducing antigens. Some protein cross-reactions were also apparent in immunoelectrophoretic studies between *L. purpureus* proteins and antibodies to proteins I and II of *P. vulgaris*.

The distribution of specific organic compounds in legume seeds is often of biosystematic interest. Bell (1971) and Bell *et al.* (1978) reported that canavanine, which is present in many species of papilionoid legumes, is absent in the Phaseolineae including *Lablab*, *Macrotyloma* and *Dolichos*. An interesting difference between the hyacinth bean and horse gram is in the occurrence of arginase in their seeds, present in the hyacinth bean, absent in the horse gram; urease, on the other hand, is found in the seed of both species. Comparisons of amino acid profiles of hyacinth bean and *P. vulgaris* showed broad similarity (Bell, 1971).

The occurrence of compounds of little biosystematic significance but of nutritional interest has been widely studied. The nature and distribution of carbohydrate material has been reviewed by Bailey (1971) and Arora (1982), of lipids by Wolff and Kwolek (1971) and Salunkhe *et al.* (1982), and of protein content and composition by Bell (1971), Mossé and Pernollet (1982) and Gupta (1982). A point of some nutritional significance which emerges is that, relative to amino acid profiles in other pulses, contents of tryptophan and sulphur amino acids are low (Boulter and Derbyshire, 1971).

The occurrence of toxic materials and anti-metabolites in the seed has also been investigated. The presence of cyanogenic glycosides has been established, as has that of protease inhibitors and lectins (Toms and Western, 1971; Liener, 1982). Some evidence for presence–absence polymorphisms has also emerged. Although toxic material and anti-metabolites have been identified from the hyacinth bean these do not pose serious difficulties in its utilisation.

Domestication and evolution

Phytogeography and ecology

The hyacinth bean has been regarded as an essentially Asiatic species, perhaps on the grounds that it is most widely cultivated there, covering virtually the whole of tropical Asia (Purseglove, 1974). However, more recently Verdcourt (1970*a*) and Zeven and Zhukhovsky (1975) have postulated an African origin. It is cultivated in north Africa as well as in tropical sub-Saharan Africa. It is effectively absent from extensive cultivation in southern Africa, although it can be found in central Africa. A good rainfall is required to establish the crop (2–3 months) after which, as noted, it tolerates drought well; it is, however, not tolerant of water-logging but tolerates a wide range of temperature régimes from warm temperate to humid tropical (Duke, 1981).

The question of the geographical origin of this bean has perhaps been confused by the recognition in *Flora Indica* by Roxburgh (1874) of seven varieties, five cultivated and two wild. The status of Indian wild forms has not been confirmed as such; it is possible that these might represent early escapes from cultivation. The accurate location of the centre of origin for a cultigen depends critically on the location of the natural area of distribution of conspecific wild relatives. In Verdcourt's view (1980) this is clearly East Africa.

Evolution of form

The domestication and evolution of the hyacinth bean has scarcely been considered and not studied in detail. There is a general belief that it has been in cultivation for a long time. Certainly it shows strong development of characters which in legumes are indicative of domestication. Variants differing in growth habit are known and variation in pod shape and seed number per pod is well established. Parallels with other pulse crops are found in the reduction of stem, flower and testa pigmentation. The range of colours is from black to cream; some variants are self-coloured, others are speckled. There is a broad tendency for the darker flowers, seeds and pigmented stems to be correlated. It is reasonable to presume that reduced pigmentation can be related to the improved cooking quality this seems generally to confer.

9.2. Horse gram (*Macrotyloma uniflorum* (Lam.) Verdc.)

Introduction

The horse gram is a relatively low-grade pulse cultivated most extensively in southern India. The seeds are not very palatable and tend to have an earthy aroma (Duke, 1981). It persists in Indian cultivation because it is hardy, drought-tolerant and can be grown on a range of soil types. It does have some potential as a forage and can be used as a green manure.

Classification and biosystematics

The horse gram was formerly included in the genus *Dolichos* but has been transferred to *Macrotyloma* leaving no (significant) cultigens in the genus *Dolichos*. The genus *Macrotyloma* now comprises some 25 species (Verdcourt, 1980). The genus *Dolichos* as recognised by Linnaeus was an extremely heterogeneous group and it is only recently that the biosystematic relationships of its former constituents have been satisfactorily clarified. The yellow-flowered species formerly assigned to *Dolichos* (including horse gram) were recognised as belonging properly to their own genus, *Macrotyloma*. There is a report in the literature (Stanton, 1966) of hybridisation between the hyacinth bean and horse gram; this would imply a closer affinity than recognised by the new nomenclature. In the absence of confirmation of this report it seems reasonable to accept the revised nomenclature, especially as a difference in chromosome number is established between the hyacinth bean ($2n = 22$) and horse gram ($2n = 20$). As Verdcourt (1980) has observed, many of the problems of classification here arise from nomenclatural difficulties rather than intrinsic biosystematic problems.

Domestication and evolution

There has been remarkably little incentive to study domestication and evolution of this crop. Polymorphism for testa colours has been observed; non-cryptic testa colours and patterns occur in addition to the cryptic mottled primitive form. There appears to be some genetic diversity within the species; the prospects for increased exploitation probably lie in use as forage or in livestock feeds. The possibility of developing a suitable fermentation technology for processing the seed and also perhaps for protein extraction and TVP production, may in the future open up prospects for further exploitation.

9.3. Kersting's (Hausa) groundnut (*Macrotyloma geocarpum* (Harms) Maréchal & Baudet

Introduction

This is a rare and interesting crop cultivated in parts of West Africa. It is probably better known from the literature than in the flesh. It is undoubtedly an under-exploited legume and the main reason for this state of affairs is clearly apparent. Although it produces a very palatable and nutritious bean-like seed, the yield is quite consistently low. Its status as an agricultural crop is therefore decidedly precarious.

Classification and biosystematics

The general morphology is typical of the *Phaseoleae* and can be described simply as bean-like (i.e. like *Phaseolus* species) with an important difference. It produces geocarpic fruit, which ties it to a prostrate growth habit. The sequence of events after fertilisation of the ovules is remarkably similar to that in the groundnut (*Arachis hypogaea*). The fertilised ovules are pushed into the soil by geotropic growth of a carpophore produced by meristematic growth at the base of the ovary and the fruit matures underground. The mature pods are relatively short and contain 1–3 (commonly 2) seeds and do not commonly exceed 30 mm in length. The testa colour in cultigens is variable, white, brown or black, self-coloured or speckled. The flower colour of pale-seeded genotypes tends to be white while those with dark seed tend towards bluish-purple. The mature seed is susceptible to attack by weevils in storage and in generally bean-like in form.

Hepper (1963) collected the wild plant in the Cameroons; it strongly resembled the cultivated form but was appreciably less robust and more

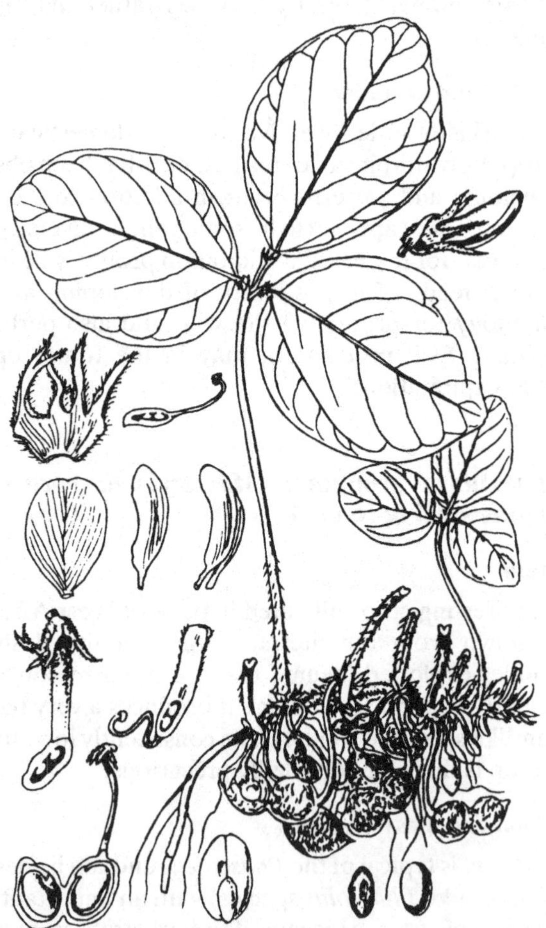

Fig. 9.2. The Hausa (Kersting's) groundnut, *Macrotyloma geocarpum* (from Duke, 1981).

delicate in vegetative structure. In general the cultigen shows gigantism of most parts, vegetative and reproductive, coupled with a more compact growth habit. It is quite clear that responses to domestication have been evoked by selection which are typical of the *Phaseoleae*. Pod and seed size have been increased, a non-cryptic testa colour and pattern polymorphism has developed, leaves and stems are larger and more robust while internodes are shorter. These effects have obviously been achieved in West Africa as its botanical origin is clearly there, but its cultivation has extended little if at all beyond its area of origin.

Little experimental study of the species has been carried out; there is perhaps some uncertainty on chromosome number. Counts of $2n = 20$ and $2n = 22$ have both been recorded. According to Goldblatt (1981) the correct count is $2n = 20$ rather than $2n = 22$ (the latter is the more typical of the *Phaseoleae*). It is only comparatively recently (Maréchal and Baudet, 1977) that this species has been assigned to the genus *Macrotyloma*. The geocarpic habit is no longer considered adequate grounds for exclusion from *Macrotyloma*, the other species of which produce normal aerial pods.

The National Academy of Sciences (1979) consider Kersting's groundnut to be an under-utilised crop, the major obvious reason for which is its poor yielding capacity. The poor storage quality, that is extreme susceptibility to damage by insects in storage, is probably a contributory cause. It is possible that some improvement might be effected by breeding and selection but unfortunately the gene pool of the species does not appear to be very large at the present time. It might usefully be enlarged by artificial mutagenesis.

9.4. The jack bean (*Canavalia ensiformis* (L.) DC.) and the sword bean (*C. gladiata* (Jacq.) DC.)

Introduction

These two very similar beans have geographically diverse origins: the jack bean is from the New World while the sword bean is from the Old. They are both commonly used for forage, green manuring and as cover crops in soil erosion control. Their use as food is restricted. The immature pods and seeds can be eaten but use of the mature seed as food requires care in preparation. The hazards can be reduced by cooking with salt in one or more changes of water. The overall appeal of these beans to the consumer is low and they are unlikely to find extensive use as a pulse when alternatives are available. Poor texture and flavour probably account for their lack of popularity.

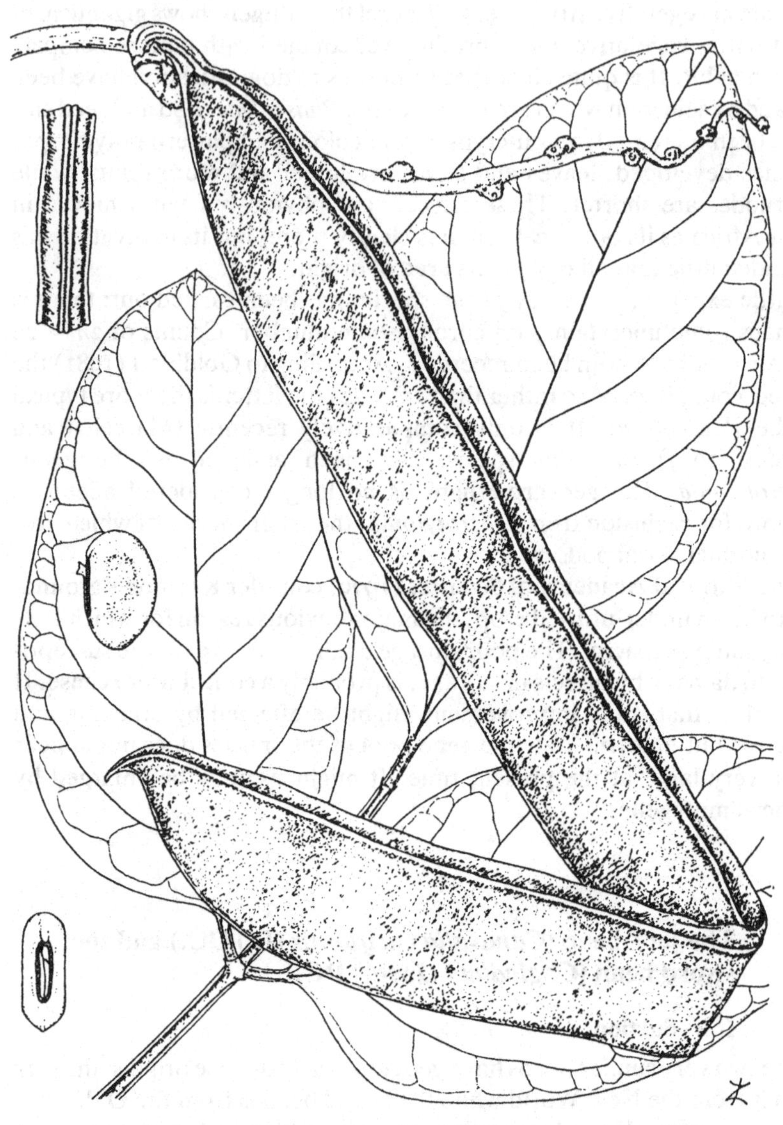

Fig. 9.3. The jack bean, *Canavalia ensiformis* (from Westphal, 1974).

Fig. 9.4. The sword bean, *Canavalia gladiata* (from Duke, 1981).

Classification and biosystematics

Taxonomy

The genus *Canavalia* comprises some 50 species (Sauer, 1964; Lackey, 1981). The status of some forms, such as *C. plagiospermus* Piper, is dubious. This is probably only a geographic race of *C. ensiformis* from S. America, the more typical form of which is found in Central America and the Caribbean. Purseglove (1974) suggested that *C. virosa* Wight & Arn., a wild bean found in tropical Asia and Africa, is the ancestral form of *C. gladiata*. Westphal (1974) favours the view that these three forms together constitute a single species.

Hybridisation

Experimental hybridisation with sword beans is difficult using conventional techniques of emasculation and pollination. The flowers are very sensitive to emasculation damage and emasculated flowers usually abscise. Bud pollination is likely to be much more successful. If interspecific crosses are to be attempted it is important to ensure that the introduced pollen has adequate time to effect fertilisation before the flowers' own pollen can act. Crosses between reputed species from cultivation and the wild could shed considerable light on biosystematic relationships within the genus and also clarify our ideas on the evolution of the cultigens.

Chemotaxonomy and biochemistry

Although *Canavalia* beans can be regarded as under-exploited in agriculture, they have provided fruitful material for study by biochemists. A very wide variety of compounds found in the seed have been extracted and investigated. This began sixty years ago when Sumner (1926) extracted and crystallised the enzyme urease, a substance of value in clinical laboratories. Boulter and Derbyshire (1971) found that the seed proteins in *C. ensiformis* and *C. gladiata* were so similar as to be identical for all practical purposes. Seed protein characters were therefore of no value in distinguishing the two species. The seed protein content is 20% (Wolff and Kwolek, 1971). The occurrence of the free amino acid canavanine is of some chemotaxonomic interest and its distribution has been studied by Lackey (1977) in the *Phaseoleae*. It is of metabolic significance and interest because of its chemical similarity to arginine (Bell, 1971). Lectins have been found in the seed and have been studied (Liener, 1982) since Sumner (1919) first isolated concanavalin A. In addition, protease inhibitors that inhibit activity of trypsin and chymotrypsin have been found. Wolff and Kwolek (1971) have placed these beans in their low lipid group (±2.6%). The content of polysaccharides (starch and amyloids) is relatively high (Bailey, 1971). The presence of flatulence factors in the seed has been noted (Courtois and Percheron, 1971); alkaloids have been reported in addition (Mears and Mabry, 1971) as also have saponins (Charavanapvan, 1943). Gibberellins in *Canavalia* spp. have also been studied (Harborne, 1971).

It is curious that in spite of all this biochemical knowledge, the relative contribution of compounds with toxic potential to the overall toxicity of the mature beans has not been satisfactorily worked out. There has, however, been rather little incentive to study in depth the biochemical problems restricting consumption and to devise solutions on account of the low appeal of the mature bean as a pulse. A somewhat similar situation

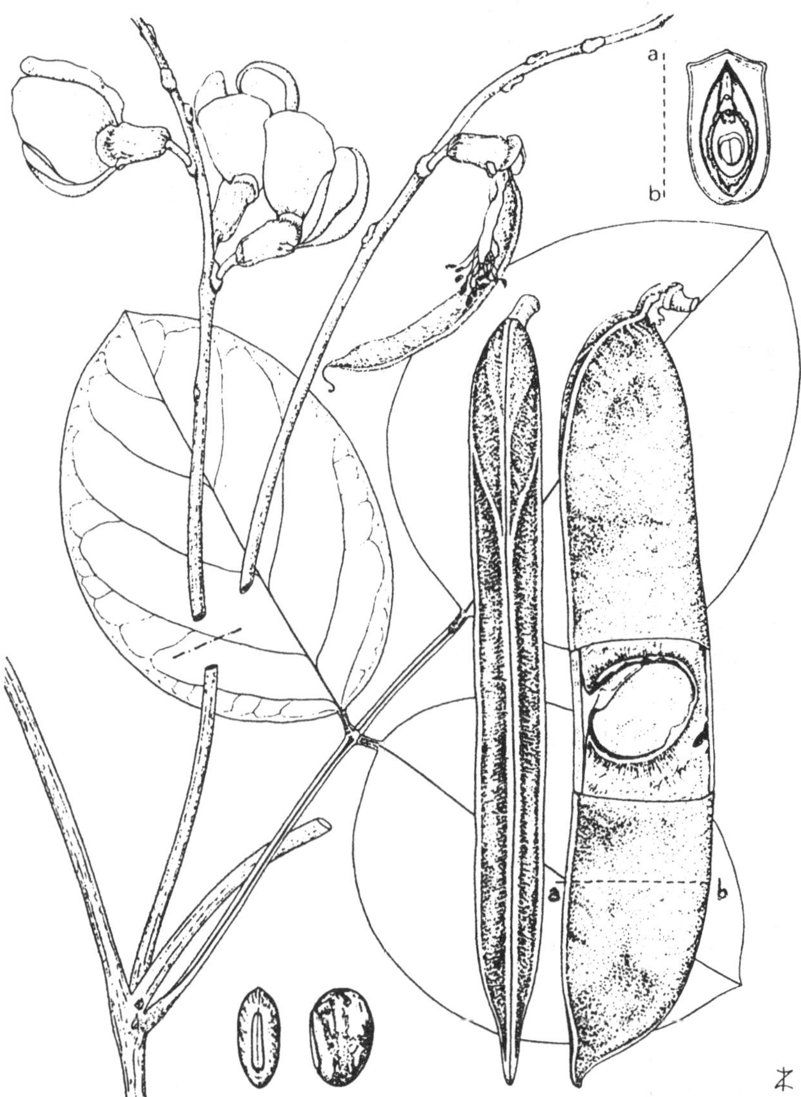

Fig. 9.5. *Canavalia virosa* (from Westphal, 1974).

Table 9.1. *Taxonomy of* Canavalia

Genus *Canavalia* Adans.

Sub-genus *Catadonia* Sauer
1. *Canavalia nitida* (Cav.) Piper
2. *C. bonariensis* Lindl.
3. *C. mandibulata* Sauer
4. *C. eurycarpa* Piper
5. *C. parviflora* Benth.
6. *C. macropleura* Piper
7. *C. sericophylla* Ducke

Sub-genus *Wenderothia* (Schlecht.) Sauer
8. *C. altipendula* (Piper) Standl.
9. *C. obidensis* Ducke
10. *C. concinna* Sauer
11. *C. bicarinata* Standl.
12. *C. septentrionalis* Sauer
13. *C. oxyphylla* Standl. & L. O. Williams
14. *C. dura* Sauer
15. *C. glabra* (Martens & Gal.) Sauer
16. *C. picta* Mart. ex Benth.
17. *C. grandiflora* Benth.
18. *C. mattogrossensis* (Rodr.) Malme
19. *C. villosa* Benth.
20. *C. hirsutissima* Sauer
21. *C. palmeri* (Piper) Standl.
22. *C. matudae* Sauer
23. *C. acuminata* Rose

Sub-genus *Canavalia*
24. *C. piperi* Killip & Macbride
25. *C. plagiosperma* Piper
26. *C. ensiformis* (L.) DC.
27. *C. brasiliensis* Mart. ex Benth.
28. *C. boliviana* Piper
29. *C. dictyota* Piper
30. *C. gladiolata* Sauer
31. *C. gladiata* (Jacq.) DC. var. *gladiata* and var. *alba* (Makino) Hisauchi
32. *C. regalis* Dunn
33. *C. virosa* Roxb.
34. *C. madagascariensis* Sauer
35. *C. papuana* Merr. & Perry
36. *C. aurita* Sauer
37. *C. ramosii* Sauer
38. *C. mollis* Wallich ex Wight & Arn.
39. *C. macrobotrys* Merr.
40. *C. cathartica* Thouars
41. *C. lineata* (Thunb.) DC.

Table 9.1. (*cont.*)

42. *C. maritima* (Aubl.) Thouars
43. *C. sericea* A. Gray
44. *C. vitiensis* Sauer
45. *C. megalantha* Merr.
46. *C. raiateensis* Moore

Sub-genus *Maunaloa*

47. *C. pubescens* Hook. & Arn.
48. *C. kauaiensis* Sauer.
49. *C. galeata* (Gaud.) Vogel
50. *C. molokaiensis* Deg.
51. *C. hawaiensis* Deg.

occurs in velvet beans, which are otherwise capable of producing very high yields but have unsolved problems of seed toxicity which preclude their use.

Domestication and evolution

Geographic distribution

Canavalia species occur, as has already been noted, in both the Old and the New Worlds. This distribution, apart from the reciprocal exchanges of *C. ensiformis* and *C. gladiata* between the hemispheres, appears to owe little or nothing to the activities of man. Westphal's (1974) suggestion, that *C. ensiformis*, *C. gladiata* and *C. virosa* are morphologically so similar as to constitute in effect a single species, raises the possibility that these are geographical or domesticated races within a single biological species. It should be a straightforward exercise to test this hypothesis experimentally.

The jack bean and sword bean are not very exacting in their cultural requirements. They are tolerant of drought when established and, unusually for legumes, of waterlogging. They are capable in good conditions of producing high yields of green matter for forage (most palatable when dried) up to 60 t ha^{-1} in Puerto Rico; corresponding seed yields of 5.4 t ha^{-1} have also been obtained. More commonly yields are of the order of 1.5 t ha^{-1} (Duke, 1981).

Archaeology

The jack bean has been recovered from archaeological sites in Mexico dated at approximately 3000 BC (Sauer and Kaplan, 1969). No comparable reports have been published of the sword bean from Old World sites (Purseglove, 1974).

Evolution of form

As the jack bean and sword bean have become widespread in both the Old and the New World tropics (although their use as both a vegetable and a pulse crop is limited), it would seem reasonable to conclude that they have achieved this wide distribution because of their value as a forage crop and perhaps later as a green manure and cover crop. The large seed and pod would seem to have been virtually pre-adaptations to cultivation as a pulse and vegetable crop but, whereas other species have responded to selection for reduction in biochemical toxins, the *Canavalia* beans appear not to have responded to such selection. The history of both beans in domestication would be clarified by further elucidation of the biosystematic relationships between *Canavalia* species. Purseglove (1974) has suggested that *C. gladiata* is related to *C. virosa* as a derived cultigen but no such progenitor type has been suggested among New World species for *C. ensiformis*. If, however, Westphal (1974) is correct in suggesting that both cultivated Canavalias constitute a single species, then both could have derived ultimately from *C. virosa*. *C. gladiata* with its perennial life form and twining growth is closer to the wild species than *C. ensiformis*, which appears to be more advanced with its annual life form and a more restrained and bushy growth habit. This is not an unattractive hypothesis but more supporting evidence is needed to give it reasonable credibility.

Further evolutionary potential

A possible future role for the sword and jack beans might develop if the mature seed were to be produced in quantity and processed on an industrial scale. Extraction of the protein and inactivation of the lectins and protease inhibitors might well produce a useful protein-rich raw material for further processing. The undesirable non-protein amino acids, alkaloids, saponins, etc. could probably be removed without great difficulty in an industrial extraction system. Whether such a capital-intensive food technology could develop in the tropics remains to be seen. The most favourable area for such a development could be the Far East, following the example of Japan in the industrialisation of traditional food fermentation technologies of the region.

9.5. Conclusions

The problem which often faces those interested in promoting the cultivation of legumes for human consumption is to determine why certain species are not more widely grown. It is important to determine precisely which factors are in fact limiting wider or more intensive exploitation.

These limiting factors can cover a wide range. The crop may have very specific climatic and cultural requirements; there may be acceptability problems ranging from palatability to taboos. There may be no specific niche for a particular crop in a local farming system, or it may not be amenable to traditional cooking methods. Whatever the reason or reasons behind the under-exploitation of particular species, it is important that these be very carefully ascertained. They should not be ignored or minimised; the potential producer and/or consumer is the last person in the world to be seduced by the public relations type of promotional hyperbole which is sometimes apparent. It seems that some promotional agencies succeed only in convincing themselves by their propaganda. Although acceptability problems are not necessarily insuperable, it can be a very slow process to bring about significant changes in taste.

From the brief consideration in this chapter, it is clearly apparent that a diversity of factors discourage more extensive production of some crops. There can be little doubt but that the Hausa groundnut would be more widely cultivated if its yields were higher. There would be a greater incentive to cultivate sword and jackbeans if they were of better flavour and texture. Future prospects for grain legumes with such acceptability problems could be greatly improved if they were found to be useable sources of protein for production of textured vegetable protein or suitable for use in production of foods produced by industrial-scale fermentation.

10 Germplasm resources and the future

Although the grain legumes have shown remarkably similar patterns of evolutionary response to the selection pressures which have operated under domestication (Smartt, 1976*a*, *b*, 1978*a*, 1980*a*), the genetic resources which are available for future conscious, man-directed evolution of pulse and legume oilseed crops are very different in both their nature and extent for the different species. The present time is crucial for collection and conservation of crop genetic resources; in the case of grain legumes no comprehensively consistent or coherent strategy has as yet evolved, but might well do so following the guidelines of Ford-Lloyd and Jackson (1986). There could be a considerable gain in the effectiveness of this effort if there were to be more overt rationalisation, co-ordination of conservation activity and the adoption of consistent procedures.

The work of Harlan and de Wet (1971) gives a very sound foundation on which to base genetic resource conservation strategies. Some modification of their approach may be necessary for grain legumes owing to the distinctly different pattern of biosystematic relationships found in the Leguminosae as compared with the Gramineae on which their work was largely based. On the whole, inter-specific and even inter-generic hybridisation is more common in the grass family than in the legumes, and development of polyploidy is much greater in the grasses, even though both major oilseed legumes (groundnut and soyabean) are in fact polyploid (tetraploid). Where inter-specific hybridisation occurs readily in the legumes, especially between autogamous species, the indication is that the evolutionary divergence of compatible species has been comparatively recent, e.g. in *Arachis* (Gregory *et al.*, 1980). The range of inter-specific hybridisation possible in many 'good' legume genera is often very restricted indeed. Inter-generic hybridisation in the legumes is negligible and where it does occur, e.g. *Cajanus* × *Atylosia*, it is generally construed as casting doubt on the validity of the genera (De, 1974). McComb (1975) has concluded that hybridisation between 'good' legume genera is in fact of very little significance. This contrasts with the Gramineae where, for example, two genera *Festuca* and *Lolium* are maintained with general agreement in spite of the occurrence of inter-generic hybrids between them. One area of incompletely resolved generic status concerns *Dolichos*, *Lablab* and *Macrotyloma*, in which an unconfirmed report of inter-specific (?inter-generic) hybridisation between *Lablab niger* and *Macrotyloma uniflorum* (both formerly included in *Dolichos*) perhaps calls into question the status of all three genera.

10.1. Hierarchy of gene pools

Harlan and de Wet (1971) considered that the gene pool concept could usefully be applied to ease the difficulties in producing satisfactory taxonomies for cultivated plants and their relatives. Their system is of equal value in consideration of genetic resources, for purposes of their classification, evaluation and documentation in particular. It has already been discussed in connection with individual crop species and genera but further discussion in some detail of this important concept is appropriate.

To recapitulate, there are three basic gene pools, namely:

(a) Primary gene pool (GP1) equivalent to the biological species between the components of which gene flow is not seriously impeded.

(b) Secondary gene pool (GP2) equivalent to the syngameon or coenospecies; this includes all species which can produce viable, partially fertile hybrids able to exchange genes to a limited extent.

(c) Tertiary gene pool (GP3) includes those species between which hybridisation is possible but hybrids are sterile, inviable or otherwise anomalous.

From the standpoint of germplasm resources in general, some amplification of this basic scheme is desirable. Harlan and de Wet (1971) have clearly recognised that the wild and domesticated components of GP1 merit distinction, a point which has been emphasised by Smartt (1981*a*). In allogamous forms, however, it may be quite difficult to maintain the desired degree of separation between GP1A (domesticated) and GP1B (wild) segments of this gene pool. This is not, however, a serious problem with the great majority of grain legumes, which are autogamous.

In terms of actual utilisation a useful additional distinction can also be drawn between domesticated and wild components of the primary gene pool. They constitute, from the point of view of the plant breeder, gene pools of first and second choice within the total gene pool of the biological species the GP1. The original definition of the secondary gene pool (GP2) is perfectly satisfactory as it stands. However, some closer definition of the tertiary gene pool (GP3) could be useful. In a group such as the legumes, where successful inter-specific hybridisation is relatively uncommon, the tertiary gene pool might well be considered to include all those taxa which can cross-fertilise the cultigen. There are arguably

Table 10.1. *A possible extension of Harlan and de Wet's gene pool system*

Gene pool	Constituents	Harlan and de Wet category	Experimental taxonomic category
1st order	(*a*) cultigen	GP1	same ecospecies (biological species)
2nd order	(*b*) (weedy form) (wild counterpart)		
3rd order	cross-compatible species producing ± fertile hybrids	GP2	same coenospecies (syngameon)
4th order	cross-compatible species producing viable but sterile hybrids	GP3	same comparium
5th order	cross-compatible species producing inviable hybrids		
6th order	incompatible related species		

grounds for subdividing this gene pool, if exploitation became feasible, on whether hybrids produced are viable or inviable. This would approximately equate the order of the range of gene pools with the degree of genetic isolation maintained by the breeder or that which is determined by naturally evolved isolating mechanisms. If more extensive exploitation of the GP3 developed or if germplasm resources were very restricted, it might be useful to extend Harlan and de Wet's scheme as in Table 10.1.

10.2. Factors controlling development of gene pools

The most important gene pool from the point of view of the plant breeder is without doubt Harlan and de Wet's GP1A. This gene pool presents fewer problems in utilisation than any other. The factors, therefore, which have influenced the extent of its development are of considerable importance. It is axiomatic that wide geographic dispersion promotes the development and maintenance of genetic diversity by a combination of founder effect, hybridisation, genetic drift and selection with mutation; these together can often produce and maintain highly distinctive genotypes in specific areas. Different modes of utilisation, and varying incidence of pathogens and pests, are two potent factors in maintaining diversity. Ecological diversity, intensity of cultivation and diversity of usage together with moderately high population densities in producing regions also serve to generate and maintain high levels of genetic variation. These factors can act and interact cumulatively to amass very extensive resources in a cultigen, to produce both rich primary and secondary centres of diversity in pulses which have extended their distributions from the Old World to the New and vice versa in post-

Columbian times. A good exemplar is the cultivated groundnut, which has developed a very extensive secondary centre of diversity in sub-Saharan Africa (Gibbons *et al.*, 1972).

The extent of variability in GP1B depends on the present range and distribution of the wild conspecific counterpart of the cultigen. This may be very restricted in the case of *Arachis hypogaea* (*A. monticola*) (i.e. north-west Argentina) (Krapovickas and Rigoni, 1957) but very extensive in *Phaseolus vulgaris* (*P. aborigineus*) (i.e. from Mexico to north Argentina) (Burkart and Brücher, 1953; Miranda Colín, 1968; Gentry, 1969). In one, perhaps exceptional, case at least there appears to be no known GP1B, namely *Vicia faba*; in virtually all other cultigens that have been adequately collected and investigated a GP1B appears to exist.

It is perhaps in development of the GP2 that the greatest variation is shown between different grain legumes. The first relevant consideration is the species richness of the genus; the second is the biosystematic relationships which exist within it. A crop which belongs to a genus or infra-generic taxon which shows considerable and comparatively recent speciation is likely to possess a more extensive GP2 than one which has developed in a small genus and/or one in which recent evolutionary diversification has not apparently occurred. The genus *Arachis* furnishes examples of both situations at the sectional level. The common cultivated groundnut is a member of section *Arachis*, which apparently is of recent evolutionary origin; extensive gene flow is possible between taxa within this section in spite of the occurrence within it of both diploid and tetraploid species. The rare diploid cultigen *A. villosulicarpa* is a member of a different and more ancient section, the *Extranervosae*. Intra-sectional hybridisation is possible but hybrids are sterile and little gene flow seems possible between species. This section at the present time comprises only four or five known species (Gregory *et al.*, 1980). The consequence of this difference in apparent evolutionary age is that *A. hypogaea* has a secondary gene pool but *A. villosulicarpa* has not. A somewhat comparable situation is found in *Phaseolus* where *P. vulgaris*, *P. coccineus* and *P. polyanthus* are related so that each species is part of the secondary gene pool for the other, while *P. lunatus* and *P. acutifolius* appear to have no GP2.

In legumes (and perhaps in other crop plants) there is some difficulty in defining the GP3 when its extent is determined by production of hybrids which may be 'sterile, inviable or anomalous'. Although it is comparatively easy to compile data on reported viable but sterile hybrids, published reports may not be sufficiently detailed regarding inviable hybrids. Early aborted inviable hybrids may not be reported on the one hand and even the occurrence of fertilisation may be inferred when it has not actually taken place on the other. G. Froussios (unpublished) observed parthenocarpic development of pods in *Phaseolus vulgaris* following emasculation. Smartt (1979) noted that the ovules in such pods remained hyaline whereas early aborted fertilised ovules were brown and autolysed. Any presumption of fertilisation in *P. vulgaris* based solely on pod production after attempted inter-specific hybridisation clearly cannot be justified.

The suggestion is advanced that in these circumstances the GP3 be subdivided into A and B components to include viable but sterile hybrids in A and inviable hybrids in B. The B division would include all species between which cross-fertilisation could be demonstrated with the appropriate cultigen.

In taxonomic terms, then, the position can be summarised that the primary gene pool approximates to the taxonomic species where it is possible to apply the biological species concept. The secondary gene pool may, in some cases at least (e.g. in *Arachis*), correspond to a division of the genus such as a section. This depends on the evolutionary state of isolating mechanisms *vis-à-vis* morphological divergence, which do not necessarily go hand in hand. It would be unwise in most grain legumes to attempt any generalisation on this point. The tertiary gene pool may well correspond in its extent to the genus as in the case of *Arachis*, which is well isolated taxonomically from its closest allies. The situation in the Vicieae could be different, where generic boundaries are much less sharp (as between *Pisum*, *Vicia* and *Lathyrus*, for example) (*vide* Gritton and Wierzbicka, 1975) and the GP3 may extend beyond generic boundaries.

10.3. Genetic resources profiles

Perhaps the most useful way of looking at the genetic resources of a crop is in terms of a genetic resources profile. This ideally would record not only the presence or absence of the three major gene pools and their subdivisions but also the magnitude of each, on a simple scoring system (0–5) covering the range from absence to the very substantial. If the groundnut were to be taken as an example, the known genetic resource constituted by the collection of cultivars, landraces and breeders' lines is very substantial and could be rated 5; in contrast the wild counterpart *A. monticola* constitutes a very small resource which could be given a rating of 1. The section *Arachis* comprises approximately ten diploid known wild species in addition to the tetraploid cultigen and its wild counterpart; thus the GP2 might be rated 3. The GP3, which is rather more extensive, might be rated 4. The genus *Arachis* thus constitutes a remarkably well-balanced genetic resource which is available for improvement of the groundnut.

Comparison with other cultigens in, for example, the genus *Phaseolus* reveals numerous differences in the profiles of gene pools produced. In

Table 10.2. *Known germplasm resources in grain legumes*

			Gene pools				
Genus	No. of species	Cultigens	GP1A	1B	2	3A	3B
Arachis	± 60 spp.	*A. hypogaea*	+	+	+	−	+
		A. villosulicarpa	+	?	−	+	+
Cajanus (*Atylosia*)	2 + 35 spp. of *Atylosia*	*C. cajan*	+	?	+	?	?
Cicer	40 spp.	*C. arietinum*	+	+	+	−	+
Glycine	9 spp.	*G. max*	+	+	−	+	+
Lablab	1 sp.	*L. niger*	+	+	?+	?+	?
Macrotyloma	24 spp.	*M. geocarpum*	+	+	?	?	?
		M. uniflorum	a	?+	?+	?+	?
Lathyrus	150 spp.	*L. sativus*	+	+	?	?	?
Lens	5 spp.	*L. culinaris*	+	+	+	?	?
Phaseolus	50 spp.	*P. acutifolius*	+	+	−	+	+
		P. coccineus	+	+	+	+	+
		P. lunatus	+	+	−	+	+
		P. vulgaris	+	+	+	+	+
Pisum	2 spp.	*P. sativum*	+	+	?+	?+	+
Psophocarpus	9 spp.	*P. tetragonolobus*	+	?+	?	?	+
Vicia	140 spp.	*V. faba*	+	−	−	−	+
Vigna	150 spp.	*V. aconitifolia*	+	+	?	+	?
		V. angularis	+	+	+	+	?
		V. mungo	+	+	+	+	?
		V. radiata	+	+	+	+?	
		V. subterranea	+	+	?	?	?
		V. umbellata	+	+	+	+	?
		V. unguiculata	+	+	−	−	?

Phaseolus vulgaris there is, as in *A. hypogaea*, an extensive GP1A but unlike the latter it has a very considerable GP1B represented by the wild prototype. This has a very extensive range from Mexico to Argentina and represents a substantial, if poorly evaluated, genetic resource. A further contrast is afforded by the GP2, which is much more substantial for *A. hypogaea* than for *P. vulgaris*. The probable extents of GP3 for the groundnut and common bean, however, are probably similar in order of magnitude.

On the basis of present biosystematic knowledge, principally derived from collection activity and experimental taxonomic studies involving inter-specific hybridisation, it should be possible to construct, for most grain legumes, at least tentative germplasm resources profiles involving the simple scoring system outlined. The basis for which could be an amplification of the kind of information presented in Table 10.2, which could be used to plan, where appropriate, further germplasm collection strategies and also strategies for exploitation of available germplasm in breeding programmes. It would help to avoid pitfalls which could lead to wasted effort in employing inappropriate strategies in the collection of germplasm resources.

10.4. Genetic resources and breeding objectives

The establishment of rational breeding objectives for grain legumes is a complex matter and the approach to it is facilitated by a consideration of current breeding problems and how these can be most effectively tackled employing known genetic resources. Since the best-balanced grain legume genetic resources collection and conservation programme which has been in operation over the past three decades has been concerned with the groundnut, this can be taken as exemplar and reference made as appropriate to other crops.

The excellent balance which has been achieved in the collection and conservation of *Arachis* germplasm has been due to the pioneering activities of two men, Gregory and Krapovickas, together with their associates (Gregory *et al.*, 1980). It is to their credit that in collection they have collected not only wild species but also South American landrace material. In addition the biosystematics of the genus has been studied intensively by experimental taxonomic methods (Gregory and Gregory, 1979). The nature and accessibility of the genetic resource represented by the wild species is now quite well understood. This is quite remarkable considering that many wild species of the critically important section *Arachis* are to be found only in some of the more remote areas of South America in hazardous terrain. Unfortunately, since so much effort has been expended in studying these collections of wild species, those of the landrace material have suffered comparative neglect. This can perhaps best be illustrated by the quest for improved resistance to the *Cercospora* leafspots in the cultivated groundnut. Abdou (1966) showed that the wild species *A. cardenasii* and *A. chacoense* had resistance to one or other of the leafspot pathogens. Although some resistance was also apparent in *A. hypogaea* landraces this was, in the greenhouse, less marked than that of the wild species. On the face of it, it appeared that a shortcut to improvement of leafspot resistance in the groundnut would be to exploit wild species resistance and this can be justified. What is less justifiable is the virtually total neglect of the landrace resistance which was observed by Abdou. This could be incorporated in new cultivars more readily than that from wild species. A further point which has been neglected is the genetic control of leafspot resistance and the investigation of resistance mechanisms in the host. It is possible that the effects of these mechanisms (if a number of mechanisms is found) might be additive and a large part of the agrichemical control costs might well be saved by their exploitation. Experience has shown that it may be unwise to rely on a single or a narrow range of resistance mechanisms in controlling the major pathogens of a crop. Accumulation and integration of several resistance mechanisms may be more effective in the long term; major gene resistance may be only a short-term palliative. Ultimately we may have to mobilise and exploit all available genetic resources in securing satisfactory control of pests and pathogens with the minimum economic and environmental penalties. It is short-sighted in the extreme to put all our disease control eggs in the one basket of major gene resistance, especially in the light of the cereal breeders' experience.

Another area in which inappropriate breeding strategies and objectives are adopted is in the production of pulse crops for the developing countries. Breeding objectives must be clearly related to an appropriate level of technology. Breeding methodology should also be the simplest consistent with efficiency. Sophisticated and advanced methodologies are quite out of place in most breeding programmes in the developing world, when the basic strategies of local collection, introduction and selection have not yet been fully exploited.

Many important pulse crops of the developing world are also grown extensively elsewhere under capital-intensive production systems. It is a tribute to the versatility of species such as *Phaseolus vulgaris* that they are capable of responding effectively to such highly divergent selection pressures. Many products of selection programmes for specialised, capital-intensive production are of little practical value in other systems. Many of their qualities are irrelevant elsewhere, since in this particular example green pod production is often the objective. Quality of mature seed is of obvious importance in the developing world and appropriate selection objectives for production would also include suitable photoperiod requirement or day-neutrality, low contents of anti-metabolites and toxic materials, high palatability and cookability (i.e. tender seeds), stable yields and tolerance of climatic variations between growing seasons, pest and pathogen resistance and general suitability for the local farming system. The available genetic variability in a pulse such as the common bean is such that considerable progress towards these objectives should be possible. However, it must be appreciated that some objectives may be incompatible. Generally speaking, for example, the most palatable pulses are those with the least pigment in testas; however, these have the lowest levels of inherent resistance to fungal pathogens. Testa pigments are phenolic compounds which are fungistatic if not fungicidal. It is thus easier to produce a satisfactory crop of a pigmented haricot bean than of a white-seeded cultivar. Harvest conditions must be near perfect to achieve a high quality, blemish-free crop of white haricots (navy beans); except under irrigation or with reliable rainfall distribution, this may be difficult to achieve in practice under erratic rainfall regimes.

The mode of exploitation of the crop must also be borne in mind. Many pulses grown in Africa, e.g. cowpeas and common beans, are exploited not only for their seeds but also for their leaves. These may be plucked and used as a spinach. Where this practice prevails, tolerance of defoliation should be assessed in breeding programmes. Traditional practices die hard and thus cannot safely be ignored. Very leafy genotypes might well be selected in such circumstances which would rightly be discarded otherwise.

It is generally accepted that pulse crops are more important in the developing world than elsewhere (Smartt, 1976*a*). They give variety to the diet as well as contributing protein, carbohydrate and other valuable nutrients. It is thus desirable that the place of legumes in the farming systems of the developing world be made secure in order to maintain nutritional standards. Equally powerful arguments can be advanced from the agronomic viewpoint. In some areas the position of legumes in the farming system is being undermined by improved cereal varieties. Landrace pulses cannot compete on economic terms with 'Green Revolution' cereals. They have often evolved under farming systems where inputs are minimal and frequently lack the capacity to respond to increased inputs. What landraces tend to achieve is a fairly stable but low yield on minimal inputs. The very real value of pulses under adverse conditions is exemplified by the grasspea (*Lathyrus sativus*), which produces yields under conditions in which other food crops, most notably cereals, fail completely. Unfortunately, excessive consumption of the grasspea leads to neuro-lathyrism. This has led to perhaps misguided attempts to ban cultivation of this crop (Rutter and Percy, 1984) (which would have promoted starvation) without stimulating effectively the initiation of programmes for selection of genotypes with acceptably low seed contents of the toxin. This has been successfully carried out on the lima bean (*Phaseolus lunatus*) in which acceptable contents of cyanogenic glycosides have been established in selected genotypes (Viehover, 1940); a similar goal might well be achieved in the grasspea by a comparable breeding programme.

The great advantage that legumes have over most other plants is their ability to benefit from the legume–*Rhizobium* symbiosis; this is all too often taken for granted. However, this advantage can be lost in part, as when the relatively primitive indeterminate climbing forms of the common bean, which nodulate more freely than the dwarf determinate cultivars, are replaced by them. Determinate growth and resulting early maturity are characteristic of many modern grain legume cultivars; this growth habit thus entails an obvious yield penalty. Delayed nodulation, which may occur rather late in the life cycle, is also apparent in many common bean dwarf cultivars. Selection for genotypes which nodulate earlier could lift this penalty somewhat.

The question of yield level in legumes is subject to inadequate presentation and even misrepresentation. There is a common tendency to ascribe a low yielding capacity to all grain legumes. Such a sweeping generalisation is quite unwarranted as very high yields of groundnuts have been produced under favourable conditions, ±6 tonnes ha^{-1} in Central Africa (Smartt, 1978*b*), and other high legume yields are also on record (Duke, 1981). Legume yields are often compared with those of cereals on a simple dry-matter basis, disregarding the very much higher calorific values of legume seeds (especially the oilseeds) *vis-à-vis* cereals. These very high groundnut yields and the unsound basis of current yield comparisons notwithstanding, the yield of many grain legumes in the developing world is none the less very low indeed. The groundnut itself has genotypes capable of producing high yields and others quite incapable of doing so. Detailed morphological, physiological, developments, and biochemical comparisons might yield some clues as to the causes of the difference. One factor too little appreciated is that success of a legume crop is largely dependent on the effectiveness of the host–*Rhizobium* symbiosis. If this is not effective the bacterium may be a virtual parasite rather than a symbiont. Bacterial and host genotypes must be compatible before effective symbiosis can occur. While superior compatible genotypes of both legume and bacterium can be selected, it is in practice easier to manipulate the host genotype than that of the soil-inhabiting bacterial population. Short of soil sterilisation and subsequent introduction of selected strains, selected *Rhizobium* strains will almost certainly be outcompeted by the pre-existing established strains. Crop genotypes should, if possible, be selected for the ability to produce effective symbiosis with local *Rhizobium* strains. The important thing to bear in mind is that an interaction between genotypes is involved and selection should take account of the nature of this interaction. When legume species such as the soyabean and clovers were introduced to new areas, spectacularly efficient nitrogen fixation was achieved. This is because matched genotype combinations of host and symbiont were customarily introduced. The success of the soyabean in North America and of clovers in Australia and New Zealand is due in part at least to this.

The occurrence of toxic materials and anti-metabolites is an impediment to easy exploitation of some grain legumes. The common bean (*Phaseolus vulgaris*) and the soyabean contain phytohaemagglutinins (lectins) and protease inhibitors. Polymorphisms for presence or absence or lectins have been established in the common bean (Klozová and

Turková, 1978) and soyabean (Orf *et al.*, 1978) and for the Kunitz trypsin inhibitor in the soyabean (Orf and Hymowitz, 1979*a*, *b*). It may well be possible to select for genotypes with significantly reduced contents of these substances. Their success in cultivation could depend on the existence and magnitude of any yield penalty entailed.

10.5. Innovative approaches to germplasm exploitation

It is commonplace in grain legumes that the genetic resources of many congeneric species are inaccessible as regards the improvement of grain legume crops. Two lines of approach to this problem are possible; the first is to attempt to produce somatic hybrids where sexual hybrids cannot be produced, and the second is to attempt to produce fertile polyploids from sterile F1 inter-specific hybrids. The prospects for successful somatic hybrids depend on the nature of the genetic isolating mechanisms between the two sexually incompatible species. If this is due to somatoplastic sterility (i.e. failure of embryo support system, endosperm, suspensor etc.) then there is some possibility of success. If on the other hand the two genomes are incompatible and incapable of working harmoniously together the end result may be the same as in sexual fusions. A further point to be borne in mind is that in transferring a gene or group of genes from the genetic background of one species to that of another (i.e. by chromosome addition or substitution), its effects may be modified to a greater or lesser degree by the change in genetic background. It must be recognised that the greater the difficulty in making such a genetic transfer, the higher the probability that the effectiveness of the transferred genes may be reduced. Such difficult transfers are in effect a last resort; they would be costly to effect and even if successful the results could be disappointing. We are certainly very far from achieving spectacular feats of genetic engineering in legumes. Some basic problems such as the induction of differentiation in tissue cultures have not been solved for the most important species.

There are two possible ways in which the products of somatic cell fusion might be exploited. The hybrid could be used directly in backcross programmes if it had appreciable fertility. If it were completely sterile some attempt should be made to double chromosome complement and restore fertility. Backcrossing could then be attempted. In addition some attempt could be made to induce inter-genomic interchange of chromosome segments by use of ionising radiation in the induced polyploid prior to such backcrossing. The ideal procedure to be developed would be to extract the desirable genes from the donor species and induce the incorporation of these in the recipient genome, by a transformation process. Such an operation would be extremely complex if more than a single or a few loci were involved.

The possibility of induced polyploidy is relevant to potential exploitation of viable but sterile hybrids. These have been induced in such inter-specific hybrids in the genera *Arachis* and *Phaseolus*, for example (Smartt and Haq, 1972; Smartt *et al.*, 1978*a*). Polypolidy is commonly induced by colchicine treatment but not all inter-specific F1 hybrids respond to this treatment. A great deal depends on the genotypes of the parental species lines used; some combinations will respond, others not (Prendota *et al.*, 1982; Thomas and Waines, 1984). The polyploids once obtained may not be immediately useful (Dhaliwal *et al.*, 1962) and may not be stable (Smartt and Haq, 1972). This may account for the relatively low frequency of polyploids among the legumes, although it is worth reminding ourselves that the two most important and successful legume grains are polyploids, the groundnut and soya bean. The possibilities of further exploitation of polyploidy among legumes should therefore be borne in mind. In general, however, it does appear that the genetic systems of the legumes as a whole do not accommodate polyploidy as readily as those of the Gramineae.

Some reference has been made to the possible development of inbred lines and the development of synthetic varieties in the more outbreeding species such as *Phaseolus coccineus*. It is possible that single-cross F1 hybrids might also be developed and offered commercially. Male sterility systems are not infrequently encountered (J. Smartt, unpublished) and it may well prove possible to develop F1 hybrid common beans which are commercially viable. It is a moot point whether the yield differentials between standard autogamous grain legume cultivars and single-cross hybrids would offer sufficient inducement either to breeders or to producers of the crop to take up F1 hybrids on a significant scale.

10.6. Future prospects: new domestications?

The present grain legume species carry with them the legacy of their past evolutionary history. Under domestication they have evolved new morphologies, and have established variant forms differing in anatomy, biochemistry and physiology from their progenitors. What may well have changed least is what might be termed their ecophysiology. This has probably remained within a relatively constant range; dissemination of crop species from their centres of origin has been determined effectively by the general ecology of the region to which they have been introduced. A good illustration has been given already by the faba bean, which as noted is a Mediterranean winter annual species. This has been introduced successfully to northern Europe and other cooler temperate areas as a summer-maturing annual crop and exploits broadly similar climatic conditions to those of its normal growing season in its own area of origin. Other suitable temperate environments have been found extensively in China and also in South America at high elevations, in Peru for example.

Although it is possible to achieve some response to selection for modified ecological tolerances, such changes are usually marginal. For example, it is highly unlikely that any amount of breeding and selection would produce a productive groundnut for cool temperate regions. The amount of potential latitude outside the present climatic tolerance range is likely to be small. Any attempt to extend cultivation of existing crop species substantially into marginal areas could carry a significant yield penalty. In addition it is quite possible that the stability of yield would also be poor; the case of the soyabean in Britain illustrates this point well.

If we cannot reasonably expect to extend effective cultivation of existing crop species into agriculturally marginal areas, is there no prospect of usefully extending cultivation into such regions? There is a possibility which is worth considering and that is to examine, carefully and critically, species in agriculturally marginal areas which are actually exploited and to explore the possibilities of domestication. The process of domestication in the past has been a lengthy one. At the present time there are considerable pressures to explore all possible means of accelerating the process. In these circumstances it is worth while to consider a model system which could test the efficacy of artificial mutagenesis in generating useful genetic variability. Such a system could be found in the legume tribe Phaseoleae.

The biosystematic background

It is suggested that, since the tribe Phaseoleae contains a large proportion of the domesticated grain legume crops grown at the present time, this could provide both a substantial range of potential new domesticates and a model system which could be used to monitor progress towards any goals which might be set. The justification for considering the Phaseoleae as a potentially valuable system for the study of accelerated domestication is readily apparent after a brief review of the biosystematics of the family. This very large family, the Leguminosae, contains three major sub-groupings: the sub-families Caesalpinioideae, Mimosoideae and Papilionoideae. These sub-families are regarded by some authorities as families in their own right; some consider that a fourth sub-family or family, the Swartzioideae, should be recognised.

The Papilionoideae

Of the species which have been domesticated, those which have been most extensively modified by selection are undoubtedly the grain legumes, which all belong to the Papilionoideae. In this sub-family according to the taxonomic scheme of Polhill and Raven (1981) there are 32 tribes; of these there are five tribes which contain species producing edible seeds, the Aeschynomeneae, Cicereae, Vicieae, Genisteae and Phaseoleae. The Aeschynomeneae contains only one species of major economic importance, *Arachis hypogaea*. The tribes Cicereae and Vicieae formerly constituted a single tribe, which included all the indigenous pulse crops of the Mediterranean and Near Eastern classical civilisations, five in number. The important grain legume members of the Genisteae are the lupins, some of which have been in cultivation for a considerable period and which have obvious potential for future development as grain legumes. However, the tribe which has contributed by far the greatest number of edible grain legumes to agriculture is undoubtedly the Phaseoleae. Within this tribe the sub-tribe Phaseolinae includes the majority of the successful grain legumes, most notably five species of *Phaseolus* and seven species of *Vigna*. The most important species of the tribe not included in this sub-section are first and foremost the soyabean and secondly the pigeonpea.

The genera *Phaseolus* and *Vigna*

This concentration of domesticated species in the Phaseoleae–Phaseolinae is of particular interest because it does not appear to be fortuitous. Their domestication has not been concentrated in a restricted geographic area but is world wide. *Phaseolus* species have apparently been domesticated in both Middle and South America from south-west Texas and north Mexico to Argentina. *Vigna* species, it is thought, have been domesticated in Africa (*V. unguiculata* and *V. subterranea*), in the Indian sub-continent (*V. radiata*, *V. mungo* and *V. aconitifolia*) and the Far East (*V. angularis* and *V. umbellata*). This pattern clearly must reflect a profound potentiality in these two genera to respond to the selection pressures which are imposed under domestication.

The actual response achieved is a reflection not only of the genetic potential of the species in question to respond but also of the range of selection pressures imposed, the range of environments under which these have operated and the size of the basic gene pool. If one considers the outcome of selection in *Phaseolus* species over the 7–9 millennia in which they are known to have been in cultivation, some very interesting points emerge. There can be no doubt but that the most successful in cultivation is *Phaseolus vulgaris*, the common bean, followed by *P. lunatus*, *P. coccineus*, *P. polyanthus* and *P. acutifolius* in approximately that order. What has contributed, then, to the greater success of *P. vulgaris* in domestication than its congeners? In the first place the wide geographic range of the wild prototype from Mexico to Argentina could be expected to generate a large gene pool. There is evidence to suggest (Gepts, 1985) that there have been two major areas of domestication (Mexico and Peru) and at least one minor centre (Colombia). In *P. lunatus* two major centres have been established, namely Peru and Mexico (Kaplan, 1965). There seems to be a consensus view that the pattern of domestication in *P. vulgaris* has been the more complex. Domestication of *P. polyanthus* is possibly similar to that of *P. lunatus* in that there are obvious Mexican and South American (Colombian) centres of diversity.

The genus *Phaseolus* has a very simple taxonomic structure (Maréchal *et al.*, 1978); it comprises three sections, *Phaseolus* (24 species), *Alepidocalyx* (2 species) and *Minkelersia* (5 species). All the cultigens are in the section *Phaseolus*. The genus in toto contains 31 species. The breakdown of the genus is fairly straightforward. This contrasts with that of *Vigna*, which is more species-rich (81 species) and has a much more complex structure, with 7 sub-genera; of these 4 are further broken down into sections which number 6 in sub-genus *Vigna*. The African cultigens belong to different sections in this sub-genus. The Asian cultigens are all members of the undivided sub-genus *Ceratotropis*, which contains only 4 totally wild species. Such a concentration of cultivated species within such a taxon is of considerable interest and suggests that it constitutes unusually good basic material for the production of domesticates. Such a group may also be relatively young in evolutionary terms; its members probably share a very similar basic genetic constitution and thus a similar capacity to respond to selection under domestication. A basically similar situation exists in *Phaseolus* where, arguably, this group is also of comparatively recent evolutionary origin and it also exhibits a comparable cluster of cultigens. A common feature of these clusters of *Vigna* and *Phaseolus* species is that they retain within each cluster some ability to produce inter-specific hybrids. This situation contrasts strongly with that of the cowpea, which is genetically isolated from its closest relatives and appears not to produce hybrids with them. The ability to produce inter-specific hybrids is markedly less in the legumes than in, say, the grasses and while it is possible to entertain the validity of claims for inter-generic hybridisation in the grasses it is not for legumes (McComb, 1975). In the legumes, then, groups such as the domesticated species clusters of *Phaseolus* and *Vigna* and their closest relatives offer very considerable scope for generation of experimental populations with high levels of genetic variability. The intrinsic variability within the species can be augmented in many cases by introgression in the first place and in all cases by artificial mutagenesis.

Process of domestication

New domesticates in *Phaseolus* and *Vigna*

The remarkable species clusters in *Phaseolus* and *Vigna* clearly indicate 'hotspots' of the kind of genetic variability which has led to successful domestication over a wide geographic range. However, by no means all species in these taxa have been exposed to the selection pressures of domestication. Domestication is by its very nature opportunistic; those plants which were domesticated were those immediately to hand. Species geographically remote from the early plant domesticators, with perhaps equal potential as domesticates, could well have been neglected subsequently because of the head start achieved by those species taken earliest into cultivation. By processes of cultural diffusion the idea of cultivating plants could have spread along with the protocrops themselves. If these could be grown in new locations they would be grown rather than initiating domestication of a new set of crop plants. Fortuitously, then, some plant species with good potential in domestication could by phytogeographic accidents have been totally neglected. However, some indications are available to pinpoint response potential and such are apparent in the Phaseoleae, i.e. taxonomic 'hotspots'. Genetic advance of domesticates under selection is, other things being equal, a function of time, and we do have to face the reality that any novel domesticate will face extremely strong competition from established crops with known histories of 9000 years of domestication and perhaps longer. Thus new domesticates will have to meet a need which no established crop can in specific and possibly marginal environments (J. M. J. de Wet, personal communication). The versatility of many of our established crop species and the proliferation of modes of exploitation over the years of the crop product has undermined the position of many long-established crops. This is noticeable especially among the oilseeds. The competition of relatively new sources of edible oil, e.g. maizegerm, has undermined the economic position of some of the classical traditional oil crops, which are increasingly being produced for specialist rather than mass markets. Such considerations could be of critical importance with respect to economic viability.

Justification for new domestication

The ever-changing economic balance between crops is of course influenced by considerations other than those of strict economics. Political, sociological and strategic pressures can be important in the evolution of new crops; the sugar beet is the prime example of this. The domestication of new species could be an important factor in meeting crises in drought-prone areas (J. M. J. de Wet, personal communiation) and more broadly in providing a base for agricultural self-sufficiency in developing economies. It may not always be possible to meet essential local needs from locally produced crops; imports may be necessary if for climatic or other reasons particular crops cannot be grown. The development of alternative crops is one way out of this particular difficulty. There are several strategies which could be employed, the actual choice being dictated by prevailing circumstances. The major possibilities are:

1 totally new domestication;
2 improvement of under-developed domesticates (second cycle domestication of A. Ashri, personal communication);
3 development of new modes of exploitation for existing cultigens.

Totally new domestication might well be indicated if there were a situation in which there was a pair of closely related species which had very different environmental requirements, one of which was domesticated and the other not. It might well be possible to produce homologous variants by selection in the wild species after mutagenesis and make progress towards establishment of a new crop. It could be assumed that appropriate adaptation to the local environment could be maintained while progress was made towards the selection goals. The latter would obviously include changes in morphology, physiology and biochemistry which are widespread if not universal among the grain legumes.

In order to produce a satisfactory domesticate, substantial changes would be necessary if the starting material was to be collections from the wild. Consideration could well be given to what might be considered under-developed domesticates which have already responded in some degree to the selection pressures of cultivation but which could provide basic materials for a second cycle of domestication (A. Ashri, personal communication). Induced genetic changes could here provide relief from the constraints which in the past have inhibited their development as successful cultigens.

The question might well be asked at this juncture as to what grounds we might have to expect that artificial mutagenesis could be an important tool in accelerating processes of domestication. In the first place, in taxonomically and genetically related organisms, homologous patterns of variation have been observed by Vavilov (1951) and subsequently. Such a pattern has been noted by Smartt (1976*a*) in *Phaseolus* among domesticated species and their wild conspecific counterparts. It is appropriate at this juncture to consider in some detail the potential predictive value of Vavilov's law of homologous series, not only in connection with potential for domestication in the tribe but also in the broader context of the whole family.

Patterns of variation in grain legumes

Vavilov's law of homologous series: its predictive value

If one considers the pattern of response to selection under domestication of the grain legumes as a whole, the responses are very closely similar. Similarities, nevertheless, are greater within genera than between them, and within tribes than between members of different tribes. Broad similarities in response within the family as a whole are none the less quite remarkable. The range of changes which are required in legumes to produce satisfactory agricultural crops are considerable, arguably more so than those involved in the production of the cereal crops. Equally profound changes have been evoked in the suppression of dispersal mechanisms in both groups; comparatively superficial changes have been necessary in growth form of the cereal grasses whereas profound changes have occurred in the grain legumes. For this reason a wide range of changes must be considered with regard to their evolution.

A potential case study: the winged bean

The consistency in the pattern of response to selection under domestication in grain legumes enables one to make predictions regarding the possible changes that might be effected in new and second cycle domesticates and to project beyond these changes in their consequences. A good example can be cited in the winged bean. This potentially valuable crop is in its present form a vining plant which typically produces very long pods. It is considered desirable to produce a dwarf determinate variant which would not require staking. Although this is a perfectly reasonable objective, the consequences of such a change have not been considered in detail. In the first place the phenotypic expression of the determinate mutant, if produced, cannot be predicted totally, in the sense that the number of nodes produced before the terminal inflorescence is an imponderable. If, say, 5–10 nodes are produced this is likely to be satisfactory (as in *Phaseolus vulgaris*); if only 2–4 nodes develop this could be unsatisfactory (as in *P. coccineus*). A better solution to the problem might be found in dwarf indeterminacy. The establishment of dwarf determinate forms would have repercussions on the desirable morphology of pods. The great advantage of the vining habit in relation to fruiting is that pods are kept out of soil contact and less liable to invasion by soil-inhabiting pathogens; there is thus little constraint on pod size. Conversely with a dwarf growth habit successful maturation of fruit implies some limit on pod size, in order to achieve avoidance of soil contact, and some constraints on shoot architecture. This must allow circulation of air to prevent excessive humidity levels building up, conducive to invasion of pods by pathogens. Substantial reduction in the innate hard-seededness of the winged bean is also desirable.

General consideration for first and second cycle domestication

The pod dehiscence mechanism disruption which produces non-dehiscent and ultimately non-fibrous pods is basically selected to favour the satisfactory harvesting of any grain legume crop. Extreme expression of this trait in total suppression of pod lignification produces pods which can be used as a green vegetable over a relatively long period: the 'mangetout' trait.

Possibly the greatest challenge in new domestication of grain legumes would be the elimination, or reduction to acceptable levels, of toxic materials and anti-metabolites found in the seed of many highly productive legume species. Species in the genus *Mucuna* spring readily to mind. As far as I have been able to ascertain the major barrier to extended cultivation of these species is the presence of non-protein amino acids, which probably act antagonistically in protein metabolism. In another legume, *Lathyrus sativus*, a useful purpose would be served by reducing content of the neuro-lathyrism-inciting principle.

Changes in photoperiodic response could also be useful in extending the possible geographic area of successful cultivation. Day-neutrality is a desirable feature. However, photoperiodic sensitivity does seem to be an entrenched feature of species such as the soyabean, but there are polymorphisms even here.

Application of Vavilov's law

The appliciation of Vavilov's law can assist us in making a more informed assessment of the boundaries of the realms of possibility in selection. Nevertheless, one must accept that in spite of broad similarities of responses to selection in the family some species will exhibit unique, idiosyncratic features. We can take encouragement none the less from the very broadly similar evolutionary patterns over the grain legumes as a whole and explore the possibilities of inducing critical, strategic changes which will remove obstacles to advance as domesticates.

Artificial mutagenesis

Basic strategies of accelerated domestication

The purpose of using mutagens in this domestication process is first of all to achieve a compression of the time scale. Measurable progress should be apparent in 5–10 years. The second purpose is to achieve a concentration of the input of human effort. Initial domestication was probably achieved by the necessarily intermittent efforts of numerous farmers over wide geographic areas on an extended time scale. This we cannot afford to do.

Feasibility models

In order to justify investment of resources in a programme of accelerated domestication, a clear demonstration of biological feasibility is required at the outset. The predictive value of Vavilov's law provides this to a very marked degree. What is desirable in addition is the provision of some concrete means of monitoring progress and the identification of characters which are recalcitrant or resistant to genetic change induced by mutagens. Such a feasibility model system could be set up in the genus *Phaseolus*. There is no doubt that the common bean has responded the most to selection pressures under domestication of virtually any grain legume. It still exists in extensive areas as a wild plant in Middle and South America as well as in a vast array of domesticated genotypes. What can be attempted, then, is in essence a recapitulation of the evolutionary processes which produced the common bean coupled with an attempt to produce domesticated forms of a totally wild species such as *Phaseolus polystachyus*.

The starting point for any such study would be selection of the most effective mutagens for use and techniques and rates of application. The genetic structure of the base population would require consideration; it might well be worth while to treat both pure lines and segregating hybrid populations. Another consideration is the possibility of exploiting both seed and vegetatively propagated generations from the M1. This should enable material of this generation to recover from the immediate transient effects of the mutagens and enable the kind of selection to be practised which has been so successful in vegetatively propagated horticultural plants (e.g. chrysanthemum). This might be particularly useful in selecting useful morphological variants produced by further vegetative segregation from the induced variants previously produced by vegetative segregation from the induced chimaera.

The role of artificial mutagenesis

The purpose of artificial mutagenesis is to compress the time scale so that the necessary genetic variability essential for the achievement of the necessary advance under selection can be generated. It should also enable effort to be concentrated efficiently on the achievement of closely defined objectives. In domestication mankind can no longer afford to spend millennia in waiting for useful mutations to accumulate, nor can this effort afford to be diffuse and widely dispersed. Effective means also need to be devised for evaluating the progress which is achieved, so that efforts can be concentrated selectively on those areas where advances are actually being made.

The possible role of new domesticates

The main justification for contemplating domestication with a view to the development of new crops can be found in the developing economies. These crops could serve to broaden the base of the agricultural economy, improve efficiency of production and generate productive capacity surplus to subsistence requirements. This could then be devoted to production of cash crops for local use and export. In this way economic vulnerability, arising from over-dependence on cash crops to generate income to purchase food, could be avoided. Another equally important incentive is to attempt to alleviate famine arising from drought by improvement of drought-resistant plants which produce seeds, tubers or foliage with good nutritional value and developing these as crops.

The unexploited potential for domestication

A consideration of the facts of domestication leads to the conclusion that it is highly improbable that the potential for domestication in the plant world has been totally exhausted by our ancestors. Only those plants which were domesticated in areas which subsequently became cradles of agriculture have been adequately evaluated for agricultural potential. It seems highly probable that plants with potential for domestication which occurred outside these areas have been totally neglected. They would have been uncompetitive with protocrop species which had already had the benefit of selection.

The selection of potential domesticates

The questions then arise of where to search and how to make selection of new protocrop species. At least two guiding principles can be used: firstly Vavilov's law of homologous series can indicate potentially useful taxa to explore in the first place, and secondly the use which is already being made of wild plants can be examined critically. Plants which are extensively resorted to in times of shortage and famine could be evaluated as potential domesticates.

It is in this context that the potential for second cycle domestication (*sensu* Ashri) can be considered. Here attempts could be made to remove identified constraints on the production of crops with undoubted but unrealised potential.

10.7. General conclusions

The selection processes operating on plants taken into cultivation which evolved into our crop plants have resulted, progressively, in our dependence on an ever-narrowing range of crop species and within these a diminishing array of genotypes. As a result of man's other activities similar effects are apparent in natural ecosystems; these are under similar pressures, being reduced in number and containing fewer species with depleted levels of genetic variation. Loss of genetic variation in crop species may be such that reversal of selection pressures may be impossible; equally, forward progression may also be impossible with the result that a genetic standstill or freeze may occur. This could mean, for example, that disease resistance which had inadvertently been lost in the process of cultivar development could not be re-instated while at the same time the possibility of introducing novel but currently desirable characters might no longer exist. In a situation of nearly total genetic erosion, restoration of genetic variation would become directly dependent on mutation, natural or induced, which would not necessarily re-instate prior levels of genetic variation on an acceptable time scale. It clearly behoves us to collect, conserve, evaluate, document and exploit our plant genetic resources as fully as we can.

In any such conservation programme priorities must be established even between crop species. The setting of priorities would depend on the importance of individual species in supporting human populations and the current perception of the extent to which specific crops' genetic resources are at risk. The maintenance of human society depends on the maintenance of balanced and sustainable farming systems, which necessitates a broad approach and not one concentrating exclusively on an overly narrow range of crops.

The situation foremost in the minds of agriculturalists at large is that where agriculture as a way of life is accepted as the norm. However, in some areas where this once was the case, the insidious process of desertification has prompted some radical re-thinking. Human populations now exist in the Sahel, for example, which are clearly in excess of the carrying capacity of the land. If we assume that some agriculture is possible, how then do we determine agricultural production strategy? Do we seek to produce drought-tolerant genotypes of crops formerly produced in the area or do we have to consider producing entirely new crops for these agriculturally marginal areas? Production of established crops has in many cases precipitated the current crisis and so alternative strategies are worth considering, especially since the success of selecting established crop species for production in marginal environments can only be considered as modest. Although responses to selection in breeding programmes for readily identifiable quality characters, morphological features or simple physiological attributes may be highly satisfactory, the responses to improvement of tolerance to extremes of temperature, water and soil conditions can only be considered modest. In this context, the search for plants potentially exploitable as crops from agriculturally marginal habitats makes very good sense. The prime essential is obvious, that it gives an acceptable product. From this point on the urgent concern is to devise an appropriate production strategy. This may very well involve discarding many cherished current practices of modern intensive agriculture and devising efficient systems of extensive agriculture with crop stand densities the environment can support, for example.

The horizons of the present day student of crop evolution have broadened. Whereas only a decade or so ago there was an implied acceptance of the inevitability of mankind becoming dependent on an ever-narrowing range of highly versatile crop species, this point of view is now being questioned. It makes better political and economic sense to encourage basic agricultural self-sufficiency in the developing world as a foundation for future broadly based agricultural and industrial development. This obviously cannot be done on the basis of poorly adapted crop species, so the possibility of developing new crops must be explored. It cannot be too strongly stressed that the crop product must be acceptable; the prime candidates for new crop status could well be wild plants which produce material collected for food at the present time. Effective cultivation might well depend on removal of a genetically simply controlled constraint, such as seed dormancy. A simple irradiation and selection programme might well bring this about.

An imaginative approach, as free from pre-conceived ideas as possible, is needed to reverse the trend of the human race's over-dependence on a narrowing range of crop species. There is a broadening field of activity for crop plant evolutionists; not only can they profitably unravel the past historical complexities of the evolution of our current crops, but there is also scope for putting the domesticates which have fallen by the wayside back into the evolutionary race as well as initiating domestication of crops *de novo* and directing their evolution consciously.

References

Abdou, Y. A. M. 1966. The source and nature of resistance in *Arachis* L. species to *Mycosphaerella arachidicola* Jenk. and *Mycosphaerella berkeleyii* Jenk. and factors influencing sporulation of these fungi. Ph.D. thesis, North Carolina State University, Raleigh, North Carolina.

Abdou, Y. A. M., Gregory, W. C. and Cooper, W. E. 1974. Sources and nature of resistance to *Cercospora arachidicola* Hori and *Cercosporidium personatum* (Beck. et Curtis) Deighton in *Arachis* species. *Peanut Science* **1**: 6–11.

Abu-Shakara, S. and Tannous, R. I. 1981. Nutritional value and quality of lentils. In *Lentils*, pp. 191–202 (eds. C. Webb and G. Hawtin). Farnham Royal: Commonwealth Agricultural Bureaux (CAB).

Adams, M. W. 1982. Plant architecture and yield breeding. *Iowa State Journal of Research* **56**: 225–254.

Adams, M. W., Coyne, D. P., Davis, J. H. C., Graham, P. H. and Francis, C. A. 1985. Common Bean (*Phaseolus vulgaris* L.). In *Grain Legume Crops*, pp. 433–476 (eds. R. J. Summerfield and E. H. Roberts). London: Collins.

Adiga, P. R., Padmanaban, G., Rao, S. L. N. and Sarma, P. S. 1962. The isolation of a toxic principle from *Lathyrus sativus* seeds. *Journal of Scientific and Industrial Research, New Delhi* **21** C: 284–286.

Ahmad, Q. N., Britten, E. J. and Byth, D. E. 1979. Inversion heterozygosity in the hybrid soyabean *Glycine soja*. Evidence from a pachytene loop configuration and other meiotic irregularities. *Journal of Heredity* **70**: 358–364.

Ahn, C. S. and Hartman, R. W. 1978. Interspecific hybridization between mung bean (*Vigna radiata* (L.) Wilczek) and adzuki bean (*V. angularis* (Willd.) Ohwi et Ohashi). *Journal of the American Society for Horticultural Science* **103**: 3–6.

Allchin, F. R. 1969. Early cultivated plants in India and Pakistan. In *The Domestication and Evolution of Plants and Animals*, pp. 323–329 (eds. P. J. Ucko and G. W. Dimbleby). London: Duckworth.

Alvarez, M. N., Ascher, P. D. and Davis, D. W. 1981. Interspecific hybridization in *Euphaseolus* through embryo rescue. *HortScience* **16**: 541–543.

Amaya, F., Young, C. T. Hammons, R. O. and Martin, G. 1977. The tryptophan content of the U.S. commercial and some South American wild genotypes of the genus *Arachis*. A survey. *Oléagineux* **32**: 225–229.

Amin, P. W. 1985. Resistance of wild species of groundnut to insect and mite pests. In *Cytogenetics of Arachis*, pp. 57–61 (ed. J. P. Moss). Pantancheru,

India: International Crops Research Institute for the Semi-Arid Tropics (ICRISAT).

Anson, M. L. 1958. Potential uses of isolated oilseed protein in foodstuffs. In *Processed Plant Protein Foodstuffs*, pp. 277–289 (ed. A. M. Altschull). New York: Academic Press.

Arora, S. K. 1982. Legume carbohydrates. In *Chemistry and Biochemistry of Legumes*, pp. 1–50 (ed. S. K. Arora). New Delhi: Oxford and IBH Publishing Co. (London: Edward Arnold, 1983.)

Arora, R. K., Chandel, K. P. S. and Ioshi, B. S. 1973. Morphological diversity in *Phaseolus sublobatus* Roxb. *Current Science* **42**: 359–361.

Ashri, A. 1976*a*. Frequencies of plasmon types in a germplasm sample of peanuts, *Arachis hypogaea*. *Euphytica* **25**: 777–785.

Ashri, A. 1976*b*. Plasmon divergence in peanuts (*Arachis hypogaea*): a third plasmon and locus affecting growth habit. *Theoretical and Applied Genetics* **48**: 17–21.

Atreya, C. D., Rama Krishna, T., Pandit, M. W. and Subramanyam, N. C. 1985. Molecular approaches to genome analysis in *Arachis* species. In *Cytogenetics of Arachis*, pp. 87–92 (ed. P. J. Moss). Patancheru, India: International Crops Research Institute for the Semi-Arid Tropics (ICRISAT).

Atkin, J. D. 1972. Nature of the stringy pod rogue of snap beans, *Phaseolus vulgaris*. *Search Agriculture* **2**: 1–3.

AVRDC (The Asian Vegetable Research and Development Center) *Progress Reports* 1976 *et seq*. Taiwan.

Aykroyd, W. R. and Doughty, J. 1964. *Legumes in Human Nutrition*. FAO Nutritional Studies no. 19. Rome: Food and Agriculture Organization of the United Nations. (2nd edition 1982, revised by J. Doughty and A. Walker as FAO Food and Nutrition Paper no. 20.)

Babu, C. N. 1955. Cytogenetical investigations on groundnuts. I. The somatic chromosomes. *Indian Journal of Agricultural Science* **25**: 41–46.

Badami, V. K. 1923. Hybridization work on groundnut, pp. 29–30. *Mysore Agricultural Department (India) Report for 1922–23*.

Badami, V. K. 1928. Ph.D. thesis (unpublished) (Cambridge: University Library), cited in H. Hunter and H. M. Leake 1933, *Recent Advances in Agricultural Plant Breeding*, Philadelphia: Blakiston.

Bailey, C. J. and Boulter, D. 1971. Urease, a typical seed protein of the Leguminosae. In *Chemotaxonomy of the Leguminosae*, pp. 485–502 (eds. J. B. Harborne, D. Boulter and B. L. Turner). London: Academic Press.

Bailey, R. W. 1971. Polysaccharides in the Leguminosae. In *Chemotaxonomy of the Leguminosae*, pp. 503–541 (eds. J. B. Harborne, D. Boulter and B. L. Turner). London: Academic Press.

Bajaj, Y. P. S., Kumar, P., Singh, M. M. and Labana, K. S. 1982. Interspecific hybridization in the genus *Arachis* through embryo culture. *Euphytica* **31**: 365–370.

Balaiah, C., Reddy, P. S. and Reddy, M. V. 1977. Genic analysis in groundnut. I. Inheritance studies on 18 morphological characters in crosses with Gujarat narrow leaf mutant. *Proceedings of the Indian Academy of Sciences* B **85**: 340–350.

Ballon, F. B. and York, T. L. 1959. Crossing the common and scarlet bean with *Vigna* species. *Philippine Agriculturalist* **42**: 454–455.

Banks, D. J. 1974. Interspecific hybridization, p. 8. *Oklahoma Agricultural Experiment Station Progress Report P-702*. Stillwater, Oklahoma, USA.

Banks, D. J. 1976. Peanuts: germplasm resources. *Crop Science* **16**: 499–502.

Banks, D. J. 1986. Origin and characteristics of a primitive peanut (*Arachis hypogaea* L. var. *hirsuta* Kohler). (Abstract.) *American Journal of Botany* **73**: 750.

Barulina, H. 1930. Lentils of the U.S.S.R. and other countries (Russian with English summary), 40th supplement (pp. 265–304) to the *Bulletin of Applied Botany, Genetics and Plant Breeding, Leningrad.*

Baudet, J. C. 1977. Origine et classification des espèces cultivées du genre *Phaseolus*. *Bulletin Société Royale Botanique Belge* **110**: 65–76.

Baudoin, J. P. 1981. Observations sur quelques hybrides interspécifiques avec *Phaseolus lunatus* L. *Bulletin des Recherches Agronomiques de Gembloux* **16**: 273–286.

Bauhin, J., Cherlero, J. H., Chabraeus, D. and Agraffenried, F. L. 1651. *Historia Plantarum Universalis (Eboraduni)*.

Belivanis, T. and Doré, C. 1986. Interspecific hybridization of *Phaseolus vulgaris* L. and *Phaseolus angustissimus* A. Gray using *in vitro* embryo culture. *Plant Cell Reports* **5**: 329–331.

Bell, E. A. 1964. Relevance of biochemical taxonomy to the problem of lathyrism. *Nature* **203**: 378–380.

Bell, E. A. 1971. Comparative biochemistry of non-protein amino acids. In *Chemotaxonomy of the Leguminosae*, pp. 179–206 (eds. J. B. Harborne, D. Boulter and B. L. Turner). London: Academic Press.

Bell, E. A., Lackey, J. A. and Polhill, R. M. 1978. Systematic significance of canavanine in the Papilionoideae (Faboideae). *Biochemical Systematics and Ecology* **6**: 201–212.

Bentham, G. 1841. On the structure and affinities of *Arachis* and *Voandzeia*. *Transactions of the Linnean Society of London* **18**: 155–162.

Ben-Ze'ev, N. and Zohary, D. 1973. Species relationships in the genus *Pisum*. *Israel Journal of Botany* **22**: 73–91.

Bergland-Brücher, O. and Brücher, H. 1976. The South American wild bean (*Phaseolus aborigineus* Burk.) as ancestor of the common bean. *Economic Botany* **30**: 257–272.

Bernard, R. L. and Weiss, M. G. 1980. Qualitative genetics. In *Soyabeans: Improvement, Production and Uses*, pp. 117–154 (ed. B. E. Caldwell). Madison, Wisconsin: American Society of Agronomy.

Bhapkar, D. G., Patil, P. S. and Patil, V. A. 1986. Dormancy in groundnut – a review. *Journal of Maharashtra Agricultural Universities* **11**: 68–71.

Bharathi, M. and Murty, U. R. 1984. Comparative embryology of wild and cultivated species of *Arachis*. *Phytomorphology* **34**: 48–56.

Bisby, F. A. 1981. Tribe 32. Genisteae. In *Advances in Legume Systematics*, pp. 409–423 (eds. R. M. Polhill and P. H. Raven). Kew: Royal Botanical Gardens.

Biswas, M. R. and Dana, S. 1975. Interchange heterozygosity in a triploid species hybrid of *Phaseolus*. *Indian Agriculture* **19**: 273–274.

Biswas, M. R. and Dana, S. 1976*a*. *Phaseolus aconitifolius* × *Phaseolus trilobus*. *Indian Journal of Genetics and Plant Breeding* **36**: 125–131.

Biswa, M. R. and Dana, S. 1976*b*. Meiosis in amphidiploid of *Phaseolus aureus*

Roxb. × *P. calcaratus* Roxb. *Journal of the Society of Experimental Agriculturists* **1**: 43–44.

Bliss, F. A. 1986. Quantitative expression of bean seed proteins. (Abstract.) *Journal of the American Oil Chemists' Society* **63**: 457–458.

Boissier, E. 1872. *Flora Orientalis*. Geneva.

Bolkhovskikh, Z., Grif, Y., Matvejeva, T. and Zakharyeva, O. 1969. Chromosome numbers of flowering plants. Leningrad: Academy of Sciences of the USSR.

Bond, D. A. 1976. Field Bean. In *Evolution of Crop Plants*, pp. 179–182 (ed. N. W. Simmonds). London: Longman.

Bond, D. A. and Poulsen, M. H. 1983. Pollination. In *The Faba Bean (Vicia faba L.). A Basis for Improvement*, pp. 77–101 (ed. P. D. Hebblethwaite). London: Butterworths.

Bose, R. D. 1939. Studies in Indian pulses. IV. Mung or green grams (*Phaseolus aureus* Roxb.). *Indian Journal of Agricultural Science* **2**: 607–624.

Boulter, D. and Derbyshire, E. 1971. Taxonomic aspects of the structure of legume proteins. In *Chemotaxonomy of the Leguminosae*, pp. 285–308 (eds. J. B. Harborne, D. Boulter and B. L. Turner). London: Academic Press.

Bowdidge, E. 1935. *The Soya Bean, its History, Cultivation [in England] and Uses*. Oxford University Press.

Braak, J. P. and Kooistra, E. 1975. A successful cross between *Phaseolus vulgaris* L. and *Phaseolus ritensis* Jones with the aid of embryo culture. *Euphytica* **24**: 669–679.

Bressani, R. 1985. Nutritive value of cowpea. In *Cowpea Research, Production and Utilization*, pp. 353–359 (eds. S. R. Singh and K. O. Rachie). Chichester: J. Wiley.

Brim, C. A. 1980. Quantitative genetics and breeding. In *soybeans: Improvement, Production and Uses*, pp. 155–186 (ed. B. E. Caldwell). Madison, Wisconsin: American Society of Agronomy.

Burkart, A. and Brücher, H. 1953. *Phaseolus aborigineus* Burkart, die mutmassliche andine Stammform der Kulturbohne. *Züchter* **23**: 65–72.

Burkill, I. H. 1906. *Psophocarpus tetragonolobus* (Goa bean) in India. *The Agricultural Ledger* **4**: 51–64.

Burkill, I. H. 1935. *A Dictionary of the Economic Products of the Malay Peninsula*. London: Crown Agents.

Charavanapvan, C. 1943. The utilization of sword bean and jack bean as food. *Tropical Agriculturist and Magazine of the Ceylon Agricultural Society* **99**: 157–159.

Cheah, C. H. 1973. Evaluation of genetic diversity in varieties of *Phaseolus vulgaris* (L.) Savi. Ph.D. thesis, Cambridge University.

Cherry, J. P. 1975. Comparative studies of seed proteins and enzymes of species and collections of *Arachis* by gel electrophoresis. *Peanut Science* **2**: 57–65.

Cherry, J. P. 1977. Potential sources of peanut seed proteins and oil in the genus *Arachis*. *Journal of Agricultural and Food Chemistry* **25**: 186–193.

Chevalier, A. 1933, 1934, 1936. Monographie de l'Arachide. *Revue de Botanique Appliquée et d'Agriculture Tropicale* **13**: 689–789; **14**: 565–632, 709–755, 834–864; **16**: 673–871.

Chooi, W. Y. 1971*a*. Variation in the nuclear content in the genus *Vicia*. *Genetics* **68**: 195–211.

Chooi, W. Y. 1971*b*. Comparison of the DNA of six *Vicia* species by the method of DNA–DNA hybridization. *Genetics* **68**: 213–230.

Chowdhury, K. A., Saraswat, K. S., Hasan, S. N. and Gaur, R. C. 1971. 4,000–3,500 year old barley, rice and pulses from Atranjikhera. *Science and Culture* **37**: 531–532.

Clausen, R. T. 1944. *A Botanical Study of the Yam Beans* (Pachyrrhizus). Memoir 264, Cornell University Agricultural Experiment Station, Ithaca, New York.

Company, M., Stalker, H. T. and Wynne, J. C. 1982. Cytology and leafspot resistance in *Arachis hypogaea* × wild species hybrids. *Euphytica* **31**: 885–893.

Conagin, C. H. T. M. 1962. Especies selvagems do gênero *Arachis*. Observacôes sobre os exemplares de colecâo da Secâo de Citologia. *Bragantia* **21**: 341–374.

Conagin, C. H. T. M. 1963. Número de cromosomas das especies selvagems de *Arachis*. *Bragantia* **22**: 125–129.

Conagin, C. H. T. M. 1964. Número de cromosomas em *Arachis* selvagem. *Bragantia* **23**: XXV–XXVII.

Courtois, J. E. and Percheron, F. 1971. Distribution des monosaccharides, oligosaccharides et polyols. In *Chemotaxonomy of the Leguminosae*, pp. 207–229 (eds. J. B. Harborne, D. Boulter and B. L. Turner). London: Academic Press.

Coyne, D. P. 1964. Species hybridization in *Phaseolus*. *Journal of Heredity* **55**: 5–6.

Cubero, J. I. 1974. On the evolution of *Vicia faba* L. *Theoretical and Applied Genetics* **45**: 47–51.

Cubero, J. I. 1981. Origin, taxonomy and domestication. In *Lentils*, pp. 15–38 (eds. C. Webb and G. J. Hawtin). Farnham Royal: Commonwealth Agricultural Bureaux (CAB).

Dalziel, J. M. 1937. *The Useful Plants of West Tropical Africa*. London: Crown Agents.

Dana, S. 1964. Interspecific cross between tetraploid *Phaseolus* species and *P. riccardianus* Ten. *Nucleus* **7**: 1–10.

Dana, S. 1966*a*. Species cross between *Phaseolus aureus* Roxb. and *P. trilobus* Ait. *Cytologia* **31**: 176–187.

Dana, S. 1966*b*. Interspecific hybrid between *Phaseolus mungo* L. × *P. trilobus* Ait. *Journal of Cytology and Cytogenetics* **1**: 61–66.

Dana, S. 1966*c*. Cross between *Phaseolus aureus* Roxb. and *P. riccardianus* Ten. *Genetica Iberica* **18**: 141–156.

Dana, S. 1966*d*. Chromosome differentiation in tetraploid *Phaseolus* species and *P. riccardianus* Ten. *Nucleus* **9**: 97–101.

Darlington, C. D. 1963. *Chromosome Botany and the Origins of Cultivated Plants*. 2nd edition. London: George Allen and Unwin.

Darlington, C. D. and Wylie, A. P. 1955. *Chromosome Atlas of Flowering Plants*. London: George Allen and Unwin.

Daussant, J., Neucere, N. J. and Conkerton, E. J. 1969. Immunochemical studies on *Arachis hypogaea* proteins with particular reference to the reserve proteins. II. Protein modification during germination. *Plant Physiology* **44**: 480–484.

Daussant, J., Neucere, N. J. and Yatsu, L. 1969. Immunochemical studies on

Arachis hypogaea proteins with particular reference to the reserve proteins. I. Characterization distribution and properties of α arachin and α conarachin. *Plant Physiology* **44**: 471–479.

Davies, A. J. S. 1958. A cytotaxonomic study in the genus *Lathyrus*. Ph.D. thesis, Manchester University.

Davies, D. R., Berry, G. J., Heath, M. C. and Dawkins, T. C. K. 1985. Pea (*Pisum sativum* L.). In *Grain Legume Crops*, pp. 147–198 (eds. R. J. Summerfield and E. H. Roberts). London: Collins.

Davis, P. H. 1970. *Pisum* L. *Flora of Turkey* **3**: 370–372. Edinburgh University Press.

Davis, K. S. and Simpson, C. E. 1976. Variable chromosome numbers in two 'amphidiploid' populations of *Arachis*. *Proceedings of the American Peanut Research and Education Association* **8**: 93.

D'Cruz, R. and Chakravarty, K. 1961. Spontaneous allopolyloidy in *Arachis*. *Indian Oilseeds Journal* **5**: 55–57.

D'Cruz, R. and Tankasale, M. P. 1961. A note on chromosome complement of four groundnut varieties. *Indian Oilseeds Journal* **5**: 58–59.

D'Cruz, R. and Upadhyaya, B. R. 1962. Allopolyploidy in *Arachis*. *Indian Oilseeds Journal* **6**: 33–37.

De, D. N. 1974. Pigeon pea. In *Evolutionary Studies in World Crops*, pp. 79–87 (ed. J. B. Hutchinson). Cambridge University Press.

De, D. N. and Krishnan, R. 1966. Cytological studies of the hybrid *Phaseolus aureus* × *P. mungo*. *Genetica* **37**: 588–600.

Debouck, D. G. 1986. Primary diversification of *Phaseolus* in the Americas: three centres? *Plant Genetic Resources Newsletter* **67**: 2–8.

Debouck, D. and Hidalgo, R. 1984. Morfologia de la planta de fríjol común (*Phaseolus vulgaris* L.) Publication series; 04*SB*-09.01. Cali Colombia: CIAT.

de Candolle, A. P. 1882. *Origines des Plantes Cultivées*, Paris. English translation 1886. London: Kegan Paul.

Deodikar, G. B. and Thakar, C. V. 1956. Cyto-taxonomic evidence for the affinity between *Cajanus indicus* Spreng. and certain erect species of *Atylosia* W. and A. *Proceedings of the Indian Academy of Sciences* B **43**: 37–45.

de Tau, E. W., Baudoin, J. P. and Maréchal, R. 1986. Obtention d'allopolyploides fertiles chez le croisement entre *Phaseolus vulgaris* et *Phaseolus filiformis*. *Bulletin des Recherches Agronomiques de Gembloux* **21**: 35–46.

Dhaliwal, A. S., Pollard, L. H. and Lorz, A. P. 1962. Cytological behaviour of an F1 species cross (*Phaseolus lunatus* L. var. Fordhook × *Phaseolus polystachyus* L.). *Cytologia* **27**: 369–374.

Drayner, J. M. 1956. Regulation of outbreeding in field beans. *Nature* **177**: 489–490.

Drayner, J. M. 1959. Self and cross-fertility in field beans (*Vicia faba* Linn.). *Journal of Agricultural Science* **53**: 387–404.

Duke, J. A. 1981. *Handbook of Legumes of World Economic Importance*. New York: Plenum.

Duncan, W. G., McCloud, D. E., McGraw, R. L. and Boote, K. J. 1978. Physiological aspects of peanut yield improvement. *Crop Science* **18**: 1015–1020.

East African Agriculture and Forestry Research Organization. *Annual Reports 1954–1956*. Nairobi, Kenya: Government Printer.

El-Sherbeeny, M. H., Mytton, L. R. and Lawes, D. A. 1977*a*. Symbiotic variability in *Vicia faba*. 1. Genetic variation in the *Rhizobium leguminosarum* population. *Euphytica* **26**: 149–156.

El-Sherbeeny, M. H., Lawes, D. A. and Mytton, L. R. 1977*b*. Symbiotic variability in *Vicia faba*. 2. Genetic variation in *Vicia faba*. *Euphytica* **26**: 377–383.

Erskine, W. 1978. The genetics of winged beans. In *The Winged Bean*, pp. 29–35. Los Baños, Philippines: Philippine Council for Agriculture and Resources Research.

Evans, A. M. 1976. Species hybridization in the genus *Vigna*. In *Proceedings of IITA Collaborators' Meeting on Grain Legume Improvement*, pp. 31–34 (eds. R. A. Luse and K. O. Rachie). Ibadan, Nigeria: Plant Improvement, International Institute of Tropical Agriculture (IITA).

Evans, A. M. 1980. Structure, variation, evolution and classification in *Phaseolus*. In *Advances in Legume Science*, pp. 337–347 (eds. R. J. Summerfield and A. H. Bunting). Kew: Royal Botanic Gardens.

Evans, A. M. and Davis, J. H. C. 1978. Breeding *Phaseolus* beans as grain legumes for Britain. *Applied Biology* **3**: 1–43.

FAO 1981. *FAO Production Year Book*. Rome: Food and Agriculture Organization of the United Nations.

FAO/WHO 1973. *World Health Organization Technical Report Series No. 522*. Geneva: World Health Organization.

Fahn, A. and Zohary, M. 1955. On the peripheral structure of the legumen, its evolution and relation to dehiscence. *Phytomorphology* **5**: 99–111.

Faris, D. G. 1964. The chromosome number of *Vigna sinensis* (L.) Savi. *Canadian Journal of Genetics and Cytology* **6**: 255–258.

Faris, D. G. 1965. The origin and evolution of the cultivated forms of *Vigna sinensis*. *Canadian Journal of Genetics and Cytology* **7**: 433–452.

Feenstra, W. J. 1960. *Biochemical aspects of seedcoat colour inheritance in Phaseolus vulgaris L. Wageningen: Medelingen Landbouwhogeschool.*

Fernandez, A. 1973. El acido lactico como fijador cromosomico. *Boletin de la Sociedad Argentina de Botanica* **15**: 287–290.

Ford-Lloyd, B. and Jackson, M. T. 1986. *Plant Genetic Resources*. London: Edward Arnold.

Foster, D. J. 1979. Evaluation of *Arachis* germplasm for resistance to early leafspot [*Mycosphaerella arachidis* Deighton (Conidial state: *Cercospora arachidicola* Hori)]. M.S. thesis, North Carolina State University, Raleigh.

Foster, D. J., Wynne, J. C., Stalker, H. T. and Beute, M. K. 1979. Resistance of *Arachis hypogaea* and wild relatives to *Cercospora arachidicola* Hori. *Agronomy Abstracts* (71st Annual Meeting Report), p. 61.

Fox, D. J., Thurman, D. A. and Boulter, D. 1964. Studies on the proteins of seeds of the Leguminosae I. Albumins. *Phytochemistry* **3**: 417–419.

Fozdar, B. S. 1963. Cytological investigation of parents, offspring and backcross derivatives involved in the interspecific cross *Phaseolus lunatus* L. × *Ph. polystachyus* (L.) BSP. *Dissertation Abstracts* **24**: 480–481.

Frahm-Leliveld, J. A. 1960. Chromosome numbers in leguminous plants. *Acta Botanica Neerlandica* **9**: 327–329.

Freeman, G. F. 1912. South Western beans and teparies. *University of Arizona Experiment Station Bulletin No. 68* (Revised 1918). Tucson, Arizona.

Froussios, G. 1970. Genetic diversity and agricultural potential in *Phaseolus vulgaris*. *Experimental Agriculture* **6**: 129–141.

Gardner, M. E. B. and Stalker, H. T. 1983. Cytology and leafspot resistance of section *Arachis* amphidiploids and their hybrids with *Arachis hypogaea*. *Crop Science* **23**: 1069–1074.

Gentry, H. S. 1969. Origin of the common bean *Phaseolus vulgaris*. *Economic Botany* **23**: 55–69.

Gentry, H. S. 1971. *Pisum* resources, a preliminary survey. *Plant Genetic Resources Newsletter* **25**: 3–13.

Gepts, P. L. 1985. Nutritional and evolutionary implications of phaseolin seed protein variability in common bean (*Phaseolus vulgaris* L.). *Dissertation Abstracts International (Sciences and Engineering)* **45**: 3715B–3716B.

Gepts, P. and Bliss, F. A. 1985. F1 hybrid weakness in the common bean: differential geographic origin suggests two gene pools in cultivated bean germplasm. *Journal of Heredity* **76**: 447–450.

Gepts, P. and Bliss, F. A. 1986. Phaseolin variability among wild and cultivated common beans (*Phaseolus vulgaris*) from Colombia. *Economic Botany* **40**: 469–478.

Gepts, P., Osborn, T. C., Rashka, K. and Bliss, F. A. 1976. Phaseolin-protein variability in wild forms and landraces of the common bean (*Phaseolus vulgaris*): evidence for multiple centers of domestication. *Economic Botany* **40**: 451–468.

Ghimpu, V. 1930. Recherches cytologiques sur les genres: *Hordeum*, *Acacia*, *Medicago*, *Vitis* et *Quercus*. *Archives d'Anatomie Microscopique* **26**: 136–234.

Gibbons, R. W. 1966. The branching habit of *Arachis monticola*. *Rhodesia, Zambia and Malawi Journal of Agricultural Research* **4**: 9–11.

Gibbons, R. W. and Bailey, B. E. 1967. Resistance to *Cercospora arachidicola* in some species of *Arachis*. *Rhodesia, Zambia and Malawi Journal of Agricultural Research* **5**: 57.

Gibbons, R. W., Bunting, A. H. and Smartt, J. 1972. The classification of varieties of groundnut (*Arachis hypogaea* L.). *Euphytica* **21**: 78–85.

Gibbons, R. W. and Turley, A. C. 1967. ARC Grain Legume Pathology Research Team. Botany and Plant Breeding, pp. 86–90. *Annual Report of the Agricultural Research Council of Central Africa*.

Gladstones, J. S. 1970. Lupins as crop plants. *Field Crop Abstracts* **23**: 123–148.

Gladstones, J. S. 1974. *Lupins of the Mediterranean Region and Africa*. Technical Bulletin No. 26. Department of Agriculture, Western Australia.

Gladstones, J. S. 1980. Recent developments in the understanding, improvement and use of lupins. In *Advances in Legume Science*, pp. 603–611 (eds. R. J. Summerfield and A. H. Bunting). Kew: Royal Botanic Gardens.

Goldblatt, P. 1981. Cytology and the phylogeny of Leguminosae. In *Advances in Legume Systematics*, pp. 427–463 (eds. R. M. Polhill and P. H. Raven). Kew: Royal Botanic Gardens.

Gopinathan Nair, P., Ponnaiya, B. W. X. and Raman, V. S. 1964. Breeding behaviour of interspecific hybrids in *Arachis*. (Abstract.) *Madras Agricultural Journal* **51**: 360.

Goshen, D., Ladizinsky, G. and Muehlbauer, R. J. 1982. Restoration of meiotic regularity and fertility among derivatives of *Lens culinaris* × *L. nigricans* hybrids. *Euphytica* **31**: 795–799.

Grant, V. 1963. *The Origin of Adaptations*. New York: Columbia University Press.

Gray, A., Parker, J. P. and Michaels, T. E. 1985. Hybrid viability in interspecific crosses between *Phaseolus vulgaris* L. and *P. acutifolius*. (Abstract.) *Canadian Journal of Plant Sciences* **65**: 1113.

Gregory, M. P. and Gregory, W. C. 1979. Exotic germplasm of *Arachis* L. interspecific hybrids. *Journal of Heredity* **70**: 185–193.

Gregory, W. C. 1946. Peanut breeding program under way. *69th Annual Report North Carolina Agricultural Experiment Station*, pp. 42–44.

Gregory, W. C. (ed.) 1968. A radiation breeding experiment with peanuts. *Radiation Botany* **8**: 88–147.

Gregory, W. C. and Gregory, M. P. 1967. Induced mutation and species hybridization in the de-speciation of *Arachis*. *Ciência et Cultura* **19**: 166–174.

Gregory, W. C. and Gregory, M. P. 1976. Groundnut. In *Evolution of Crop Plants*, pp. 151–154 (ed. N. W. Simmonds). London: Longman.

Gregory, W. C., Gregory, M. P., Krapovickas, A., Smith, B. W. and Yarbrough, J. A. 1973. Structures and genetic resources of peanuts. In *Peanuts – Culture and Uses*, pp. 47–133. Stillwater, Oklahoma: American Peanut Research and Education Association Inc.

Gregory, W. C., Krapovickas, A. and Gregory, M. P. 1980. Structures, variation, evolution and classification in *Arachis*. In *Advances in Legume Science*, pp. 469–481 (eds. R. J. Summerfield and A. H. Bunting). Kew: Royal Botanic Gardens.

Gritton, E. T. and Wierzbicka, B. 1975. An embryological study of a *Pisum sativum* × *Vicia faba* cross. *Euphytica* **24**: 277–284.

Gupta, Y. P. 1982. Nutritive value of food legumes. In *Chemistry and Biochemistry of Legumes*, pp. 287–327 (ed. S. K. Arora). New Delhi: Oxford and IBH Publishing Co. (London: Edward Arnold, 1983).

Hadley, H. H. and Hymowitz, T. 1973. Speciation and cytogenetics. in *Soybeans: Improvement, Production and Uses*, pp. 97–116 (ed. B. E. Caldwell). Madison, Wisconsin: American Society of Agronomy.

Halliday, J. 1976. An interpretation of seasonal and short term fluctuations in nitrogen fixation. Ph.D. thesis, University of Western Australia.

Hammons, R. O. 1970. Registration of Spancross peanuts. *Crop Science* **10**: 459.

Hammons, R. O. 1971. Inheritance of inflorescences in mainstem leaf axils in *Arachis hypogaea* L. *Crop Science* **11**: 570–571.

Hammons, R. O. 1973*a*. Early history and origin of the peanut. In *Peanuts – Culture and Uses*, pp. 17–45. Stillwater, Oklahoma: American Peanut Research and Education Association, Inc.

Hammons, R. O. 1973*b*. Genetics of *Arachis hypogaea*. In *Peanuts – Culture and Uses*, pp. 135–173. Stillwater, Oklahoma: American Peanut Research and Education Association Inc.

Hammons, R. O. 1982. Origin and early history of the peanut. In *Peanut Science and Technology*, pp. 1–20 (eds. H. E. Pattee and C. T. Young). Yoakum, Texas: American Peanut Research and Education Association Inc.

Hanelt, P. 1972*a*. Die infraspezifische Variabilität von *Vicia faba* L. und ihre Gliederung. *Kulturpflanze* **20**: 75–128.

Hanelt, P. 1972*b*. Zur Geschichte des Anbaues von *Vicia faba* L. und ihrer verschieden Formen. *Kulturpflanze* **20**: 209–223.

Haq, N. and Smartt, J. 1977. Chromosome complements in *Psophocarpus* spp. *Tropical Grain Legume Bulletin* **10**: 16–19.

Harborne, J. B. 1971. Terpenoids and other low molecular weight substances of systematic interest in the Leguminosae. In *Chemotaxonomy of the Leguminosae*, pp. 257–283 (eds. J. B. Harborne, D. Boulter and B. L. Turner). London: Academic Press.

Harborne, J. B., Boulter, D. and Turner, B. L. (eds.) 1971. *Chemotaxonomy of the Leguminosae*. London: Academic Press.

Hardy, R. W. E., Burns, R. C., Herbert, R. R., Holsten, R. D. and Jackson, E. K. 1971. Biological nitrogen fixation, a key to world protein. *Plant and Soil (Special Volume)*, pp. 561–590.

Harlan, J. R. 1975. *Crops and Man*. Madison, Wisconsin: American Society of Agronomy.

Harlan, J. R. and de Wet, J. M. J. 1971. Toward a rational classification of cultivated plants. *Taxon* **20**: 509–517.

Hassan, H. and Beute, M. K. 1977. Evaluation of resistance to *Cercospora* leafspot in peanut germplasm potentially useful in a breeding program. *Peanut Science* **4**: 78–83.

Hawkes, J. G. 1983. *The Diversity of Crop Plants*. Cambridge, Massachusetts: Harvard University Press.

Hawtin, G. C. and Hebblethwaite, P. D. 1983. Background and history of faba bean production. In *The Faba Bean* (*Vicia faba* L.) *A Basis for Improvement*, pp. 3–22 (ed. P. D. Hebblethwaite). London: Butterworths.

Helbaek, H. 1965. Isin-Larsan and Horian food remains at Tell Bazmosian in the Dokan Valley. *Sumer* **19**: 27–35.

Helbaek, H. 1970. Plant husbandry of Hacilar. In *Excavation in Hacilar*, pp. 189–191 (ed. J. Mellaart). Edinburgh University Press.

Hepper, F. N. 1963. Plants of the 1957–58 West African Expedition. II. The Bambara groundnut (*Voandzeia subterranea*) and Kersting's groundnut (*Kerstingiella geocarpa*) wild in West Africa. *Kew Bulletin* **16**: 395–407.

Herklots, G. A. C. 1972. *Vegetables in South-East Asia*. London: George Allen and Unwin.

Hermann, F. J. 1954. A synopsis of the genus *Arachis*. *Agricultural Monograph No. 19*. Beltsville, Maryland: United States Department of Agriculture.

Hermann, F. J. 1962. A revision of the genus *Glycine* and its immediate allies. *Technical Bulletin 1268*. Beltsville, Maryland: United States Department of Agriculture.

Hernandez Xolocotzi, E., Miranda Colín, S. and Prywer, C. 1959. El origen de *Phaseolus coccineus* L. *darwinianus* Hdz. et Miranda C. subspecies nova. *Revista de la Sociedad Mexicana de Historia Natural* **20**: 99–121.

Hildebrand, D. F. and Hymowitz, T. 1981. Two soyabean genotypes lacking lipoxygenase-1. *Journal of the American Oil Chemists' Society* **58**: 583–586.

Hildebrand, G. L. and Smartt, J. 1980. The utilization of Bolivian groundnut (*Arachis hypogaea* L.) germplasm in Central Africa. *Zimbabwe Journal of Agricultural Research* **18**: 39–48.

Hill, G. D. 1977. The composition and nutritive value of lupin seed. *Nutrition Abstracts and Reviews, B. Livestock Feeds and Feeding* **47**: 511–529.
Hitchcock, C. L. 1952. A revision of the N. American species of *Lathyrus*. *University of Washington Publications in Biology* 5.
Hoehne, F. C. 1940. Leguminosas – Papilionadas, Gênero *Arachis*. *Flora Brasilica* **25** (2) 122: 1–20.
Honma, S. 1956. A bean interspecific hybrid. *Journal of Heredity* **47**: 217–220.
Honma, S. and Heeckt, O. 1958. Bean interspecific hybrid involving *Phaseolus coccineus* × *P. lunatus*. *Proceedings, American Society for Horticultural Science* **72**: 360–364.
Honma, S. and Heeckt, O. 1959. Interspecific hybrid between *Phaseolus vulgaris* and *P. lunatus*. *Journal of Heredity* **50**: 233–237.,
Hoover, E. E., Brenner, M. L. and Ascher, P. D. 1985*a*. Comparison of development of two bean crosses. *HortScience* **20**: 884–886.
Hoover, E. E., Brenner, M. L. and Brun, W. A. 1985*b*. Partitioning of 14C-photoassimilates within seeds of *Phaseolus coccineus* (Lam.) and *Phaseolus coccinus* × *Phaseolus vulgaris* L. *Plant Growth Regulation* **3**: 63–69.
Hopf, M. 1969. Plant remains and early farming in Jericho. In *The Domestication and Exploitation of Plants and Animals*, pp. 355–359 (eds. P. J. Ucko and G. W. Dimbleby). London: Duckworth.
Hu, T. C. 1963. Discourse on the character shu (soybeans). *Essays on Chinese Literature and History, 3rd Series* (Shanghai), pp. 111–15. (In Chinese, cited in Hymowitz and Newell, 1981.)
Hucl, P. and Scoles, G. J. 1985. Interspecific hybridization in the common bean. *HortScience* **20**: 352–357.
Hull, F. H. 1937. Inheritance of rest period of seeds and certain other characters in the peanut. *Bulletin 314, University of Florida Agricultural Experiment Station, Gainesville, Florida*.
Hull, F. H. and Carver, W. A. 1938. Peanut improvement. *Annual Report, Florida Agricultural Experiment Station*, pp. 39–40.
Hume, D. J., Shanmugasundaram, S. and Beversdorf, W. D. 1985. Soyabean. In *Grain Legume Crops*, pp. 391–432 (eds. R. J. Summerfield and E. H. Roberts). London: Collins.
Husted, L. 1931. Chromosome numbers in species of peanut, *Arachis*. *American Naturalist* **65**: 476–477.
Husted, L. 1933. Cytological studies of the peanut *Arachis*. I. Chromosome number and morphology. *Cytologia* **5**: 109–117.
Husted, L. 1936. Cytological studies of the peanut *Arachis*. II. Chromosome number, morphology and behaviour and their application to the origin of cultivated forms. *Cytologia* **7**: 396–423.
Hwang, J. K. 1979. Interspecific hybridization between *Phaseolus vulgaris* L. and *Phaseolus acutifolius* A. Gray. *Dissertation Abstracts International B.* **39**: 4672B.
Hymowitz, T. 1970. On the domestication of the soybean. *Economic Botany* **24**: 408–421.
Hymowitz, T. 1972. The trans-domestication concept as applied to guar. *Economic Botany* **26**: 49–60.
Hymowitz, T. 1976. Soybeans. In *Evolution of Crop Plants*, pp. 159–162 (ed. N. W. Simmonds). London: Longman.

Hymowitz, T. and Boyd, J. 1977. Origin, ethnobotany and agricultural potential of the winged bean – *Psophocarpus tetragonolobus*. *Economic Botany* **31**: 180–188.

Hymowitz, T. and Harlan, J. R. 1983. Introduction of the soybean to North America by Samuuel Bowen in 1765. *Economic Botany* **37**: 371–379.

Hymowitz, T. and Newell, C. A. 1980. Taxonomy, speciation, domestication, dissemination, germplasm resources and variation in the genus *Glycine*. In *Advances in Legume Science*, pp. 251–264 (eds. R. J. Summerfield and A. H. Bunting). Kew: Royal Botanic Gardens.

Hymowitz, T. and Newell, C. A. 1981. Taxonomy of the genus *Glycine*, domestication and uses of soybeans. *Economic Botany* **35**: 272–288.

IBPGR (International Board for Plant Genetic Resources)
1982 *Lima bean descriptors*
1982 Phaseolus vulgaris *descriptors*
1983 Phaseolus coccineus *descriptors*
1985 Phaseolus acutifolius *descriptors*
1985 *Groundnut descriptors (revised)*
Rome: Italy: IBPGR Secretariat, FAO.

ICRISAT (International Crops Research Institute for the Semi Arid Tropics) *1982 Annual Report*. Patancheru, India.

ICRISAT 1985. *Proceedings of an International Workshop on Cytogenetics of* Arachis. *31 October–2 November 1983*. Patancheru, India: ICRISAT Center.

Jackson, M. T. and Yunus, A. G. 1984. Variation in the grass pea (*Lathyrus sativus* L.) and wild species. *Euphytica* **33**: 549–559.

Jaffé, W. G., Brücher, O. and Palozza, A. 1972. Detection of four types of specific phytohaemagglutinins in different lines of beans (*Phaseolus vulgaris*). *Zeitschrift für Immunitätsforschung und experimentelle Therapie* **142**: 439–447.

Jahnavi, M. R. and Murty, U. R. 1985. Interspecific gene transfer in *Arachis hypogaea* L., in relation to the behaviour of triploid, pentaploid and hexaploid derivatives. *Cytologia* **50**: 865–878.

Jain, H. K. and Mehra, K. L. 1980. Evolution, adaptation, relationships and uses of the species of *Vigna* cultivated in India. In *Advances in Legume Science*, pp. 459–468 (eds. R. J. Summerfield and A. H. Bunting). Kew: Royal Botanic Gardens.

Janzen, D. H. 1969. Seed-eaters versus seed size, number toxicity and dispersal. *Evolution* **23**: 1–27.

Johansen, E. L. and Smith, B. W. 1956. *Arachis hypogaea* × *Arachis diogoi*. Embryo and seed failure. *American Journal of Botany* **43**: 250–258.

Johnson, D. R., Wynne, J. C. and Campbell, W. V. 1977. Resistance of wild species of *Arachis* to the two-spotted spider mite, *Tetranychus urticae*. *Peanut Science* **4**: 9–11.

Jolivet, E. and Mossé, J. 1982. Non protein nitrogenous compounds with particular reference to ureides. In *Chemistry and Biochemistry of Legumes*, pp. 111– 193 (ed. S. K. Arora). New Delhi, Oxford and IBH Publishing Co. (London: Edward Arnold, 1983.)

Kajale, M. D. 1974. Plant economy at Bhokardan. In *Excavations at Bohkardan (Bhogavadhana)* 1973, vol. 2, pp. 7–224 (eds. S. B. Deu and R. S. Gupta). India: Nagpur University and Maharashtra University.

Kamal, S. S. 1976. Resistance of species of *Arachis* to lesser cornstalk borer. (Abstract.) *Oléagineux* **33**: 78.

Kaplan, L. 1965. Archeology and domestication in American *Phaseolus*. *Economic Botany* **19**: 358–368.

Kaplan, L. 1981. What is the origin of the common bean? *Economic Botany* **35**: 240–254.

Kaplan, L., Lynch, T. F. and Smith, C. E. Jr. 1973. Early cultivated beans (*Phaseolus vulgaris*) from an intermontane Peruvian valley. *Science* **179**: 76–77.

Kawakami, J. 1930. Chromosome numbers in Leguminosae. *Botanical Magazine (Tokyo)* **44**: 319–328. (Cited by Husted, 1933.)

Khan, T. N. 1976. Papua New Guinea: a centre of genetic diversity in winged bean (*Psophocarpus tetragonolobus* (L.) DC.). *Euphytica* **25**: 693–706.

Khan, T. N. and Eagleton, G. E. 1980. The winged bean, *Psophocarpus tetragonolobus*. In *Advances in Legume Science*, pp. 383–392 (eds. R. J. Summerfield and A. H. Bunting). Kew: Royal Botanic Gardens.

Khvostova, V. V. 1983. *Genetics and Breeding of Peas*. New Delhi: Oxonian Press.

Kirti, P. B., Murty, U. R., Bharati, M. and Rao, N. G. P. 1982. Chromosome pairing in F1 hybrid *Arachis hypogaea* L. × *A. monticola* Krap. et. Rig. *Theoretical and Applied Genetics* **62**: 139–144.

Kloz, J. 1971. Serology of the Leguminosae. In *Chemotaxonomy of the Leguminosae*, pp. 309–365 (eds. J. Harborne, D. Boulter and B. L. Turner). London: Academic Press.

Kloz, J. and Klozová, E. 1968. Variability of proteins I and II in the seeds of species of the genus *Phaseolus*. In *Chemotaxonomy and Serotaxonomy*, pp. 93–102 (ed. J. G. Hawkes). London: Academic Press.

Klozová, E. 1965. Interrelations among several Asiatic species of the genus *Phaseolus* studied by immunochemical methods. In *Proceedings of the Symposium on the Mutational Process, Prague, August 9–11, 1965*. Prague: Academia.

Klozová, E., Kloz, J. and Winfield, P. 1976. Atypical composition of seed proteins in cultivars of *Phaseolus vulgaris* L. *Biologia Plantarum* **18**: 200–205.

Klozová, E., Švachulová, J., Smartt, J., Hadač, E., Turková, V. and Hadačová, V. 1983*a*. The comparison of seed protein patterns within the genus *Arachis* by polyacrylamide gel electrophoresis. *Biologia Plantarum* **25**: 266–273.

Klozová, E. and Turková, V. 1978. The polymorphism of a seed protein with phytohaemagglutinating activity in the cultivar of *Phaseolus vulgaris* L. *Biologia Plantarum* **20**: 373–376.

Klozvoá, E., Turková, V., Smartt, J., Pitterová, K. and Švachulová, J. 1983*b*. Immunochemical characterization of seed proteins of some species of the genus *Arachis* L. *Biologia Plantarum* **25**: 201–208.

Kousalya, G., Ayyavoo, R., Muthuswamy, M. and Kandaswamy, T. K. 1972. Reaction of wild species of *Arachis* to groundnut viruses. *Madras Agricultural Journal* **59**: 563.

Krapovickas, A. 1968. Origen, variabilidad y diffusion del maní (*Arachis hypogaea*). *Actas y Memorias del XXVII Congreso Internacional de Americanistas* **2**: 517–534.

Krapovickas, A. 1969. The origin, variability and spread of the groundnut

(*Arachis hypogaea*). In *The Domestication and Exploitation of Plants and Animals*, pp. 427–441 (eds. P. J. Ucko and G. W. Dimbleby). London: Duckworth.

Krapovickas, A. 1973. Evolution of the genus *Arachis*. In *Agricultural Genetics – Selected Topics*, pp. 131–151 (ed. R. Moav). New York: J. Wiley.

Krapovickas, A., Fernandez, A. and Seeligman, P. 1974. Recuperación de la fertilidad en un híbrido interspecifico estéril de *Arachis* (Leguminosae). *Bonplandia* **3**: 129–142.

Krapovickas, A. and Gregory, W. C. 1960. *Arachis rigonii* nueva especie silvestre de maní. *Revista de Investigaciones Agricolas* **14**: 157–160.

Krapovickas, A. and Rigoni, V. A. 1949. Cromosomas de un especie silvestre de *Arachis*. *Idia* **2**: 23–24.

Krapovickas, A. and Rigoni, V. A. 1950. Observaciones citologicas y geneticas en *Arachis*. *Memoria de la primera reunion de maní y girasol*. Ministerio de Agricultura y Ganaderia Nacion Argentina.

Krapovickas, A. and Rigoni, V. A. 1951. Estudios citologicos en el genero *Arachis*. *Revista de Investigaciones Agricolas* **5**: 289–293.

Krapovickas, A. and Rigoni, V. A. 1953. Cromosomas de una especie silvestre de *Arachis*. *Archivo Fitotecnico del Uruguay* **5**: 282–420.

Krapovickas, A. and Rigoni, V. A. 1954. Progreso realizado en las investigaciones agricolas durante el año 1953. *Idia* **73–75**: 1–159.

Krapovickas, A. and Rigoni, V. A. 1956. Noroeste argentino y Bolivia probable centro de origen del maní. *Darwiniana* **11**: 197–228.

Krapovickas, A. and Rigoni, V. A. 1957. Nuevas especies de *Arachis* vinculadas al problema del origen del maní. *Darwiniana* **11**: 431–455.

Krapovickas, A. and Rigoni, V. A. 1960. La nomenclature de la subespecies y variedades de *Arachis hypogaea* L. *Revista de Investigaciones Agricolas* **14**: 198–228.

Krishna, T. G. and Reddy, L. J. 1982. Species affinities between *Cajanus cajan* and some *Atylosia* species based on esterase isoenzymes. *Euphytica* **31**: 709–713.

Krishnan, R. and De, D. N. 1965. Studies on pachytene and somatic chromosomes of *Phaseolus aureus*. *Nucleus* **8**: 7–16.

Krishnan, R. and De, D. N. 1968*a*. Cytogenetical studies in *Phaseolus*. I. Autotetraploid *Phaseolus aureus* × a tetraploid species of *Phaseolus* and the backcrosses. *Indian Journal of Plant Breeding and Genetics* **28**: 12–22.

Krishnan, R. and De, D. N. 1968*b*. Cytogenetical studies in *Phaseolus*. II. *Phaseolus mungo* × tetraploid *Phaseolus* species and the amphidiploid. *Indian Journal of Plant Breeding and Genetics* **28**: 23–30.

Kumar, L. S. S. and D'Cruz, R. 1957. Aneuploidy in species hybrid of *Arachis*. *Journal of the Indian Botanical Society* **36**: 545–547.

Kumar, L. S. S., D'Cruz, R. and Oke, J. G. 1957. A synthetic allohexaploid in *Arachis*. *Current Science* **26**: 121–122.

Kumar, L. S. S., Thombre, M. V. and D'Cruz, R. 1958. Cytological studies of an intergeneric hybrid of *Cajanus cajan* (Linn.) Millsp. and *Atylosia lineata* W. and A. *Proceedings of the Indian Academy of Sciences* B **47**: 252–261.

Kupicha, F. K. 1976. The infrageneric structure of *Vicia*. *Notes from the Royal Botanic Garden, Edinburgh* **34**: 287–326.

Kupicha, F. K. 1977. The delimitation of the tribe Vicieae (Leguminosae) and the relationships of *Cicer* L. *Botanical Journal of the Linnean Society* **74**: 131–162.

Kupicha, F. K. 1981. Tribe 21 – Vicieae. In *Advances in Legume Systematics*, pp. 377–381 (eds. R. M. Polhill and P. H. Raven). Kew: Royal Botanic Gardens.

Kupicha, F. K. 1983. The intrageneric structure of *Lathyrus*. *Notes from the Royal Botanic Garden, Edinburgh* **41**: 209–244.

Kyle, J. H. 1959. Factors influencing water entry through the micropyle in *Phaseolus vulgaris* L. and their significance in inheritance studies of hard seeds. *Dissertation Abstracts* **20**: 840.

Kyle, S. H. and Randall, T. E. 1963. A new concept of the hard seed character in *Ph. vulgaris* L. and its use in breeding and inheritance studies. *Proceedings of the American Society for Horticultural Science* **83**: 461–475.

Lackey, J. A. 1977. A revised classification of the tribe Phaseoleae (Leguminosae, Papilionoideae) and its relation to canavanine distribution. *Botanical Journal of the Linnean Society* **74**: 163–178.

Lackey, J. A. 1981. Tribe 10 Phaseoleae. In *Advances in Legume Systematics*, pp. 301–327 (eds. R. M. Polhill and P. H. Raven). Kew: Royal Botanic Gardens.

Ladizinsky, G. 1975*a*. A new *Cicer* from Turkey. *Notes from the Royal Botanic Garden, Edinburgh* **34**: 201–202.

Ladizinsky, G. 1975*b*. On the origin of the broad bean *Vicia faba* L. *Israel Journal of Botany* **24**: 80–88.

Ladizinsky, G. 1975*c*. Seed protein electrophoresis of the wild and cultivated species of section *Faba* of *Vicia*. *Euphytica* **24**: 785–788.

Ladizinsky, G. 1979*a*. The origin of the lentil and its wild gene pool. *Euphytica* **28**: 179–187.

Ladizinsky, G. 1979*b*. The genetics of several morphological traits in the lentil. *Journal of Heredity* **70**: 135–137.

Ladizinsky, G. 1979*c*. Species relationships in the genus *Lens* as indicated by seed protein electrophoresis. *Botanical Gazette* **140**: 449–451.

Ladizinsky, G. 1985. The genetics of hard seed coat in the genus *Lens*. *Euphytica* **34**: 339–543.

Ladizinsky, G. 1986. A new *Lens* from the Middle East. *Notes from the Royal Botanic Garden, Edinburgh* **43**: 489–492.

Ladizinsky, G. and Adler, A. 1975. The origin of chickpea as indicated by seed protein electrophoresis. *Israel Journal of Botany* **24**: 183–189.

Ladizinsky, G. and Adler, A. 1976*a*. The origin of chickpea *Cicer arietinum* L. *Euphytica* **25**: 211–217.

Ladizinsky, G. and Adler, A. 1976*b*. Genetic relationships among the annual species of *Cicer* L. *Theoretical and Applied Genetics* **48**: 197–203.

Ladizinsky, G. and Hamel, A. 1980. Seed protein profiles of pigeon pea (*Cajanus cajan*) and some *Atylosia* species. *Euphytica* **19**: 313–317.

Lamprecht, H. 1966. *Die Entstehung der Arten*. Vienna: Springer.

Larsen, A. L. 1967. Electrophoretic differences in seed proteins among varieties of soybean. *Crop Science* **7**: 311–313.

Larsen, A. L. and Caldwell, B. E. 1968. Inheritance of certain proteins in soybean seed. *Crop Science* **8**: 474–476.

Larsen, A. L. and Caldwell, B. E. 1969. Sources of protein variants in soybean seeds. *Crop Science* **9**: 385–386.

Lawes, D. A., Bond, D. A. and Poulsen, M. H. 1983. Classification, origin, breeding methods and objectives. In *The Faba Bean (Vicia faba L.). A Basis for Improvement*, pp. 23–76 (ed. P. D. Hebblethwaite). London: Butterworths.

Le Marchand, G. and Maréchal, R. 1977. Wide crossing with lima beans. *Tropical Grain Legume Bulletin* **8**: 7–11.

Le Marchand, G., Maréchal, R. and Baudet, J. C. 1976. Observations sur quelques hybrides dans le genre *Phaseolus*. III. *P. lunatus*: nouveaux hybrides et considérations sur les affinités interspéficiques. *Bulletin des Recherches Agronomiques de Gembloux* **11**: 183–200.

Leuck, D. B. and Hammons, R. O. 1968. Resistance of wild peanut plants to the mite *Tetranychus tumidellus*. *Journal of Economic Entomology* **61**: 687–688.

Liener, I. E. 1978. Protease inhibitors and other toxic factors in seeds. In *Plant Proteins*, pp. 117–140 (ed. G. Norton). London: Butterworths.

Liener, I. E. 1982. Toxic constituents in legumes. In *Chemistry and Biochemistry of Legumes*, pp. 217–257 (ed. S. K. Arora). New Delhi: Oxford and IBH Publishing Co. (London: Edward Arnold, 1983.)

Linnaeus, C. 1737. *Hortus Cliffortianus, Historiae Naturalis Classica*, p. 499 (eds. J. Cramert and H. K. Swann). Reprinted 1968, New York: Stochert-Hafner Service Agency.

Linnaeus, C. 1753. *Species Plantarum*. Stockholm.

Linnaeus, C. 1754. *Genera Plantarum*. Stockholm.

Lorz, A. P. 1952. An interspecific cross involving the lima bean *Phaseolus lunatus* L. *Science* **115**: 702–703.

Lukoki, L., Maréchal, R. and Otoul, E. 1980. Les ancêtres sauvages des haricots cultivées: *Vigna radiata* (L.) Wilczek et *V. mungo* (L.) Hepper. *Bulletin du Jardin Botanique de Belgique* **50**: 385–391.

Lush, W. M. 1979. Floral morphology of wild and cultivated cowpeas. *Economic Botany* **33**: 442–447.

Lush, W. M. and Evans, L. T. 1980. The seed coats of cowpeas and other grain legumes: structure in relation to function. *Field Crops Research* **3**: 267–286.

Lush, W. M. and Evans, L. T. 1981. The domestication and improvement of cowpeas (*Vigna unguiculata* (L.) Walp.). *Euphytica* **30**: 579–587.

McComb, J. A. 1975. Is intergeneric hybridization in the Leguminosae possible? *Euphytica* **24**: 497–502.

Machado, M., Tai, W. and Baker, L. R. 1982. Cytogenetic analysis of the interspecific hybrid *Vigna radiata* × *V. umbellata*. *Journal of Heredity* **73**: 205–208.

Madhava Menon, P., Raman, V. S. and Krishanswami, S. 1970. An X-ray induced monosomic in groundnut. *Madras Agricultural Journal* **57**: 80–82.

Makasheva, R. Kh. 1983. *The Pea* (translated from the Russian edition, 1973). New Delhi: Oxonian Press.

Mallikarjuna, N. and Sastri, D. C. 1985. In vitro culture of ovules and embryos from some incompatible interspecific crosses in the genus *Arachis*. In *Cytogenetics of Arachis*, pp. 147–151 (ed. J. P. Moss). Patancheru, India: International Crops Research Institute for the Semi-Arid Tropics.

Maréchal, R. 1969. Données cytologiques sur les espèces de la sous-tribu des Papilionaceae – Phaseoleae – Phaseolinae. Première serie. *Bulletin du Jardin Botanique National de Belgique* **39**: 125–165.

Maréchal, R. 1982. Arguments for a global conception of the genus *Vigna*. *Taxon* **31**: 280–283.

Maréchal, R. and Baudet, J. C. 1977. Transfert du genre africain *Kerstingiella* Harms à *Macrotyloma* (Wight et Arn.) Verdc. (Papilionaceae). *Bulletin du Jardin Botanique National de Belgique* **47**: 49–52.

Maréchal, R. and Baudoin, J. P. 1978. Observations sur quelques hybrides dans le genre *Phaseolus*. IV. L'hybride *Phaseolus vulgaris* × *Phaseolus filiformis*. *Bulletin des Recherches Agronomiques de Gembloux* **13**: 233–240.

Maréchal, R., Mascherpa, J. M. and Stainier, F. 1978. Etude taxonomique d'un groupe complexe d'espèces des genres *Phaseolus* et *Vigna* (Papilionaceae) sur la base de données morphologiques et polliniques, traitées par l'analyse informatique. *Boissiera, Genève* **28**: 1–273.

Mears, J. A. and Mabry, T. J. 1971. Alkaloids in the Leguminosae. In *Chemotaxonomy of the Leguminosae*, pp. 73–178 (eds. J. B. Harborne, D. Boulter and B. L. Turner). London: Academic Press.

Meikle, R. D. 1969. *Pisum* L. *Notes from the Royal Botanic Garden, Edinburgh* **29**: 320.

Mendel, G. 1986. Versuche über Pflanzenhybriden. *Verhandlungen des Naturforschenden Vereins in Brunn* **4**: 3–47. (English translation in *Experiments in Plant Hybridization*, 1965 (ed. J. H. Bennett). Edinburgh: Oliver and Boyd.)

Mendes, A. J. T. 1947. Estudos citologicos no gênero *Arachis*. *Bragantia* **7**: 257–267.

Miège, J. 1960. Troisième liste de nombres chromosomiques d'Afrique occidentale. *Annales de la Faculté des Sciences, Université de Dakar* **5**: 75–85.

Miller, J. C. Jr., Zary, K. W. and Fernandez, G. C. J. 1986. Inheritance of N_2 fixation efficiency in cowpea. *Euphytica* **35**: 551–560.

Miranda Colín, S. 1968. Origen de *Phaseolus vulgaris* L. (Frijol común). *Agronomia Tropical* **18**: 191–205.

Miranda Colín, S. 1969. Estudio sobre la herencia de tres caracteres de frijol. *Agrociencia* **4**: 115–122.

Miranda Colín, S. 1974. Evolutionary genetics of wild and cultivated *Phaseolus vulgaris* L. and *Phaseolus coccineus* L. Ph.D. thesis, Cambridge University.

Mok, D. W. S., Mok, M. C. and Rabakoarihanta, A. 1978. Interspecific hybridization of *Phaseolus vulgaris* with *P. lunatus* and *P. acutifolius*. *Theoretical and Applied Genetics* **52**: 209–215.

Montgomery, R. D. 1964. Observations on the cyanide content and toxicity of tropical pulses. *West Indian Medical Journal* **13**: 1–11.

Moreno, M. T. and Cubero, J. I. 1978. Variation in *Cicer arietinum* L. *Euphytica* **27**: 465–485.

Moreno, M. T. and Martinez, A. 1980. The divided world of *V. faba*. *FABIS* **2**: 18–19.

Moss, J. P. 1980. Wild species in the improvement of groundnuts. In *Advances in Legume Science*, pp. 525–535 (eds. R. J. Summerfield and A. H. Bunting). Kew: Royal Botanic Gardens.

Mossé, J. and Pernollet, J. C. 1982. Storage proteins of legume seeds. In *Chemistry and Biochemistry of Legumes*, pp. 111–193 (ed. S. K. Arora). New Delhi: Oxford and IBH Publishing Co. (London: Edward Arnold, 1983.)

Mukherjee, P. 1968. Pachytene analysis in *Vigna*. Chromosome morphology in *Vigna sinensis* (cultivated). *Science and Culture* **34**: 252–253.

Murty, U. R. and Jahnavi, M. R. 1986. The 'A' genome of *Arachis hypogaea* L. *Cytologia* **51**: 241–250.

Nabhan, G. P. 1979. Tepary beans, the effects of domestication on adaptations to arid environments. *Arid Lands Newsletter* **10**: 11–16.

NAS (National Academy of Sciences) 1975. *Underexploited Tropical Plants with Promising Economic Value*. Washington, DC: National Academy of Sciences.

NAS (National Academy of Sciences) 1979. *Tropical Legumes – Resources for the Future*. Washington, DC: National Academy of Sciences.

NRC (National Research Council) 1981. *The Winged Bean – a High Protein Crop for the Tropics*. (2nd edition). Washington, DC: National Academy Press.

Neucere, N. J. and Cherry, J. P. 1975. An immunochemical survey of proteins in species of *Arachis*. *Peanut Science* **2**: 66–72.

Nevill, D. J. 1978. Breeding groundnuts (*Arachis hypogaea* L.) for resistance to foliar pathogens. Ph.D. thesis. Cambridge: University Library.

Newell, C. A. and Hymowitz, T. 1982. Successful wide hybridization between the soybean and a wild perennial relative. *G. tomentella* Hayata. *Crop Science* **22**: 1062–1065.

Newell, C. A. and Hymowitz, T. 1983. Hybridization in the genus *Glycine* subgenus *Glycine* Willd. (Leguminosae – Papilionoideae). *American Journal of Botany* **70**: 334–348.

Nicholls, L., Sinclair, H. M. and Jelliffe, D. B. 1961. *Tropical Nutrition and Dietetics* (4th edition). London: Bailliere, Tindall and Cox.

Ng, N. Q. and Maréchal, R. 1985. Cowpea taxonomy, origin and germplasm. In *Cowpea – Research, Production and Utilization*, pp. 11–21 (eds. S. R. Singh and K. O. Rachie). Chichester: J. Wiley.

Ohwi, J. 1965. *Flora of Japan*. Washington, DC: Smithsonian Institution.

Orf, J. H. and Hymowitz, T. 1979*a*. Inheritance of the absence of the Kunitz trypsin inhibitor in seed protein of soybeans. *Crop Science* **19**: 107–109.

Orf, J. H. and Hymowitz, T. 1979*b*. Genetics of the Kunitz trypsin inhibitor: an antinutritional factor in soybeans. *Journal of the American Oil Chemists' Society* **56**: 722–726.

Orf, J. H., Hymowitz, T., Pull, S. P. and Pueppke, S. G. 1978. Inheritance of a soybean seed lectin. *Crop Science* **18**: 899–900.

Osborn, T. C. and Bliss, F. A. 1985. Effects of genetically removing lectin seed protein on horticultural and seed characteristics of common bean. *Journal of the American Society for Horticultural Science* **110**: 484–488.

PAG (Protein Advisory Group of the United Nations System) 1973. *Nutritional Improvement of Food Legumes by Breeding*. New York: United Nations Organization.

Parker, P. F. 1978. The classification of crop plants. In *Essays in Plant Taxonomy*, pp. 97–124 (ed. H. E. Street). London: Academic Press.

Patil, S. H. 1968. Cytogenetics of X-ray induced aneuploids in *Arachis hypogaea* L. *Canadian Journal of Genetics and Cytology* **10**: 545–550.

Patil, S. H. and Bora, K. C. 1961. Meiotic abnormalities induced by X-rays in *Arachis hypogaea*. *Indian Journal of Genetics and Plant Breeding* **21**: 59–74.

Paul, C., Gates, P., Harris, N. and Boulter, D. 1978. Asynchronous sexual development determines the breeding system in field beans. *Nature, London* **275**: 54–55.

Payne, P. R. 1978. Human protein requirements. In *Plant Proteins*, pp. 247–263 (ed. G. Norton). London: Butterworths.

Pickersgill, B. 1980. Cytology of two species of winged bean, *Psophocarpus tetragonolobus* (L.) DC. and *P. scandens* (Endl.) Verdc. (Leguminosae). *Botanical Journal of the Linnean Society* **80**: 279–291.

Pickersgill, B., Jones, J. K., Ramsay, G. and Stewart, H. 1983. Problems and prospects of wide crossing in the genus *Vicia* for the improvement of faba bean. In *Faba Beans, Kabuli Chickpeas and Lentils in the 1980s*, pp. 57–70 (eds. M. C. Saxena and S. Varma). Aleppo: The International Center for Agricultural Research in the Dry Areas (ICARDA).

Pinkas, R., Zamir, D. and Ladizinsky, G. 1985. Allozyme divergence and evolution in the genus *Lens*. *Plant Systematics and Evolution* **151**: 131–140.

Piper, C. V. and Morse, W. J. 1910. *The Soybean: History, Varieties and Field Studies*. United States Department of Agriculture, Bureau of Plant Industry Bulletin No. 197.

Plitmann, U. 1970. Vicia. In *Flora of Turkey*, vol. 3 (ed. P. H. Davies). Edinburgh University Press.

Plukenet, L. 1692. *Phytographia* 3. London (cited in van der Maesen, 1985.)

Polhill, R. M. and Raven, P. J. (eds.) 1981. *Advances in Legume Systematics*. Kew: Royal Botanic Gardens.

Polhill, R. M. and van der Maesen, L. J. G. 1985. Taxonomy of grain legumes. In *Grain Legume Crops*, pp. 3–36 (eds. R. J. Summerfield and E. H. Roberts). London: Collins.

Pompeu, A. S. 1977. Cruzamentos entre *Arachis hypogaea* e as espécies *A. villosa* var. *correntina*, *A. diogoi* e *A. villosulicarpa*. *Ciência e Cultura* **29**: 319–321.

Pompeu, A. S. 1979. Cruzamento interspecífico entre *Arachis hypogaea* et *A. diogoi* (Abstract). *Ciência e Cultura* **31** (Supplement 7) 17.

Pompeu, A. S. 1983. Cruzamentos entre *Arachis hypogaea* e as espécies *A. diogoi* e *A. spp. (30006, 30035)*. *Bragantia* **31**: 261–265.

Pratt, R. C. and Bressan, R. A. 1983. Fertility of *Phaseolus vulgaris* × *Phaseolus acutifolius* hybrids and backcross generations (Abstract). *HortScience* **18**: 600.

Pratt, R. C., Bressan, R. A. and Hasegawa, P. M. 1975. Genotypic diversity enhances recovery of hybrids and fertile backcrosses of *Phaseolus vulgaris* L. × *P. acutifolius* A. Gray. *Euphytica* **34**: 329–344.

Prendota, K., Baudoin, J. P. and Maréchal, R. 1982. Fertile allopolyploids from the cross *Phaseolus acutifolius* × *Phaseolus vulgaris*. *Bulletin des Recherches Agronomiques de Gembloux* **17**: 177–189.

Probst, A. H. and Judd, R. W. 1973. Origin, U.S. history and development, and world distribution. In *Soybeans: Improvement, Production and Uses*, pp. 1–15 (ed. B. E. Caldwell). Madison, Wisconsin: American Society of Agronomy.

Pundir, R. P. S. 1981. Relationships among *Cajanus*, *Atylosia* and *Rhynchosia* species. Ph.D. thesis, Varanasi, India: Banaras Hindu University (cited in van der Maesen, 1985).

Pundir, R. P. S. and Sing, R. B. 1983. Cross-compatibility among *Cajanus*, *Atylosia* and *Rhynchosia* species. *International Pigeonpea Newsletter* No. 2: 12–14.

Purseglove, J. W. 1974. *Tropical Crops: Dicotyledons*. London: Longman.

Rachie, K. O. 1985. Introduction. In *Cowpea – Research, Production and Utilization*, XXI–XXVII (eds. S. R. Singh and K. O. Rachie). Chichester: J. Wiley.

Rajendra, B. R. 1979. Genetic studies of micromorphological traits in some interspecific and intergeneric crosses, I. Triticeae and II. Leguminosae (Abstract). *Dissertation Abstracts International* B **40**: 76B.

Raman, V. S. 1957. Studies of the genus *Arachis*. I. Observations on the morphological characters of certain species of *Arachis*. *Indian Oilseeds Journal* **1**: 235–246.

Raman, V. S. 1958*a*. Studies of the genus *Arachis*. II. Chromosome numbers of certain species of *Arachis*. *Indian Oilseeds Journal* **2**: 72–73.

Raman, V. S. 1958*b*. Studies of the genus *Arachis*. IV. Hybrid between *A. hypogaea* and *A. monticola*. *Indian Oilseeds Journal* **2**: 20–23.

Raman, V. S. 1959*a*. Studies in the genus *Arachis*. V. Note on 40-chromosomed interspecific hybrids. *Indian Oilseeds Journal* **3**: 46–48.

Raman, V. S. 1959*b*. Studies in the genus *Arachis*. VI. Investigation on 30-chromosomed interspecific hybrids. *Indian Oilseeds Journal* **3**: 157–161.

Raman, V. S. 1959*c*. Studies in the genus *Arachis*. VII. A natural interspecific hybrid. *Indian Oilseeds Journal* **3**: 226–228.

Raman, V. S. 1960. Studies in the genus *Arachis*. IX. A fertile synthetic tetraploid groundnut from the interspecific backcross *A. hypogaea* × (*A. hypogaea* × *A. villosa*). *Indian Oilseeds Journal* **4**: 90–92.

Raman, V. S. 1973*a*. A form of genic sterility in an allotriploid of *Arachis*. *Oléagineux* **28**: 299–300.

Raman, V. S. 1973*b*. Genome relationships in *Arachis*. *Oléagineux* **28**: 137–140.

Raman, V. S. 1976. *Cytogenetics and Breeding in Arachis*. New Delhi: Today and Tomorrow's Printers and Publishers.

Raman, V. S. and Kesavan, P. C. 1962. Studies on a diploid interspecific hybrid in *Arachis*. *Nucleus* **5**: 123–126.

Ramanujam, S. 1976. Chickpea. In *Evolution of Crop Plants*, pp. 157–158 (ed. N. W. Simmonds). London: Longman.

Ramirez, D. A. 1960. Cytology of Philippines Plants. V. *Psophocarpus tetragonolobus* (Linn.) DC. *Philippine Agriculturist* **43**: 533–534.

Ramsay, G. 1984. C-banding in *Vicia* species. In *Systems for Cytogenetic Analysis in Vicia faba L.*, vol. 11, pp. 28–39 (eds. G. P. Chapman and S. A. Tarawali). Dordrecht, Netherlands: Martinus Nijhof/Dr. W. Junk, Publishers.

Rathnaswamy, R., Sundaram, N., Vindhiyavarman, P., Muthusamy, M., Ramalingam, R. S. and Vamanbhat, M. 1986. Groundnut lines resistant to late leafspot and rust developed through hybridization of 'Mutant 1' *Arachis hypogaea* Linn. with *A. villosa* Benth. *Indian Journal of Agricultural Sciences* **56**: 537–539.

Raunkiaer, C. 1934. *The Life Forms of Plants and Statistical Plant Geography*. Oxford: Clarendon Press.

Rawal, K. M. 1975. Natural hybridization among wild, weedy and cultivated *Vigna unguiculata* (L.) Walp. *Euphytica* **24**: 699–707.

Reddy, L. J. 1973. Interrelationships of *Cajanus* and *Atylosia* species as revealed by hybridization and pachytene analysis. Ph.D. thesis, Kharagpur (India) cited in van der Maesen, 1980.

Reddy, L. J., Green, J. M. and Sharma, D. 1981. Genetics of *Cajanus cajan* (L.) Millsp. × *Atylosia* spp. In *Proceedings of the International Workshop on Pigeonpeas, 1980*, vol. 2, pp. 39–50. Patancheru, AP, India: International Crops Research Institute for the Semi-Arid Tropics.

Renfrew, J. M. 1969. The archaeological evidence for the domestication of plants: methods and problems. In *The Domestication and Exploitation of Plants and Animals*, pp. 149–172 (eds. P. J. Ucko and G. W. Dimbleby). London: Duckworth.

Renfrew, J. M. 1973. *Palaeoethnobotany*. London: Methuen.

Resslar, P. M. 1979. A cytotaxonomic study of the species of *Arachis* L. section Arachis (Leguminosae). Ph.D. thesis, Raleigh: North Carolina State University. *Dissertation Abstracts International* B **40**: 1498B.

Resslar, P. M. 1980. A review of the nomenclature of the genus *Arachis* L. *Euphytica* **29**: 813–817.

Resslar, P. M. and Gregory, W. C. 1979. A cytological study of three diploid species of the genus *Arachis* L. *Journal of Heredity* **60**: 13–16.

Roth, I. 1977. *Fruits of Angiosperms*. Berlin: Gebrüder Borntraeger.

Roxburgh, W. 1874. *Flora Indica*. Reprint of Carey's edition of 1832, Calcutta.

Roy, A. and De, D. N. 1965. Intergeneric hybridization of *Cajanus* and *Atylosia*. *Science and Culture* **31**: 93–95.

Royes, W. V. 1976. Pigeon pea. In *Evolution of Crop Plants*, pp. 154–156 (ed. N. W. Simmonds). London: Longman.

Rugman, E. E. and Cocking, E. C. 1985. The development of somatic hybridization techniques for groundnut improvement. In *Cytogenetics of Arachis*, pp. 167–174 (ed. J. P. Moss). Patancheru, India: International Crops Research Institute for the Semi-Arid Tropics.

Rutger, J. N. 1970. Variation in protein content and its relation to other characters in beans, *Phaseolus vulgaris* L. *Report Dry Bean Research Conference, Davis, California* **10**: 59–69.

Ruthenberg, H. 1980. *Farming Systems in the Tropics* (3rd edition). Oxford University Press.

Rutter, J. and Percy, S. 1984. The pulse that maims. *New Scientist* **103**, no. 1418: 22–23.

Salunkhe, D. K., Sathe, S. K. and Reddy, N. R. 1982. Legume lipids. In *Chemistry and Biochemistry of Legumes*, pp. 51–109 (ed. S. K. Arora). New Delhi: Oxford and IBH Publishing Co. (London: Edward Arnold, 1983.)

Saraswat, K. S. 1980. The ancient remains of the crop plants at Atranjikhera (c. 2,000–2,500 B.C.). *Journal of the Indian Botanical Society* **59**: 306–319.

Sastri, D. C. and Moss, J. P. 1982. Effect of growth regulators on incompatible crosses in the genus *Arachis* L. *Journal of Experimental Botany* **33**: 1293–1301.

Satyan, B. A., Mahishi, D. M. and Shivashankar, G. 1982. Meiosis in the hybrid between green gram and rice bean. *Indian Journal of Genetics and Plant Breeding* **42**: 356–359.

Sauer, J. 1964. Revision of *Canavalia*. *Brittonia* **16**: 108–181.

Sauer, J. and Kaplan, L. 1969. *Canavalia* beans in American prehistory. *American Antiquity* **34**: 417–423.
Savi, G. 1798. *Flora Pisana*. Pisa: P. Giacomelli.
Saw Lwin, 1956. Studies in the genus *Lathyrus*. M.Sc. thesis. Manchester: University Library.
Saxena, M. C. 1981. Agronomy of lentils. In *Lentils*, pp. 111–129 (eds. C. Webb and G. Hawtin). Farnham Royal, Commonwealth Agricultural Bureaux.
Schäfer, H. I. 1973. Taxonomie der *Vicia narbonensis*-Gruppe. *Kulturpflanze* **21**: 211–273.
Schultze-Motel, J. 1972. Die archäologischen Reste der Ackerbohne, *Vicia faba* L.; und die Genese der Art. *Kulturpflanze* **19**: 321–358.
Schwanitz, F. 1966. *The Origin of Cultivated Plants*. Cambridge, Massachusetts: Harvard University Press.
Seetharam, A., Muraleedharan Nyar, K., Achar, D. K. T. and Hamananthappa, H. S. 1974. Interspecific hybridization in groundnut to transfer resistance to tikka leafspot disease. *Current Research, Bangalore, India* **3**: 98–99.
Sen, N. K., and Bhowal, J. G. 1960. Colchicine-induced tetraploids of six varieties of *Vigna sinensis*. *Indian Journal of Agricultural Science* **30**: 149–161.
Sen, N. K. and Ghosh, A. K. 1961. Genetic studies in green gram. *Indian Journal of Genetics and Plant Breeding* **19**: 210–227.
Sen, N. K. and Hari, M. N. 1956. Comparative study of diploid and tetraploid cowpea (Abstract). *Proceedings of the 43rd Indian Science Congress* **3**: 258.
Sene, O. 1967. Determinisme genetique de la precocité chez *Vigna unguiculata* (L.) Walp. *Agronomie Tropicale* **22**: 309–318.
Senn, H. A. 1938*a*. Chromosome number relationships in the Leguminosae. *Bibliographia Genetica* **21**: 175–336.
Senn, H. A. 1938*b*. Experimental data for a revision of the genus *Lathyrus*. *American Journal of Botany* **25**: 67–78.
Shanmugam, A. S., Rangasamy, S. R. S. and Rathnasamy, R. 1985. Study of segregating generations of the interspecific hybrids of the genus *Vigna*. *Genetica Agraria* **39**: 387–400.
Shanmugasundaram, S., Kuo, G. C. and Nalampang, A. 1980. Adaptation and utilisation of soyabeans in different environments and agricultural systems. In *Advances in Legume Science*, pp. 265–277 (eds. R. J. Summerfield and A. H. Bunting). Kew: Royal Botanic Gardens.
Sharief, Y., Rawlings, J. O. and Gregory, W. C. 1978. Estimates of leafspot resistance in three interspecific hybrids of *Arachis*. *Euphytica* **27**: 741–751.
Sikdar, A. K. and De, D. N. 1967. Cytological studies of two species of *Atylosia*. *Bulletin of the Botanical Society of Bengal* **1**: 25–28.
Simpson, C. E. 1976. Peanut breeding strategy to exploit sources of variability from wild *Arachis* species (Abstract). *Proceedings of the American Peanut Research and Education Association* **8**: 87.
Simpson, C. E. 1984. Plant exploration: planning, organization and implementation with special emphasis on *Arachis*. In *Conservation of Crop Germplasm – An International Perspective*, chapter 1, pp. 1–20 (eds. W. L. Brown, T. T. Chang *et al.*). Madison, Wisconsin: Crop Science Society of America.
Simpson, C. E. and Higgins, D. L. 1984. Catalog of *Arachis* germplasm collec-

tions in South America, 1976–1983. College Station, Texas: Texas Agricultural Experiment Station; Rome, Italy: International Board for Plant Genetic Resources.

Sinclair, T. R. and de Wit, C. T. 1975. Photosynthate and nitrogen requirements for seed production by various crops. *Science* **189**: 565–567.

Singh, A. K. 1985. Genetic introgression from compatible wild species into cultivated groundnut. In *Cytogenetics of Arachis*, pp. 107–117 (ed. J. P. Moss). Patancheru, India: International Crops Researich Institute for the Semi-Arid Tropics.

Singh, A. K. 1986*a*. Utilization of wild relatives in the genetic improvement of *Arachis hypogaea* L. 7. Autotetraploid production and the prospects in interspecific breeding. *Theoretical and Applied Genetics* **72**: 164–169.

Singh, A. K. 1986*b*. Utilization of wild relatives in the genetic improvement of *Arachis hypogaea* L. 8. Synthetic amphiploids and their importance in interspecific breeding. *Theoretical and Applied Genetics* **72**: 433–439.

Singh, A. K. and Moss, J. P. 1982. Utilization of wild relatives in genetic improvement of *Arachis hypogaea* L. 2. Chromosome complements of species in section *Arachis*. *Theoretical and Applied Genetics* **61**: 305–314.

Singh, B., Khanna, A. N. and Vaidyn, S. M. 1964. Crossability studies in the genus *Phaseolus*. *Journal of Postgraduate School (India)* **2**: 47–50.

Singh, D. and Mehta, T. R. 1953. Inheritance of lobed leaf character in mung. *Current Science* **22**: 248.

Singh, K. P., Sharma, S. K., Singh, R. K. and Sood, D. R. 1986. Performance of amphidiploid derivatives of green gram × black gram crosses. *Indian Journal of Agricultural Sciences* **56**: 390–392.

Singh, R. J., Kollipara, K. P. and Hymowitz, T. 1987. Intersubgeneric hybridization of soybeans with a wild perennial species, *Glycine clandestina* Wendl. *Theoretical and Applied Genetics* **74**: 391–396.

Singh, S. P. and Gutierrez, J. A. 1986. Geographical distribution of the DL 1 and DL 2 genes causing hybrid dwarfism in *Phaseolus vulgaris* L., their association with seed size and their significance in breeding. *Euphytica* **33**: 337–345.

Singh, U., Jambunathan, R. and Gurtu, S. 1981. Seed protein fractions and aminoacid composition of some wild species of pigeon pea. *Journal of Food Science Technology* **18**: 83–85.

Smartt, J. 1960. A guide to soya bean cultivation in Northern Rhodesia. *Rhodesia Agricultural Journal* **57**: 459–463.

Smartt, J. 1964. Interspecific hybridization in relation to peanut improvement. *Proceedings Third National Peanut Research Conference*, pp. 53–56. Peanut Improvement Working Group.

Smartt, J. 1965. Cross-compatibility relationships between the cultivated peanut *Arachis hypogaea* L. and other species of the genus *Arachis*. Ph.D. thesis, Raleigh, North Carolina State University.

Smartt, J. 1969. Evolution of American *Phaseolus* beans under domestication. In *The Domestication and Exploitation of Plants and Animals*, pp. 451–462 (eds. P. J. Ucko and G. W. Dimbleby). London: Duckworth.

Smartt, J. 1970. Interspecific hybridization between cultivated American species of the genus *Phaseolus*. *Euphytica* **19**: 480–489.

Smartt, J. 1973. The possible status of *Phaseolus coccineus* L. ssp. *darwinianus* Hdz. X. et Miranda C. as a distinct species and cultigen of the genus *Phaseolus*. *Euphytica* **22**: 424–426.

Smartt, J. 1976*a*. *Tropical Pulses*. London: Longman.

Smartt, J. 1976*b*. Comparative evolution of the pulses. *Euphytica* **25**: 139–143.

Smartt, J. 1978*a*. The evolution of pulse crops. *Economic Botany* **32**: 185–198.

Smartt, J. 1978*b*. Makulu Red – a 'Green Revolution' groundnut variety. *Euphytica* **27**: 605–608.

Smartt, J. 1979. Interspecific hybridization in the grain legumes – a review. *Economic Botany* **33**: 329–337.

Smartt, J. 1980*a*. Evolution and evolutionary problems in grain legumes. *Economic Botany* **34**: 219–235.

Smartt, J. 1980*b*. Some observations on the origin and evolution of the winged bean (*Psophocarpus tetragonolobus*). *Euphytica* **29**: 121–123.

Smartt, J. 1980*c*. The *Phaseolus vulgaris* syngameon. *Annual Report of the Bean Improvement Co-operative* **23**: 19–20.

Smartt, J. 1981*a*. Evolving gene pools in crop plants. *Euphytica* **30**: 415–418.

Smartt, J. 1981*b*. Gene pools in *Phaseolus* and *Vigna* cultigens. *Euphytica* **30**: 445–449.

Smartt, J. 1984. Gene pools in grain legumes. *Economic Botany* **38**: 24–35.

Smartt, J. 1986. Evolution of grain legumes. VI. The future – the exploitation of evolutionary knowledge. *Experimental Agriculture* **22**: 39–58.

Smartt, J. and Gregory, W. C. 1967. Interspecific cross-compatibility between the cultivated peanut *Arachis hypogaea* L. and other members of the genus *Arachis*. *Oléagineux* **22**: 455–459.

Smartt, J., Gregory, W. C. and Gregory, M. P. 1978*a*. The genomes of *Arachis hypogaea*. 1. Cytogenetic studies of putative genome donors. *Euphytica* **27**: 665–675.

Smartt, J., Gregory, W. C. and Gregory, M. P. 1978*b*. The genomes of *Arachis hypogaea*. 2. The implications in interspecific breeding. *Euphytica* **27**: 677–680.

Smartt, J. and Haq, N. 1972. Fertility and segregation of the amphidiploid *Phaseolus vulgaris* L. × *P. coccineus* L. and its behaviour in backcrosses. *Euphytica* **21**: 496–501.

Smartt, J. and Hymowitz, T. 1985. Domestication and evolution of grain legumes. In *Grain Legume Crops*, pp. 37–72 (eds. R. J. Summerfield and E. H. Roberts). London: Collins.

Smartt, J. and Stalker, H. T. 1982. Speciation and cytogenetics in *Arachis*. In *Peanut Science and Technology*, pp. 21–49 (eds. H. E. Pattee and C. T. Young). Yoakum, Texas: American Peanut Research and Education Society Inc.

Smith, P. M. 1976. Minor crops. In *Evolution of Crop Plants*, pp. 301–324 (ed. N. W. Simmonds). London: Longman.

Smithson, J. B., Thompson, J. A. and Summerfield, R. J. 1985. Chickpea (*Cicer arietinum* L.). In *Grain Legume Crops*, pp. 312–390 (eds. R. J. Summerfield and E. H. Roberts). London: Collins.

Spielman, I. V. and Moss, J. P. 1976. Techniques for chromosome doubling in interspecific hybrids of *Arachis*. *Oléagineux* **31**: 491–494.

Sprent, J. I. 1979. *The Biology of Nitrogen-Fixing Organisms*. London; McGraw Hill.

Stalker, H. T. 1978. Cytological analysis of *Erectoides* × *Arachis* intersectional hybrids. *Proceedings of the American Peanut Research and Education Association Inc.* **10**: 62.

Stalker, H. T. 1981. Hybrids in the genus *Arachis* between sections *Erectoides* and *Arachis*. *Crop Science* **21**: 359–362.

Stalker, H. T. 1984. Utilizing *Arachis cardenasii* as a source of *Cercospora* leafspot resistance for peanut improvement. *Euphytica* **33**: 529–538.

Stalker, H. T. and Dalmacio, R. D. 1981. Chromosomes of *Arachis* species, section *Arachis*. *Journal of Heredity* **72**: 403–408.

Stalker, H. T. and Dalmacio, R. D. 1986. Karyotype analysis and relationships among varieties of *Arachis hypogaea* L. *Cytologia* **51**: 617–629.

Stalker, H. T. and Wynne, J. C. 1979. Cytology of interspecific hybrids in section *Arachis* of peanuts. *Peanut Science* **6**: 110–114.

Stalker, H. T., Wynne, J. C. and Company, M. 1979. Variation in progenies of an *Arachis hypogaea* × diploid wild species hybrid. *Euphytica* **28**: 675–684.

Stanton, W. R. 1966. *Grain Legumes in Africa*. Rome: Food and Agriculture Organization of the United Nations.

Stanton, W. R. 1969. Some domesticated lower plants in south-east Asian food technology. In *The Domestication and Exploitation of Plants and Animals*, pp. 463–469 (eds. P. J. Ucko and G. W. Dimbleby). London: Duckworth.

Stebbins, G. L. 1950. *Variation and Evolution in Plants*. New York: Columbia University Press.

Steele, W. M. 1976. Cowpeas. In *Evolution of Crop Plants*, pp. 183–185 (ed. N. W. Simmonds). London: Longman.

Steele, W. M. and Mehra, K. L. 1980. Structure, evolution and adaptation to farming systems and environments in *Vigna*. In *Advances in Legume Science*, pp. 393–404 (eds. R. J. Summerfield and A. H. Bunting). Kew: Royal Botanic Gardens.

Subrahmanyam, P., Ghanekar, A. M., Nolt, B. L., Reddy, D. V. R. and McDonald, D. 1985*a*. Resistance to groundnut diseases in wild *Arachis* species. In *Cytogenetics of Arachis*, pp. 49–55 (ed. J. P. Moss). Patancheru, India: International Crops Research Institute for the Semi-Arid Tropics.

Subrahmanyam, P., Moss, J. P., McDonald, D., Rao, P. V. S. and Rao, V. R. 1985*b*. Resistance to leafspot caused by *Cercosporidium personatum* in wild *Arachis* species. *Plant Disease* **69**: 951–954.

Summerfield, R. J. 1981. Adaptation to environments. In *Lentils*, pp. 91–120 (eds. C. Webb and G. Hawtin). Farnham Royal: Commonwealth Agricultural Bureaux.

Summerfield, R. J. and Roberts, E. H. (eds.) 1985. *Grain Legume Crops*. London: Collins.

Sumner, J. B. 1919. The globulins of the jack bean *Canavalia ensiformis* L. *Biological Chemistry* **37**: 137–144.

Sumner, J. B. 1926. The recrystallization of urease. *Journal of Biological Chemistry* **70**: 97–98.

Sutton, A. W. 1914. Results obtained by crossing a wild pea from Palestine with commercial types (and with *Pisum sativum umbellatum*). *Linnean Society's Journal – Botany* **62**: 427–434.

Taubert, P. 1894. Leguminosae. In *Die naturlichen Pflanzenfamilien*, III Teil, Abt. 3: 70–388 (*Arachis*), pp. 322 and 324–325 (eds. A. Engler and K. Prantl). Leipzig: Verlag von Wilhelm Engelman.

The Peanut – the Unpredictable Legume 1951. Washington, DC: The National Fertilizer Association.

Thomas, C. V. and Waines, J. G. 1984. Fertile backcross and allotetraploid plants from crosses between tepary beans and common beans. *Journal of Heredity* **75**: 93–98.

Tindall, H. D. 1983. *Vegetables in the Tropics*. London: Macmillan.

Tixier, P. 1965. Données cytologiques sur quelques Legumineuses cultivées ou spontanées du Vietnam et du Laos. *Revue de Cytologie et de Biologie Vegetales* **28**: 133–163.

Tombs, M. D. and Lowe, M. 1967. A determination of the sub-units of arachin by osmometry. *Biochemical Journal* **181**: 181–187.

Toms, G. C. and Western, A. 1971. *Phytohaemaglutinins*. In *Chemotaxonomy of the Leguminosae*, pp. 367–462 (eds. J. B. Harborne, D. Boulter and B. L. Turner). London: Academic Press.

Townsend, C. C. and Guest, E. 1974. *Flora of Iraq*, vol. 3. Leguminales. Baghdad: Ministry of Agriculture and Agrarian Reform.

Trapnell, C. G. 1953. *The Soils, Vegetation and Agriculture of North-Eastern Rhodesia*. Lusaka (Zambia): Government Printer.

Trapnell, C. G. and Clothier, J. N. 1957. *The Soils, Vegetation and Agricultural Systems of North-Western Rhodesia*. Lusaka (Zambia): Government Printer.

Tschermark-Seysenegg, E. 1942. Über Bastarde zwischen Fisole (*Phaseolus vulgaris* L.) und Feuerbone (*Phaseolus multiflorus* Lam.) und ihre eventuelle praktische Verwertbarkeit. *Züchter* **14**: 153–164.

Tuchlenski, H. 1958. Groundnut breeding with special reference to production of mutations. *Proceedings of the First Congress of the South African Genetic Society 1958*: 107–109.

Turková, V. and Klozová, E. 1985. Comparison of seed proteins in some representatives of the genus *Vigna*. *Biologia Plantarum* **27**: 70–73.

Turková, V., Klozová, E. and Hadačová, V. 1980. The comparison of seed proteins of several representatives of the genus *Pisum* with respect to their relationship. An immunological comparison. *Biologia Plantarum* **22**: 17–24.

Vanderborght, T. 1979. Le dosage de l'acide cyanohydrique chez *Phaseolus lunatus* L. *Annales de Gembloux* **85**: 29–41.

van der Maesen, L. J. G. 1972 *Cicer L. A Monograph of the Genus with Special Reference to the Chickpea (*Cicer arietinum*), its Ecology and Cultivation.* Wageningen: Mededelingen Landbouwhogeschool 72-10 (1972).

van der Maesen, L. J. G. 1980. India is the native home of the pigeon pea. *Miscellaneous Papers* **19**: 257–263. Wageningen: Mededelingen Landbouwhogeschool.

van der Maesen, L. J. G. 1984. Taxonomy, distribution and evolution of the chickpea and its wild relatives. In *Genetic Resources and their Exploitation – Chickpea, Faba Beans and Lentils*, pp. 95–104 (eds. J. R. Witcombe and W. Erskine). The Hague: Martinus Nijhoff/Dr. W. Junk Publishers.

van der Maesen, L. J. G. 1985. *Cajanus* D.C. and *Atylosia* W. & A. (Leguminosae). *Agricultural University Wageningen Papers* 85-4 (1985).

van Zeist, W. 1970. Prehistoric and early historic food plants in the Netherlands. *Palaeohistoria* **14**: 42–173.

van Zeist, W. 1976. On macroscopic traces of food plants in south western Asia. *Philosophical Transactions of the Royal Society of London* B **275**: 27–41.

Varisai Muhammad, S. 1973*a*. Cytogenetical investigations in the genus *Arachis* L. I. Interspecific hybrids between diploids. *Madras Agricultural Journal* **60**: 323–327.

Varisai Muhammad, S. 1973*b*. Cytogenetical investigations in the genus *Arachis* L. II. Triploid hybrids and their derivatives. *Madras Agricultural Journal* **60**: 1414–1427.

Varisai Muhammad, S. 1973*c*. Cytogenetical investigations in the genus *Arachis* L. III. Tetraploid interspecific hybrids and their derivatives. *Madras Agricultural Journal* **60**: 1428–1432.

Varisai Muhammad, S. 1973*d*. Cytogenetical investigations in the genus *Arachis* L. IV. Chiasma frequency in interspecific hybrids and their derivatives. *Madras Agricultural Journal* **60**: 1433–1437.

Vavilov, N. I. 1951. The origin, variation, immunity and breeding of cultivated plants. *Chronica Botanica* **13**: 1–364.

Verdcourt, B. 1970*a*. Studies in the Leguminosae – Papilionoideae for the 'Flora of Tropical East Africa' III. *Kew Bulletin* **24**: 379–447.

Verdcourt, B. 1970*b*. Studies in the Leguminosae – Papilionoideae for the 'Flora of Tropical East Africa' IV. *Kew Bulletin* **24**: 507–569.

Verdcourt, B. 1971. Studies in the Leguminosae – Papilionoideae for the 'Flora of Tropical East Africa' V. *Kew Bulletin* **25**: 65–169.

Verdcourt, B. 1978. The demise of two geocarpic legume genera. *Taxon* **27**: 219–222.

Verdcourt, B. 1980. The classification of *Dolichos* L. emend. Verdc., *Lablab* Adans., *Phaseolus* L., *Vigna* Savi and their allies. In *Advances in Legume Science*, pp. 45–48 (eds. R. J. Summerfield and A. H. Bunting). Kew: Royal Botanic Gardens.

Verdcourt, B. 1981. The correct name for the Bambara groundnut. *Kew Bulletin* **35**: 374.

Verdcourt, B. 1982. A revision of *Macrotyloma* (Leguminosae). *Hooker's Icones Plantarum*: 1–138.

Verdcourt, B. and Halliday, P. 1978. A revision of *Psophocarpus* (Leguminosae – Papilionoideae – Phaseoleae). *Kew Bulletin* **33**: 191–227.

Viehover, A. 1940. Edible and poisonous beans of the lima type (*Phaseolus lunatus* L.). *Thai Scientific Bulletin* **2**: 1–99.

Waines, J. G. 1975. The biosystematics and domestication of peas (*Pisum* L.). *Bulletin of the Torrey Botanical Club* **102**: 385–395.

Westphal, E. 1974. *Pulses in Ethiopia, their Taxonomy and Ecological Significance*. Wageningen: Centre for Agricultural Publishing and Documentation (PUDOC).

Williams, J. T., Sanchez, A. M. C. and Jackson, M. T. 1974. Studies on lentils and their variation. I. The taxonomy of the species. *SABRAO Journal* **6**: 133–145.

Witcombe, J. R. and Erskine, W. (eds.) 1984. *Genetic Resources and their Exploitation – Chickpeas, Faba Beans and Lentils*. The Hague: Martinus Nijhoff/Dr. W. Junk Publishers.

Wolff, I. A. and Kwolek, W. F. 1971. Lipids of the Leguminosae. In *Chemotaxonomy of the Leguminosae*, pp. 231–255 (eds. J. B. Harborne, D. Boulter and B. L. Turner). London: Academic Press.

Wynne, J. C. and Coffelt, R. A. 1982. Genetics of *Arachis hypogaea* L. In *Peanut Science and Technology*, pp. 50–94 (eds. H. E. Pattee and C. T. Young). Yoakum, Texas: American Peanut Reseaırch and Education Society Inc.

Yarnell, S. H. 1965. Cytogenetics of the vegetable crops. IV. Legumes. *Botanical Review* **31**: 247–330.

Zeven, A. and Zhukovsky, P. M. 1975. *Dictionary of Cultivated Plants and their Centres of Diversity: excluding Ornamentals, Forest Trees and Lower Plants*. Wageningen: Centre for Agricultural Publishing and Documentation (PUDOC).

Zhukovsky, P. M. 1929. The genus *Lupinus* Tourn. (Russian). *Bulletin of Applied Botany, of Genetics and Plant Breeding* (Leningrad) **21**: 241–292. (English translation No. 2839, Commonwealth Scientific and Industrial Research Organization, Australia.)

Zhukovsky, P. M. 1975. *World Gene Pool of Plants for Breeding*. Leningrad: USSR Academy of Sciences.

Zohary, D. 1972. The wild progenitor and the place of origin of the cultivated lentil *Lens culinaris*. *Economic Botany* **26**: 326–332.

Zohary, D. 1976. Lentils. In *Evolution of Crop Plants*, pp. 163–164 (ed. N. W. Simmonds). London: Longman.

Zohary, D. 1977. Comments on the origin of cultivated broad bean, *Vicia faba* L. *Israel Journal of Botany* **26**: 39–40.

Zohary, D. and Hopf, M. 1973. Domestication of pulses in the Old World. *Science* **182**: 887–894.

Postscript

In writing a book the decision on when to stop writing in a difficult one to take. Inevitably there is important work in the offing which would be worthy of note, but a halt has to be called somewhere. The present brief addendum is an attempt to indicate trends in work published since the main text was written and to draw attention to some major works published since that time.

Informal personal communications suggest that the long-awaited taxonomic monograph of the genus *Arachis* may soon appear. This is a consummation devoutly to be wished. Although it has been possible in the past to make do with nomina nuda to designate specific groups, this has not been satisfactory. In addition the more recently collected material has not yet been accommodated in the informal systematic arrangement. Many other studies have been carried out in parallel with the production of the monograph which should enable a definitive view to be formulated with regard to the sources of the *A. hypogaea* genomes.

Studies of inter-specific hybridization and inter-specific gene transfer are continuing at ICRISAT and elsewhere (Singh, 1986, 1988; Zhou and Shou, 1987; Murthy and Tiwari, 1987; Halward and Stalker, 1987; and Moss *in* Ketchum, 1988). Some potentially valuable chemotaxonomic study has also been carried out (Krishna and Mitra, 1988).

Continuing research on *Cajanus–Atylosia* inter-specific hybridization continues to yield significant information. It would appear that the pigeon pea (*Cajanus cajan*) has a very substantial secondary gene pool (GP2) among the Asiatic wild species of *Cajanus–Atylosia*. The observation by Dundas *et al.* (1987) of hybridization between pigeon pea and two native Australian species of *Atylosia* (*Cajanus*) is of considerable interest. The sterility of the inter-specific hybrids produced contrasts strongly with the relatively high fertility of hybrids with some Asiatic species (Dhanju and Gill, 1985; Pundir and Singh, 1987 Tripathi and Patil, 1987).

Wild relatives of the soyabean have continued to receive attention in the United States and elsewhere (Grant, 1986; Grant *et al.*, 1986; Newell *et al.*, 1987; Singh and Hymowitz, 1987; Singh *et al.*, 1987, 1988). This work tends to confirm the genetic isolation of the soyabean in the genus *Glycine* and is effectively an exploration of its tertiary gene pool.

The publication in 1988 of *Genetic Resources of* Phaseolus *Beans* (P. Gepts, ed.) is an event of the greatest significance to those interested in the genetic resources, evolution and breeding of *Phaseolus* species. It is a landmark publication and comprehensive in its coverage. There have been several other publications worthy of note on inter-specific hybrids and their derivatives. Much of this work has been carried out by Maréchal and co-workers at Gembloux (Belgium). The production of interspecific hybrids *Phaseolus polyanthus* × *P. vulgaris* is of interest; this has been achieved by embryo rescue techniques although the reciprocal hybrid can be obtained without them. The pollen stainability of the resultant hybrids is remarkably high (±90.0%) (Camarena and Baudoin, 1987).

The allotetraploid produced from the inter-specific hybridization of *P. vulgaris* and *P. filiformis* is worthy of note (de Tau *et al.*, 1987), as are a number of interspecific F_1 hybrids involving *P. lunatus* (Katanga and Baudoin, 1987*a*). Not the least interesting feature is that the production of some F_2 seed is reported, suggesting the possible existence of a secondary gene pool for the lima bean. The same authors report a hybrid between *P. filiformis* and *P. angustissimus* (Katanga and Baudoin, 1987*b*). Some F_2 seed has also been obtained from this cross.

A comprehensive inter-specific hybridization programme in *Vigna* is reported by Rashid *et al.* (1988). Those reported recently include the cross between the rice bean (*V. umbellata*) and *V. minima* (Gopinathan *et al.*, 1986). The hybrid produced was sterile with an irregular meiosis. Of even greater interest is the reported production of an inter-specific hybrid *V. pubescens* × *V. unguiculata* through embryo rescue (Fatokum and Singh, 1987). This is in all probability the first successful attempt at producing a viable interspecific hybrid with the cowpea. The very substantial mungbean symposium (AVRDC, 1988) is something of a milestone. The publication in 1988 of Zohary and Hopf's *Domestication of Plants in the Old World* has undoubtedly been welcomed by many besides students of grain legume evolution. The treatment of Mediterranean pulse crop domestication is magisterial and is likely to become a classic. Its production is simple, restrained but elegant.

Some interesting chemotaxonomic work on cultivated members of the Vicieae has been reported by Rougé *et al.* (1987*a*, *b*) on the use of the properties and rections of lectins in assessing phylogenetic relationships.

In the genus *Cicer* some useful basic studies on inter-specific hybridization have been carried out by Slinkard and his associates in Canada (Basiri *et al.*, 1987; Ahmad *et al.*, 1988). In these, useful studies of barriers to successful interspecific hybridization have been carried out. The publication of the monograph (Saxena and Singh, 1987) on the chickpea is another very significant event.

Recent published work on the lentil and its allies by Ladizinsky and co-workers includes the description of a new species of *Lens* (Ladizinsky, 1986), possible changes induced in *Lens culinaris* prior to domestication by selection (Ladizinsky, 1987), and a study of morphological variation in *Lens* (Hoffman *et al.*, 1988).

Reports of inter-specific hybridization in *Lupinus* are few and far between, therefore the studies of Roy and Gladstones (1988) of interspecific hybridization among Mediterranean–African lupin species are of real interest, especially as the progeny produced are somewhat fertile. The species concerned are *L. atlanticus*, *L. digitatus* and *L. cosentinii*.

Perhaps the most perplexing problem besetting those concerned with germplasm collection and conservation is the fact that so many areas of critical importance are inaccessible because of unstable or troubled political situations. The Middle East and the Horn of Africa are two good cases in point. While there are reasonable grounds for optimism here in the long term, in other areas such as the USSR and China hopeful political developments have been followed by ethnic and political unrest which have raised some uncertainty regarding the prospects for plant genetic resource collection in these areas. It is to be hoped that the present period of political volatility will be short-lived and followed by one of sufficient stability to enable the tasks of collection and conservation of these vital resources to proceed to a satisfactory conclusion.

Supplementary references

Ahmad, F., Slinkard, A. E. and Scoles, G. J. 1987. The cytogenetic relationships between *Cicer judaicum* Bois. and *Cicer chorassicum* (Bge.) M.P. *Genome* **29**: 883–886.

Ahmad, F., Slinkard, A. E. and Scoles, G. J. 1988. Investigations into the barrier(s) to interspecific hybridization between *Cicer arietinum* and eight other annual *Cicer* species. *Plant Breeding* **100**: 193–198.

Asian Vegetable Research and Development Centre, 1988. *Mungbean. Proceedings of the Second International Symposium*. Shanhua, Taiwan: AVRDC.

Bassiri, A., Ahmad, F. and Slinkard, A. E. 1987. Pollen grain germination and pollen tube growth following *in vivo* and *in vitro* self and interspecific pollinations in *Cicer* species. *Euphytica* **36**: 667–675.

Camarena, F. and Baudoin, J. P. 1987. Obtention des premiers hybrides interspécifiques entre *Phaseolus vulgaris* et *Phaseolus polyanthus* avec le cytoplasme de cette dernière forme. *Bulletin des Recherches Agronomiques de Gembloux* **22**: 43–55.

de Tau, E. W., Mathieu, A., Maréchal, R. and Baudoin, J. P. 1987. Observation par fluorescence de la croissance du gamétophyte mâle chez les allotétraploides d'un hybride interspécifique entre *Phaseolus vulgaris* L. et *Phaseolus filiformis* Benth. *Bulletin des Recherches Agronomiques de Gembloux* **22**: 143–151.

Dhanju, M. S. and Gill, B. S. 1985. Intergeneric hybridization between *Cajanus cajan* and *Atylosia platycarpa*. *Annals of Biology* **1**: 229–231.

Dundas, I. S., Britten, E. J., Byth, D. E. and Gordon, G. H. 1987. Meiotic behaviour of hybrids of pigeon pea and two Australian native *Atylosia* species. *Journal of Heredity* **78**: 261–265.

Fatokun, C. A. and Singh, B. B. 1987. Interspecific hybridization between *Vigna pubescens* and *V. unguiculata* (L.) Walp. through embryo rescue. *Plant Cell, Tissue and Organ Culture* **9**: 229–233.

Gepts, P. (ed.) *Genetic Resources of* Phaseolus *Beans*. Dordrecht: Kluwer Academic Publishers.

Gopinathan, M. C., Babu, C. R. and Shivanna, K. R. 1986. Interspecific hybridization between rice bean (*Vigna umbellata*) and its wild relative (*V. minima*): fertility–sterility relationships. *Euphytica* **35**: 1017–1022.

Grant, J. E. 1986. Hybridisation of soybean with its diploid wild perennial relatives. Special Publication no. 5 (1986;, Agronomy Society of New Zealand: 27–29.

Grant, J. E., Pullen, R., Brown, A. H. D., Grace, J. P. and Gresshoff, P. M.

1986. Cytogenetic affinity between the new species *Glycine argyrea* and its congeners. *Journal of Heredity* **77**: 423–426.

Halward, T. M. and Stalker, H. T. 1987. Incompatibility mechanisms in interspecific peanut hybrids. *Crop Science* **17**: 456–460.

Hoffmann, D. L., Muehlbauer, F. J. and Ladizinsky, G. 1988. Morphological variation in *Lens* (Leguminosae). *Systematic Botany* **13**: 87–96.

Katanga, K. and Baudoin, J. P. 1987*b*. Obtention et observations d'un nouvel hybride interspécifique entre deux espèces sauvages: *Phaseolus filiformis* Benth. × *Phaseolus angustissimus* A. Gray. *Bulletin des Recherches Agronomiques de Gembloux* **22**: 152–160.

Ketchum, J. L. F. (ed.) 1988. Contribution from J. P. Moss, ICRISAT; Newsletter, International Plant Biotechnology Network no. 9.

Krishna, T. G. and Mitra, K. 1988. The probable genome donors to *Arachis hypogaea* L. based on arachin seed storage protein. *Euphytica* **37**: 47–52.

Ladizinsky, G. 1986. A new *Lens* from the Middle-East. *Notes from the Royal Botanic Garden, Edinburgh* **43**: 489–492.

Ladizinsky, G. 1987. Pulse domestication before cultivation. *Economic Botany* **41**: 60–65.

Murthy, T. G. K. and Tiwari, S. P. 1987. Second division restitution in a fertile interspecific triploid hybrid of groundnut. *Cytologia* **52**: 667–670.

Newell, C. A., Delannay, X. and Edge, M. E. 1987. Interspecific hybrids between the soybean and wild perennial relatives. *Journal of Heredity* **78**: 301–306.

Pundir, R. P. S. and Singh, R. B. 1987. Possibility of genetic improvement of pigeon pea (*Cajanus cajan* (L.) Millsp.) utilizing wild gene sources. *Euphytica* **36**: 33–37.

Rashid, K. A., Smartt, J. and Haq, N. 1988. Hybridization in the genus *Vigna*. In *Mungbean: Proceedings of the Second International Symposium*, pp. 205–214. Shanhua, Taiwan, AVRDC.

Rougé, P., Richardson, M., Ranfaing, P., Yarwood, A. and Sousa-Cavada, B. 1987*a*. Single and two chain legume lectins as phylogenetic markers of speciation. *Biochemical Systematics and Ecology* **11**: 342–348.

Rougé, P., Garcia, M. L., Boisseau, C. and Causse, H. 1987*b*. ELISA measurement of immunochemical cross reactions among *Lathyrus* lectins as an assessment of their phylogenetic relationship. *Biochemical Systematics and Ecology* **15**: 349–353.

Roy, N. N. and Gladstones, J. S. 1988. Further studies with interspecific hybridization among Mediterranean/African lupin species. *Theoretical and Applied Genetics* **75**: 606–609.

Saxena, M. C. and Singh, K. B. (eds) 1987. *The Chickpea*. Farnham Royal: C.A.B. International; Aleppo: ICARDA.

Singh, A. K. 1986. Alien gene transfer in groundnut by ploidy and genome manipulation. In *Genetic manipulation in plant breeding. Proceedings of Eucarpia International Symposium, West Berlin, Sept. 8–13, 1985* (ed. W. Horn, C. J. Jensen, W. Oldenbach and O. Schieder). Berlin, Walter de Gruyter.

Singh, A. K. 1988. Putative genome donors of *Arachis hypogaea* (Fabaceae) evidence from crosses with synthetic amphidiploids. *Plant Systematics and Evolution* **160**: 143*f*152.

Singh, R. J. and Hymowitz, T. 1987. Inter-subgeneric crossability in the genus *Glycine* Willd. *Plant Breeding* **89**: 171–173.

Singh, R. J., Kollipara, K. P. and Hymowitz, T. 1987. Intersubgeneric hybridization of soybeans with a wild perennial species, *Glycine clandestina* Wendl. *Theoretical and Applies Genetics* **74**: 391–396.

Singh, R. J., Kollipara, K. P. and Hymowitz, T. 1988. Further data on the genomic relationships among wild perennial species ($2n$=40) of the genus *Glycine* Willd. *Genome* **30**: 166–176.

Tripathi, S. N. and Patil, B. D. 1987. Trispecific cross in the genus *Atylosia* W. and A. *Cytologia* **52**: 657–660.

Zhou, Y. L. and Zhou, R. 1987. Studies on interspecific crosses between cultivated and wild species in *Arachis* III. Autotetrapolid *A. stenosperma* and its utilization. *Oil Crops of China* **2**: 4–10. (In Chinese.)

Zohary, D. and Hopf, M. 1988. *Domestication of Plants in the Old World*. Oxford: Clarendon Press.

Author index

General index